Perilous Progress
Managing the Hazards of Technology

D1432926

Westview Special Studies

The concept of Westview Special Studies is a response to the continuing crisis in academic and informational publishing. Library budgets are being diverted from the purchase of books and used for data banks, computers, micromedia, and other methods of information retrieval. Interlibrary loan structures further reduce the edition sizes required to satisfy the needs of the scholarly community. Economic pressures on university presses and the few private scholarly publishing companies have greatly limited the capacity of the industry to properly serve the academic and research communities. As a result, many manuscripts dealing with important subjects, often representing the highest level of scholarship, are no longer economically viable publishing projects--or, if accepted for publication, are typically subject to lead times ranging from one to three years.

Westview Special Studies are our practical solution to the problem. As always, the selection criteria include the importance of the subject, the work's contribution to scholarship, and its insight, originality of thought, and excellence of exposition. We accept manuscripts in camera-ready form, typed, set, or word processed according to specifications laid out in our comprehensive manual, which contains straightforward instructions and sample pages. The responsibility for editing and proofreading lies with the author or sponsoring institution, but our editorial staff is always available to answer questions and provide guidance.

The result is a book printed on acid-free paper and bound in sturdy, library-quality soft covers. We manufacture these books ourselves using equipment that does not require a lengthy make-ready process and that allows us to publish first editions of 300 to 1000 copies and to reprint even smaller quantities as needed. Thus, we can produce Special Studies quickly and can keep even very specialized books in print as long as there is a demand for them.

About the Book and Editors

Tapping a unique data base of information on ninety-three hazards, this book offers a comprehensive and comparative perspective on a broad range of technological risks confronting contemporary society. The fourteen authors, specialists with backgrounds in geochemistry, geography, law, physics, and psychology, draw on six years of interdisciplinary research to provide a framework for thinking about hazards and hazard management, for measuring the consequences of technological risk for society, and for assessing the appropriateness of various hazard management techniques.

Robert W. Kates is professor at the Graduate School of Geography and research professor at the Center for Technology, Environment, and Development (CENTED), Clark University, Worcester, Massachusetts, where Jeanne X. Kasperson is research librarian. Christoph Hohenemser is professor of physics and chair of the Environment, Technology, and Society Program at Clark University.

Perilous Progress
Managing the Hazards of Technology

edited by
Robert W. Kates,
Christoph Hohenemser,
and Jeanne X. Kasperson

Westview Press / Boulder and London

Westview Special Studies in Science, Technology, and Public Policy

Copyright © 1985 by Clark University, Worcester, Massachusetts; except for
Chapter 10, which is © by the American Association for the Advancement of
Science

Published in 1985 in the United States of America by Westview Press, Inc.;
Frederick A. Praeger, Publisher; 5500 Central Avenue, Boulder, Colorado 80301

Library of Congress Cataloging in Publication Data
Main entry under title:
Perilous progress.
    (Westview special studies in science, technology,
and public policy)
    1. Technology assessment. 2. Risk. I. Kates, Robert
William. II. Hohenemser, Christoph. III. Kasperson,
Jeanne X.
T174.5.P47 1984   363.1'0028'7   84-25623
ISBN 0-8133-7025-6

Composition for this book was provided by the editors
Printed and bound in the United States of America

10 9 8 7 6 5 4 3 2 1

Contents

Acknowledgments

The progress of this volume has sometimes been perilous. We appreciate the patience and good humor of the contributors who bore with us as the book emerged, ever so slowly, from the realm of the perpetually forthcoming. Although we take full responsibility for the finished product, we wish to acknowledge the editorial insights of Wendy Goble, the artistic talents of Lisa Hohenemser, the typing skill of Joan McGrath, and the masterful photocopying abilities of Glenda Alin. To Donna Law, who continued to smile as she delivered yet another professional rendering of this or that figure, we owe special gratitude. Above all, we salute the incomparable perfectionism of Lu Ann Renzoni (who is chuckling as she flawlessly types this), who now holds the world's record for moonlighting, for producing sterling camera-ready copy.

Robert W. Kates
Christoph Hohenemser
Jeanne X. Kasperson
Worcester, Massachusetts

1
Introduction: Coping with Technological Hazards

Robert W. Kates, Christoph Hohenemser,
and Jeanne X. Kasperson

The realization that technological progress may be perilous is scarcely unique to our time. Even the ancient Greeks and Romans recognized that the same lead that improved their plumbing, architecture, ships, weapons, and jewelry also poisoned their miners, contaminated their wine, and polluted their water supplies (Nriagu 1983). Agricola in 1556 lamented the environmental depredation of mining regions in Europe:

> ...fields are devastated by mining operations...woods and groves are cut down...for timbers, machines, and the smelting of metals...then are exterminated the beasts and the birds...when the ores are washed, the water... poisons the brooks and streams,...destroys the fish, therefore the inhabitants of these regions on account of the devastation of their fields, woods, groves, brooks, and rivers,... find great difficulty in procuring the necessaries of life...it is clear to all that there is greater detriment to mining than the value of the metals which the mining produces (Agricola [1556]1950,8).

Similarly well-understood in its own time was the social devastation of the Industrial Revolution, including the expulsion of rural folk from the land, the replacement of natural rhythms with the mechanical discipline of endless belts, the exploitation of children in mine and factory, and the recurrent economic crises of widespread unemployment. The novels of Dickens and Zola graphically depict these horrors.

Until recently the voices of demographers, reformers, and muckrakers who chronicled this toll remained a distinct minority, whereas the majority viewed technology with awe and hope. Now, however, an increasingly technological society betrays some measure of disenchantment. A discernible popular and scientific ambivalence toward technology has emerged in the mix of responses to surveys on technological issues (LaPorte and Metlay 1975;Marsh and McLennan 1980;Miller, Prewitt, and Pearson 1980,61-66;Mitchell 1980) and in the research literature (Kates and Kasperson 1983). This ambivalence, if it persists, may well constitute a watershed that separates our time from centuries of virtually undivided commitment to industrial growth.

1

An important aspect of this uncertainty is the dual sense of technology as progress and as hazard. René Dubos (1981) reminds us: "Modern technology has clearly improved the quality of our lives in countless ways, but it has inevitably generated new risks fundamentally different in both character and magnitude from those encountered in the past." One is hard put to deny that the production and use of pesticides have brought jobs and agricultural progress to India, but the lethal leak on 4 December 1984 of methyl isocyanate (MIC) from a Union Carbide plant at Bhopal has branded that city as the site of the worst technological disaster in history (Chemical and Engineering News 1985). As Pope John Paul II cautioned the Italian National Academy of Science: "But often technology, ever more perfect and deadly, that is derived from it [science] has been turned against man, to the point of creating appalling arsenals of conventional and nuclear arms, biological and chemical means capable of destroying a great part of humanity" (Kamm 1982). This duality draws strength from four trends in hazard evolution, perception, and management: changes in technology, in the identification of hazards, in public perceptions, expectations, and demands, and in the character of societal response.

From a review of these trends, we proceed in this introduction to take stock of the first decade of interdisciplinary research spawned by these developments. Our subsequent overview of the present volume and the research herein takes its measure in the context of the field as a whole.

Trends

Beginning with World War II, an impressive technological revolution has generated a staggering array of hazardous materials, products, and processes. Most dramatic was the development of the atomic bomb, the legacy of which still hangs heavy in the fear of nuclear destruction and in the worldwide debate on nuclear power (see chapter 10). Less dramatic, perhaps, but of major significance were the exponential increase in production of synthetic chemicals, the concentration of materials normally dispersed in nature, and the changes in energy flow and mineral cycling that accompanied order-of-magnitude changes in engineering works, transportation, and waste generation and disposal.

Commoner (1971) has argued that such changes are fundamental and disjunctive, not simply a continuation of the processes set in motion by the Industrial Revolution. But whether or not a break with the past has occurred, genuine improvements in hazard identification and monitoring have surely intensified the sense of technology as hazard. These improvements include major advances in analytic and bioassay methods that facilitate positive identification of chemical and biological hazards; new screening devices that use low-cost **in vitro** tests; computer modelling of environmental sources, pathways, and sinks; and rapid growth in the general scientific effort devoted to hazard assessment. Monitoring networks for the major pollutants of air, land, and water are now functioning on a national scale. And new methods for required reporting of product failure, for accidental releases, and for routine epidemiological observations are in place or under development.

Thus extraordinary scientific and technical advances enhance capability for identifying, measuring, and assessing hazards. Similarly, heightened public perceptions of risk, combined with increased expectations and demands for protection and safety, also make for timely recognition (or suspicion) of new hazards. One recent poll found that, even in the face of major decreases in mortality, 78 per cent of Americans polled viewed life in 1980 as riskier than life 20 years before. Moreover, a smaller albeit significant majority (55 per cent) of respondents anticipate that the risks from science and technology will increase over the next twenty years (Marsh and McLennan 1980). Findings from other polls indicate that despite recent shifts in U.S. public policy, expectation of and support for added protection and safety persist unabated among the public (Mitchell 1980;Harris 1982;Insurance Information Institute 1983).

The explanation for these changing perceptions and expectations is not obvious. Surely some of the shift must be due to the actual and identified increase in hazardousness, but other factors may be at work as well. In the wake of major advances in the extension of life, the control of diseases, the elimination of hunger, and the diminishing of insecurity from unemployment and old age, former worries have given way to new concerns. Now society can enjoy the luxury of turning to less visible hazards such as radiation and chemicals. Affluence breeds security, and increased wealth fosters an expectation that risks will decrease, if not disappear, as comforts increase.

A recent report identifies three other forces that have altered people's perceptions of risk: "the intensified reporting of risks in the media, which sometimes justified and sometimes not, magnifies people's concerns; the loss of faith in institutions ostensibly created to deal with risk; and the growth of a complex, highly technological society that is interdependent in its functioning and that reduces the perception of individuals that they can control the events" (National Research Council 1982,13). Such forces may in turn relate to broad sweeps and episodic fluctuations in moods of optimism or despair, in predilections for risk or security, or in the politics of liberalism and conservatism.

Societal response parallels trends in increased hazardousness and heightened public perception. David Bazelon (1983) points to "twin revolutions"--in science and technology and in social expectations--that have transformed American values and produced a staggering litigation crisis that threatens to overwhelm the courts. In 22 years (1957-1978) the United States Congress enacted at least 178 laws dealing with technological hazards (chapter 19, this volume; Okrent and Wilson 1982). Congress works steadily, passing laws on established hazards and adding two to three new hazards each session. The cumulative outcome is a vast legislative/regulatory domain, the full extent of which is only now beginning to surface (Greenwood, Kingsbury, and Cleland 1979).

U.S. hazard prevention, reduction, and mitigation costs for 1979 totalled between $45.7 and $54.1 billion (chapter 7). Taking together the costs of hazard management, the productivity unrealized due to injury and premature death, and the residual toll of economic losses, the overall economic burden for 1979 falls between $179.5

and \$283 billion, equivalent to 8 to 12 percent of the gross national product (GNP).

Part of the effort to prevent, reduce, and mitigate hazards has been the emergence of a new "applied" discipline and profession concerned with the assessment and management of technological hazards. And with the assessors and the managers has come the development of a research program.

Modern research on the comparative management of technological hazards dates to a seminal paper of Starr (1969) relating social benefit and technological risk. Subsequent interest in Starr's findings spurred a colloquium by the National Academy of Engineering (1972) and a workshop of the Engineering Foundation (Okrent 1975). Meanwhile, the Scientific Committee on Problems of the Environment (SCOPE) sponsored international workshops in places as diverse as Woods Hole, Massachusetts (Kates 1978) and Tihany, Hungary (Whyte and Burton 1980). The publication of book-length reviews and texts on risk assessment (Lowrance 1976;Rowe 1977;Kates 1978) and a major casebook (Lawless 1977), circulated in 1974, followed. The impetus for new research endeavors developed in part through a series of workshops sponsored by the National Science Foundation (NSF) in 1977 (Kates 1977) and the subsequent establishment by NSF of its program on Technology Assessment and Risk Analysis (TARA). Under the auspices of the TARA program a committee of the National Research Council (1982) has prepared an overview of the research field. The program provided encouragement and support for the international Society for Risk Analysis, which publishes its own journal (Risk Analysis), a newsletter (Risk Newsletter), and the proceedings of its annual meetings (Covello et al. 1983;Covello, Menkes, and Mumpower 1985).

It is possible to quibble over the precise dating of the field (Otway 1980). Well before 1969 risk analyses of various types had appeared in such specialized fields as engineering, product safety management, industrial hygiene, and occupational medicine as well as in risk markets (insurance, stocks and bonds, and gambling). And by 1969, natural-hazards management (Burton, Kates, and White 1968) had enjoyed two decades of interdisciplinary research (Burton, Kates, and White 1968,1978) that addressed many of the same issues later examined for technological hazards. Yet an extensive (over 1000 citations) topical bibliography, spanning the years 1935–1983, includes only 41 entries with publication dates prior to 1969 (Covello and Abernathy 1983). Moreover, Starr's publication in Science was the first major paper to undertake explicitly the **comparative** analysis of technological hazards. The ensuing period has witnessed an exponential growth in the literature. Anyone who questions the staying power of this quasi discipline has only to note this voluminous output.

Comparative Research on Technological Hazards

Recent bibliographic forays speak to a flourishing research effort, particularly in recent years (Covello and Abernathy 1983; Kates and Kasperson 1983;Kasperson and Kates 1984). Indeed, if a recent literature survey conducted by the Hazard Assessment Group at Clark University's Center for Technology, Environment, and Development (CENTED) is any indication, the 1980s promise an inundation.

The group tapped its extensive library on technological hazards and selected for analysis 61 major books, published between 1970 and 1984, on comparative risk analysis. Insofar as 45 of the 61 titles have appeared in the period 1980-1984 and a number of additional volumes are in the wings, the 1980s have already eclipsed previous decades.

The survey is far from comprehensive. Limited as it is to English-language monographs and book-length collections of papers, it overlooks a vast international literature in books, in journal articles, and in reports from government, industry, and public-interest groups. Absent, too, are the risk assessments or case studies of specific technologies or hazards, such as the fifty or more risk assessment reports published each year by the National Research Council (1981). Critics may well find the list too long or too short and may lobby for exclusion or inclusion of this or that title, but such dissensions do not negate the utility of the survey. The selected volumes do represent a significant portion of the **comparative** research literature and thus they provide insights into the interests and concerns of an adolescent field.

Surveys of the volumes (Kates and Kasperson 1983; Kasperson and Kates 1984) have identified six recurring themes—(1) overviews of one or more areas; (2) risk estimation; (3) discussion of acceptable or tolerable risk; (4) risk perception; (5) analysis of regulation; and (6) case studies of specific technological hazards—as well as agenda for research. Table 1 lists the books in chronological order and summarizes the incidence of the six recurring themes that thread their various ways through the 61 volumes. We use these themes to take stock of the research to date and to assess the contribution of this volume to an already crowded field.

Overviews

Though fragmented and often inconsistent, the emerging litera-ture on technological hazards contains some integrative overviews and evaluations of the field. Chapter 3 depicts the structure of technological hazards as a linked causal chain bounded by four mana-gerial activities—**hazard assessment, control analysis, strategy selection,** and **implementation and evaluation.** Few volumes cover the full range of these activities. Most overviews point to methodological and conceptual shortcomings, but some seek to evalu-ate the socioeconomic, political, and cultural contexts that may have a bearing on the practice of hazard assessment. Thus Lagadec (1982) characterizes the "challenge of major risk" as a series of clashes between reason and democracy; Douglas and Wildavsky (1982) view the very selection of risks as a basic cultural choice, a de-liberate decision to worry most about those dangers that threaten beliefs and values; and geographers assume a spatial stance and explore the **regionalization of risk** by defining **hazard zones** (Zeigler, Johnson, and Brunn 1983).

Risk Estimation

Consensus in the literature has it that risks are measures of the likelihood that particular adverse consequences will follow a hazardous event. Thus the estimated lifetime risk of an average

TABLE 1
Major book-length publications on comparative risk analysis (1970–1984)*

YEAR	AUTHOR(S) OR EDITOR(S)	OVERVIEWS	RISK ESTIMATION	ACCEPTABLE RISK	PERCEPTION	REGULATION	CASE STUDIES
1970	Calabresi					+	
1972	National Academy of Engineering		+	+	+		+
1972	Sinclair et al.		+	+	+		+
1974	Epstein and Grundy		+			+	+
1975	Chicken		+	+		+	+
1975	Environmental Studies Board		+	+			
1975	National Research Council		+	+			
1976	Ashford		+	+		+	+
1976	Lowrance			+	+		
1977	Council for Science and Society		+	+	+	+	+
1977	Kates	+				+	
1977	Lawless	+	+	+	+	+	+
1977	Rowe	+	+	+	+		
1978	Kates	+	+	+	+	+	+
1979	Goodman and Rowe	+		+		+	
1979	Hammond and Selikoff		+	+		+	
1980	Conrad	+	+	+	+		
1980	Dierkes et al.	+		+	+	+	+
1980	Dowie and Lefrere	+		+	+		+
1980	Hovden		+	+	+		
1980	The Open University	+	+	+	+		
1980	Salem et al.		+	+	+		
1980	Schwing and Albers		+	+	+		
1980	Whyte and Burton		+	+			+
1981	Baram		+			+	+
1981	Berg and Maillie		+			+	+
1981	Crandall and Lave		+			+	+

Year	Reference
1981	Ferguson and LeVeen
1981	Griffiths
1981	Haimes
1981	Lave
1981	Nicholson
1981	Richmond et al.
1981	The Royal Society
1981	Siddall
1982	Burton, Fowle, and McCullough
1982	Crouch and Wilson
1982	Douglas and Wildavsky
1982	Fischhoff et al.
1982	Green
1982	Hohenemser and Kasperson
1982	Inhaber
1982	Kunreuther
1982	Kunreuther and Ley
1982	Lagadec
1982	Lave
1982	Lind
1982	National Research Council
1982	Poole
1982	Prentice and Whittemore
1983	Covello et al.
1983	National Research Council
1983	Rescher
1983	Rogers and Bates
1983	Royal Society
1983	Viscusi
1983	Zeigler, Johnson, and Brunn
1984	Deisler
1984	Hadden
1984	Perrow
1984	Ricci, Sagan, and Whipple

American's dying in an automobile accident, for example, is 2-3 percent. Hazard assessment includes three distinct activities: (1) the identification of hazards likely to produce hazardous events, (2) the estimation of the risks of such events and their attendant consequences, and (3) the social evaluation or weighting of the risks so derived (Kates 1978). Terminology and concepts, however, differ slightly from author to author (Lowrance 1976;Rowe 1977;Kates 1978; Whyte and Burton 1980;National Research Council 1982).

The initial step, hazard identification, receives short shrift in most hazard assessment. One senses an inexplicable but undeniable confidence in a known and knowable pool of hazards that require measurement, evaluation, and management. The assessors turn at once to the business of estimating and quantifying risk and worry little about unknown hazards. Beginning with an **identified** hazard, this (now) initial task of estimation relies heavily on extrapolation from past experience, from experiments (usually with animal models), or from simulations (often with computer models). As extrapolations, and frequently imperfect ones, risk estimates inevitably entail scientific uncertainty, the handling of which is crucial. In Part 3 of this volume, we explore this problem in case studies of automobile accidents (chapter 8), airborne mercury (chapter 9), and nuclear power (chapter 10) and in an overall critique of hazard assessment methodologies (chapter 11).

Risk Acceptability or Tolerability

The third step in assessing risk is to determine which risks are tolerable to society. The frequent, and somewhat misleading question "How safe is safe enough?" has become the **acceptable risk** issue. **Acceptability,** as we suggest in chapter 3, is an unfortunate term, implying a degree of consent that rarely accompanies impositions of risk. **Tolerability** better captures most actual risk situations (Kasperson and Kasperson 1983).

Whatever the label, however, the determination of tolerability, in contrast to the estimation of risk, lacks scientific precision. The question "How safe is safe enough?" is primarily one of values. Most discussions of risk tolerability acknowledge that all human activity is inherently hazardous to someone or something, that even the absence of an activity, especially a useful one, may be hazardous. Most researchers would agree that collective efforts for managing a given hazard ought to be commensurate with the degree of threat—observed or perceived—posed by the particular activity or technology in question. Hence the priority lists, the classification schemes, and the taxonomies such as that in chapter 4. No one challenges society's need to focus on the important hazards. Disagreement rears with the inference that society should optimize (in economic terms) its investment in risk reduction via some common metric (for example, number of lives saved). Meanwhile, alternative approaches have proliferated. Initially, Starr (1969), taking into account both the benefits and the voluntary/involuntary nature of a given risk, inferred its "acceptability" as "revealed" in historical statistics of mortality. Stabilization of a level of mortality over time implies that society has accepted a certain degree of risk from a particular product or activity. Later, **risk/benefit analysis** sought to define the level of risk that is tolerable in return for a given benefit (National Academy of Engineering 1972;Environmental Studies Board 1975;Crouch and Wilson 1982). Most recently, in response to the mounting critique of revealed-preferences and risk/benefit analysis, several researchers have sought to define a "risk threshold," or **de minimis** level, below which risk should command no regulatory attention (Comar 1979;Wilson 1981;Eisenbud 1980;Okrent 1982;Starr and Whipple 1982).

The foregoing approaches entail serious methodological and ethical problems (Fischhoff et al. 1982). Because such attempts fall short when quantitative risk estimates are not available, alternatives have proliferated. One approach elicits directly from the public its preferences for various risks (this volume, chapters 5 and 12;Fischhoff et al. 1978;Slovic et al. 1979); another advocates direct public involvement in risk decisions through existing legal and political processes.

Risk Perception

Practicing risk assessors find all too often that their scientific findings diverge from popular perceptions of risk. In fact, both scientific risk assessment and popular perceptions derive from judgments, the former made with the assistance of formal and sometimes reproducible methodology, the latter elicited through more informal and perhaps broader cognitive processes. Considerable research, which has progressed from the speculative to the scientific, has gone into identifying and understanding the nature of perceived risk.

Pioneering studies by Slovic, Fischhoff, and Lichtenstein (1978) compare the perceived risk of many technologies and activities. Perhaps more than any other work, this research--represented in chapters 5 and 12 of this volume and in virtually all collections of papers on risk--has enhanced our awareness of risk perception. Other researchers (Vlek and Stallen 1981;Lee 1981) have employed comparable techniques and produced consistent findings. A major finding of most of this work is that lay people's judgments of risk are qualitatively similar to those of scientific experts but differ from the latter in many important details. Each group taps its own heuristics in making quantitative judgments, hence the discrepancies and the serious over- and under-estimates.

Two threads of evidence may explain the discrepancies in quantitative judgments about risks. An observed general tendency to underestimate the frequency of common events and to overestimate the frequency of rare events makes for a compression in the scale of probability judgments (Slovic, Fischhoff, and Lichtenstein 1979). Moreover, experts usually estimate risks in terms of mortality, whereas lay persons are more prone to consider other factors.

The structure of risk perception has also been studied via extensive attitudinal surveys (Otway, Pahner, and Linnerooth 1975; Vlek and Stallen 1981). The most widely studied single hazard is nuclear power, for which national surveys exist back to the early seventies (Louis Harris and Associates 1975 and 1976;Melber et al. 1977;Mitchell 1980). In the United States, overall risk has been subject to major broad-based surveys (Marsh and McLennan 1980), which have indicated that most Americans believe life is becoming riskier over time. A recent survey (Harris 1982) established

American concern and willingness to pay for environmental protection. Generally, most of the polls indicate an erosion of confidence in institutions charged with enforcing existing health and safety regulations. This disenchantment has produced a heightened perception of danger and a clamor for better protection—even in the face of an antiregulatory climate.

Regulation

The assessment of a new hazard and its inclusion in the repertoire of public perception demand a societal response. This response, conceptualized as **hazard management** (Kates 1977;Whyte and Burton 1980;Nicholson 1981;Hadden 1984) includes a spectrum of ways in which government, industry, private groups, and individuals control, reduce, avoid, or tolerate hazards. The literature on these actors and activities is decidely one-sided. Despite the reality that most decisions about risk are made by individuals, and many are made by corporations, the literature concentrates on the relatively few that are made by government or at its insistence. Exceptions are a book (Baram 1981) that considers the alternatives to regulation, such as legal remedies, taxation, and other incentives and a collection (Poole 1982) that comes out in favor of an extreme alternative to regulation—namely, true deregulation, or the abolition of regulatory agencies.

Case Studies

Rather frequently, experience with regulation comes through in the form of case studies. As is appropriate for a field with an indistinct and still emerging structure, the literature on technological hazards abounds with case studies conceived in varying frames of reference. Some authors use case studies to illustrate the methodology of risk estimation; others employ the framework of risk tolerability criteria; still others focus on the sequence and timing of regulation, and more generally, the structure of hazard management. Some case studies have no identifiable frame of reference and simply highlight the full range of issues in celebrated cases such as Love Canal, Three Mile Island, or DDT.

A major sourcebook for systematic comparison is that of Lawless (1977), whose analysis of 45 instances of technological shock provides a standardized look at the timing and interrelation of hazard identification, media coverage, and political, legal, and regulatory response. As in chapter 13 of the present volume, the overall record is one of failures in managing hazard after hazard.

Lawless's ambitious casebook stands unmatched in scope, but other volumes also make significant use of case studies. Crandall and Lave (1981) recruit trios of experts—a scientist, an economist, and a regulator—to pool their analyses of the scientific basis for regulating passive restraints, cotton dust, waterborne carcinogens, saccharin, and sulfur dioxide. A conference volume (Hammond and Selikoff 1979) tendered four perspectives on the management of vinyl chloride. Another conference (Nicholson 1981) invited cross-national comparisons of the handling of carcinogenic risk in Sweden, Canada, Norway, and Sweden. In a Canadian study (Burton, Fowle, and McCullough 1982), a series of case studies ranging from toxic shock syndrome to 2,4-5 T serves to clarify concepts and methodologies as well as to illustrate and inform risk management in Canada. Crouch and Wilson (1982) propose a "prescription for useful analysis" of nine cases ranging from nuclear power plant accidents to swine flu vaccinations. Parts 3 and 4 of the present volume include case studies of specific hazards--nuclear power, contraceptives, airborne mercury, polychlorinated biphenyls (PCBs), automobile accidents, and television--and certain hazard managers--the United States Congress and the Consumer Product Safety Commission--to test theory, validate a model of hazard structure, and enhance the process of managing hazards.

Analytic use of case studies contributes to the conceptual understanding of hazard and risk. Certainly, Lawless's (1977) uniform comparison of 45 cases, Lave's (1981) eight "decision frameworks," and Crouch and Wilson's (1982) "prescription" enhance the theoretical data base. Yet few of the numerous and varied case studies that pervade the literature really test hypotheses about the nature of hazard and its management.

Perilous Progress: Managing the Hazards of Technology

To a large extent, the present volume grew out of a workshop, convened at Clark University in the fall of 1976, to address research needs in a fledgling field. The book derives from a project conceived as a collaborative, interdisciplinary effort by the Clark Hazard Assessment Group and Decision Research of Eugene, Oregon. At one time or another the project commingled a biochemist, a geochemist, several geographers, two physicists, and several psychologists. Despite the disparate nature of their respective disciplines, participants worked to develop a common language. As to the extent to which they succeeded, the reader must judge!

Beyond the introduction, the book falls into into four parts, each with a brief overview to establish context and organization. Part 1, **Conceptualizing Hazards** contains four chapters that take a generic, comparative approach to the understanding, classification, and management of hazards. Part 2, **Measuring Consequences,** links two closely related chapters that measure the total burden of technological hazard in terms of human and nonhuman mortality and economic costs. Part 3, **Assessing Risks,** comprises three studies of specific hazards--automobiles, airborne mercury, and nuclear power--a critique of hazard assessment, and a discussion of risk tolerability criteria. Part 4, **Managing Technological Hazards** moves from a propositional inventory of 41 publications on hazard management to case studies of specific hazards and hazard managers. Four technological hazards--automobiles, PCBs, contraceptives, and television--take up a chapter apiece. Two additional chapters present analyses of two hazard managers--the Consumer Product Safety Commission and the United States Congress.

In the context of the comparative research the 19 chapters contribute to the major themes of the existing literature: Parts 1 and 2 in themselves constitute an **overview**; Part 3 provides cases and a critique of **risk estimation** and **risk tolerability** (or **acceptability**); chapters 5 and 12 address **risk perception,** and the **case studies** of six hazards and two hazard managers add to our stockpile of analyzed experience. Moreover, the volume enhances our understanding in a number of distinct ways.

Conceptual Development

Part 1 provides a framework for thinking about hazards and hazard management. **Hazards**, defined as "threats to humans and what they value," are causal chains linking human needs and wants to choice of technology and to threatening consequences (chapter 2). Hazard reduction and control take the form of disruptions or attenuations of causal chains. Hazard control in response to experienced or predicted harm is visualized in terms of feedback. Four managerial activities--hazard assessment, control analysis, strategy selection, and implementation and evaluation surround the causal chain (chapter 3). Judgments elicited from lay persons and experts allow for the characterization of hazard perception. Statistical analysis of these judgments suggests a reproducible structure that "explains" the sometimes unexpected facts of lay perception (chapter 5). A classification scheme that uses causal structure as its organizing principle, yields to a taxonomy of hazard that spans the full domain of "hazardousness" and through 12 causal descriptors accounts for 50-75 percent of the variance in perceptions of risk (chapter 4).

Comprehensiveness

The work described in Parts 1 and 2 is comprehensive in several respects. In contrast to much of the literature, analysis in terms of causal structure emphasizes hazard control rather than risk assessment (chapters 2 and 3). The discussion of hazard management (chapter 3) goes well beyond the paradigm of regulation and thus helps to correct an imbalance that pervades much of the literature of the 1970s. The 12-descriptor characterization of hazardousness underlying the classification of hazards considerably expands upon the conventional definition of **risk** as "probability of dying" (chapter 4). And the accounting of consequences in terms of mortality (chapter 6) and economic costs and losses (chapter 7) provides a summing that approximates the measurable burden of **all** technological hazards.

Empirical Grounding

The conceptualization of hazard and hazard management derives from an intrinsically empirical approach that taps several extensive data bases. The analysis of perception elaborates earlier work on nine cognitive dimensions and 30 hazards and now includes 18 dimensions and 90 hazards, each scaled by lay subjects (chapter 5). The causal classification of hazard is based on 12 descriptors and 93 hazards, each scaled by explicit reference to the scientific literature (chapter 4). A comparison of causal structure to perception (chapter 4) employs a separate set of 12 cognitive dimensions and 81 hazards judged by lay subjects and parallels closely the descriptors and hazards employed in the analysis of causal structure. The propositional inventory that gives way to managerial case studies (Part 4) derives from a survey of 41 studies of experience with specific hazards or hazard managers (chapter 13). Finally, nine case studies, three of which emphasize risk assessment (chapters 8-10), and five of which address broader questions of hazard management (chapters 14-19), inform the analysis throughout the generic chapters in the volume.

Management Tools

The volume offers several potential tools for improving hazard management. The diagramming of causal structure provides a systematic way of identifying opportunities for hazard control, mapping the level of control by hazard stage, and assessing the timeliness and comprehensiveness of response (chapters 2 and 3). Feedback analysis permits a systematic approach to identifying cases in which hazard consequences are inadvertently increased rather than reduced (chapter 2). The classification of hazards by causal structure suggests means for identifying hazard management priorities, making quantitative comparisons of "hazardousness," and developing protocols for managing new hazards on the basis of success or failure in managing similarly classified old hazards. The linkage between lay perceptions and causal descriptors renders possible the prediction--based solely on analysis of causal structure (chapters 2, 4)--of public response to newly discovered hazards.

Informed Public Policy

Portions of the book may shed factual and conceptual light on the formulation of public policy. The disaggregation of economic costs and losses (chapter 7) can inform assessments of the burden of technological hazards and the equity of their distribution between the public and private sector. The taxonomy of hazards (chapter 4) can aid in designating which hazards should worry society. Chapter 12 proposes criteria and methods for determining tolerable levels of risk. How well we are coping with technological hazards can be evaluated in part via the historical mortality record (chapter 6) and the review of hazard management studies (chapter 13). Finally, the biases and presuppositions of experts come into play (chapter 11).

In June, 1979, at the end of the first decade of significant comparative research on technological hazard, we made the following prognosis:

The 1970s saw the creation of vast amounts of legislation regulating the known domains of technological hazard. As the decade draws to a close, concern over such hazards continues unabated, while controversy increases over cost, adequacy, and management. Inconsistencies in law, regulation, and attitudes confront the nation with the need for continual reassessment.

The coming decade, we predict, will be a period of such reassessment as we collectively decide to reduce many risks, accommodate others, and eliminate a few. (Harvey et al. 1979,15)

Nearly six years later this prognosis appears to be in full flower, albeit with some change in direction. A national administration, actively engaged in accommodating many hazards and reducing some, stands reluctant to take decisive regulatory action if it can be avoided or achieved by voluntary means. The new era seems to be one in which both the courts and the Congress are taking the lead in coping with technological hazards.

We did not believe in 1979, nor do we believe now, that decisions about hazards are the province of legislators, regulators, scientists, industrialists, and lobbyists. Decisions about hazards are, above all, public decisions. They demand broad-based scientific understanding, an appreciation of the differences between scientific and lay assessments, and a sense of the balance between peril and progress.

REFERENCES

Agricola, Georg. [1556] 1950. De re metallica. 1st ed. Reprint, trans. H. C. Hoover and L. C. Hoover. New York: Dover Publications.

Ashford, Nicholas A. 1976. Crisis in the workplace: Occupational disease and injury. Cambridge, MA: MIT Press.

Baram, Michael S. 1981. Alternatives to regulation: Managing risks to health, safety and the environment. Lexington, MA: Lexington Books.

Bazelon, David L. 1983. Technology, litigation and justice. Bulletin of the Atomic Scientists 39 no. 9 (November):101-105.

Berg, George C., and H. D. Maillie, eds. 1981. Measurement of risks. New York: Plenum Press.

Burton, Ian, C. David Fowle, and Roger S. McCullough, eds. 1982. Living with risk: Environmental risk management in Canada. Publication EM-3 (September). Toronto: Institute for Environmental Studies, University of Toronto.

Burton, Ian, Robert W. Kates, and Gilbert F. White. 1968. The human ecology of extreme events. Natural Hazards Research Working Paper no. 1. Toronto: Department of Geography, University of Toronto.

Burton, Ian, Robert W. Kates, and Gilbert F. White. 1978. The environment as hazard. New York: Oxford University Press.

Calabresi, Guido. 1970. The costs of accidents. New Haven: Yale University Press.

Chemical and Engineering News. 1985. Bhopal report. Special issue. 63 no. 6 (11 February).

Chicken, John C. 1975. Hazard control policy in Britain. New York: Pergamon Press.

Cohen, Bernard L., and I-Sing Lee. 1979. A catalog of risks. Health Physics 36 no. 6 (June):707-722.

Comar, Cyril L. 1979. Risk: A pragmatic de minimis approach. Science 203:319.

Commoner, Barry. 1971. The closing circle. New York: Knopf.

Conrad, Jobst, ed. 1980. Society, technology and risk assessment. New York: Academic Press.

Council for Science and Society. 1977. The acceptability of risks: The logic and social dynamics of fair decisions and effective controls. London: Barry Rose for the Council.

Covello, Vincent T., and Mark Abernathy. 1984. Risk analysis and technological hazard: A policy-related bibliography. In Technological risk assessment, ed. Paolo Ricci, Leonard Sagan, and Chris Whipple, 283-363. The Hague: Martinus Nijhoff.

Covello, Vincent T., W. Gary Flamm, Joseph V. Rodricks, and Robert G. Tardiff, eds. 1983. The analysis of actual vs. perceived risks. New York: Plenum Press.

Covello, Vincent T., Joshua Menkes, and Jeryl Mumpower, eds. 1985. Risk evalution and management: The state of the art. New York: Plenum Press, forthcoming.

Crandall, Robert W., and Lester B. Lave, eds. 1981. The scientific basis of health and safety regulation. Washington: The Brookings Institution.

Crouch, Edmund A. C., and Richard Wilson. 1982. Risk/benefit analysis. Cambridge, MA: Ballinger.

Deisler, Paul F., Jr., ed. 1984. Reducing the carcinogenic risks in industry. New York: Marcel Dekker.

Dierkes, Meinolf, Sam Edwards, and Rob Coppock, eds. 1980. Technological risk: Its perception and handling in the European Community. Cambridge, MA: Oelgeschlager, Gunn and Hain for the Commission of the European Communities.

Douglas, Mary, and Aaron Wildavsky. 1982. Risk and culture: An essay on the selection of technological and environmental dangers. Berkeley and Los Angeles: University of California Press.

Dowie, Jack, and Paul Lefrere, eds. 1980. Risk and chance: Selected readings. Milton Keynes, U.K.: Open University Press.

Dubos, René. 1981. Outline of remarks: Unpublished address prepared for presentation 8 January 1981, New York Hilton [a statement by Dubos on the occasion of a tribute to him on his 80th birthday. New York: René Dubos Center for Human Environments].

Eisenbud, Merril. 1980. The concept of de minimis dose. In Quantitative risk in standard setting, 64-72. Proceedings of the Sixteenth Annual Meeting of National Council on Radiation Protection and Measurements. Washington: NCRP.

Environmental Studies Board. 1975. Committee on Principles of Decision Making for Regulating Chemicals in the Environment. Decision making for regulating chemicals in the environment. Washington: National Academy of Sciences.

Epstein, Samuel S., and Richard D. Grundy, eds. 1974. The legislation of product safety. 2 vols. Cambridge, Mass.: MIT Press.

Ferguson, Allen R., and E. Phillip LeVeen, eds. 1981. The benefits of health and safety regulation. Cambridge, Mass.: Ballinger.

Fischhoff, Baruch, Paul Slovic, Sarah Lichtenstein, Steven Read, and Barbara Combs. 1978. How safe is safe enough?: A psychometric study of attitudes towards technological risks and benefits. Policy Sciences 8:127–152.

Fischhoff, Baruch, Sarah Lichtenstein, Paul Slovic, Steven Derby, and Ralph L. Keeney. 1982. Acceptable risk. New York: Cambridge University Press.

Goodman, Gordon T., and William D. Rowe, eds. 1979. Energy risk management. New York: Academic Press.

Greenwood, D. R., G. L. Kingsbury, and Cleland, J. G. 1979. Handbook of key federal regulations for multimedia environmental control. EPA-600/7-79-179. Washington: Environmental Protection Agency.

Griffiths, Richard F., ed. 1981. Dealing with risk. New York: John Wiley and Sons.

Hadden, Susan G. 1984. Risk analysis, institutions, and public policy. Port Washington, New York: Associated Faculty Press.

Haimes, Yacov Y., ed. 1981. Risk/benefit analysis in water resources planning and management. New York: Plenum Press.

Hammond, E. Cuyler, and Irving J. Selikoff, eds. 1979. Public control of environmental health hazards. Annals of the New York Academy of Sciences, vol. 329. New York: New York Academy of Sciences.

Harris, Louis. 1983. Testimony in U.S. Congress, Senate, Committee on Environment and Public Works, Subcommittee on Environmental Pollution, American Attitudes Toward Clean Water: Hearing... Dec. 15, 1982 (Serial 97-H68, 97th Cong., 2d sess). Washington: Government Printing Office.

Harvey, Marjorie, et al. 1979. Project summary: Improving the societal management of technological hazards. [Worcester, Mass.: Center for Technology, Environment, and Development (CENTED), Clark University.]

Hohenemser, Christoph, and Jeanne X. Kasperson, eds. 1982. Risk in the technological society. AAAS Selected Symposium, 65. Boulder, Colo.: Westview Press.

Hovden, Jan. 1980. Accident risks in Norway: How do we perceive and handle risks? Oslo: Risk Research Committee, Royal Norwegian Council for Scientific and Industrial Research.

Inhaber, Herbert. 1982. Energy risk assessment. New York: Gordon and Breach.

Insurance Information Institute. 1983. Public attitudes toward risk. New York: Louis Harris and Associates for the Institute.

Kamm, Henry. 1982. Pope warns scientists about the effects of their work. New York Times, 22 September, p. A4.

Kasperson, Jeanne X., and Robert W. Kates. 1984. Comparative risk analysis of technological hazards (A review)/II: A look at the book-length literature. Risk Abstracts 1 no. 4 (October):153–163.

Kasperson, Roger E., and Jeanne X. Kasperson. 1983. Determining the acceptability of risks: Ethical and policy issues. In Risk: A symposium on the assessment and perception of risk to human health in Canada, October 18 and 19, 1982... Proceedings, ed. J. T. Rogers and D. V. Bates, 135–155. Ottawa: Royal Society of Canada.

Kates, Robert W., ed. 1977. Managing technological hazard: Research needs and opportunities. Program on Technology,

Environment and Man, Monograph 25. Boulder: Institute of Behavioral Science, University of Colorado.

Kates, Robert W. 1978. Risk assessment of environmental hazard. SCOPE 8. New York: John Wiley for Scientific Committee on Problems of the Environment (SCOPE).

Kates, Robert W., and Jeanne X. Kasperson. 1983. Comparative risk analysis of technological hazards (A review). Proceedings of the National Academy of Sciences USA 80 (November):7027-7038.

Kunreuther, Howard C., ed. 1982. Risk: A seminar series. Laxenburg, Austria: International Institute for Applied Systems Analysis (IIASA).

Kunreuther, Howard C., and Eryl W. Ley, eds. 1982. The risk analysis controversy: An institutional perspective. Proceedings of a Summer Study on Decision Processes and Institutional Aspects of Risk held at IIASA, Laxenburg, Austria, 22-26 June 1981. New York: Springer Verlag.

Lagadec, Patrick. 1982. Major technological risk: An assessment of industrial disasters, trans. from the French by H. Ostwald. New York: Pergamon Press.

La Porte, Todd R., and Daniel Metlay. 1975. Technology observed: Attitudes of a wary public. Science 188:121-127.

Lave, Lester B. 1981. The strategy of social regulation. Washington: The Brookings Institution.

Lave, Lester B., ed. 1982. Quantitative risk assessment in regulation. Washington: The Brookings Institution.

Lawless, Edward W. 1977. Technology and social shock. New Brunswick, NJ: Rutgers University Press.

Lee, T. R. 1981. The public's perception of risk and the question of irrationality. Proceedings of the Royal Society of London, Series A 376:5-16. Note: A reprint of the article appears in Royal Society (1981,5-16).

Louis Harris and Associates. 1975. A survey of public and leadership attitudes toward nuclear power development in the United States. New York: Ebasco Services.

Louis Harris and Associates. 1976. A second survey of public and leadership attitudes toward nuclear power development in the United States. New York: Ebasco Services.

Lowrance, William W. 1976. Of acceptable risk: Science and the determination of safety. Los Altos, Calif.: William Kaufmann.

Marsh and McLennan, Inc. 1980. Risk in a complex society: A Marsh and McLennan public opinion survey. Conducted by Louis Harris and Associates. New York: Marsh and McLennan.

Melber, Barbara D., Stanley M. Nealey, Joy Hammersla, and William L. Rankin. 1977. Nuclear power and the public: Analysis of collected survey research. PNL-2430. Seattle: Battelle Memorial Institute, Human Affairs Research Centers.

Miller, Jon D., Kenneth Prewitt, and Robert Pearson. 1980. The attitudes of the U.S. public toward science and technology. Chicago: National Opinion Research Center, University of Chicago.

Mitchell, Robert Cameron. 1980. Public opinion on environmental issues: Results of a national pubic opinion survey. Washington: Council on Environmental Quality.

National Academy of Engineering. 1972. Committee on Public Engineering Policy. Perspectives on benefit-risk decision making. Washington: National Academy of Engineering.

National Research Council. 1975. Committee for the Working Conference on Principles of Protocols for Evaluating Chemicals in the Environment, Environmental Studies Board, and Committee on Toxicology. Principles for evaluating chemicals in the environment. Washington: National Academy of Sciences.

National Research Council. 1981. Governing Board Committee on the Assessment of Risk. The handling of risk assessments in NRC reports. Washington: National Academy of Sciences.

National Research Council. 1982. Committee on Risk and Decision Making. Risk and decision making: Perspectives and research. Washington: National Academy Press.

National Research Council. 1983. Committee on the Institutional Means for Assessment of Risks to Public Health. Risk assessment in the federal government: Managing the process. Washington: National Academy Press.

Nicholson, William J., ed. 1981. Management of assessed risk for carcinogens. Annals of the New York Academy of Sciences, vol. 363. New York: New York Academy of Sciences.

Nriagu, Jerome O. 1983. Lead and lead poisoning in antiquity. New York: Wiley-Interscience.

Okrent, David, ed. 1975. Risk-benefit methodology and applications: Some papers presented at the Engineering Foundation Workshop, Sept. 22-26, 1975. UCLA ENG 7598. Los Angeles: University of California.

Okrent, David. 1982. Comment on societal risk. In Risk in the technological society, ed. Christoph Hohenemser and Jeanne X. Kasperson, 203-215. AAAS Selected Symposium 65. Boulder, Colo.: Westview Press. Note: This article also appears in Science 208 (1980):372-375.

Okrent, David, and Richard Wilson. 1982. Safety regulations in the USA. In High risk safety technology, ed. A. E. Green, 601-613. Chichester: Wiley.

The Open University. 1980. Risk: A second level university course. 6 vols. Milton Keynes, U.K.: Open University Press.

Otway, Harry J. 1980. Discussion paper on A. Mazur: Societal and scientific causes of the historical development of risk assessment. In Society, technology and risk assessment, ed. J. Conrad, 163-164. New York: Academic Press.

Otway, Harry, Philip D. Pahner, and Joanne Linnerooth. 1975. Social values in risk acceptance. IIASA Research Memorandum, RM 75-54. Laxenburg, Austria: International Institute for Applied Systems Analysis (IIASA).

Poole, Robert W., ed. 1982. Instead of regulation: Alternatives to federal regulatory agencies. Lexington, MA: Lexington Books.

Prentice, Ross L., and Alice S. Whittemore, eds. 1982. Environmental epidemiology: Risk assessment. Philadelphia: Society for Industrial and Applied Mathematics (SIAM).

Ricci, Paolo, Leonard Sagan, and Chris Whipple, eds. 1984. Technological risk assessment. The Hague: Martinus Nijhoff.

Richmond, Chester R., Phillip J. Walsh, and Emily D. Copenhaver, eds. 1981. Health risk analysis. Proceedings of the Third Life Sciences Symposium Gatlinburg, Tennessee, October 27-30, 1980. Philadelphia: Franklin Institute Press.

Rogers, J. T., and D. V. Bates, eds. 1983. Risk: A symposium on the assessment and perception of risk to human health in Canada, October 18 and 19, 1982... Proceedings. Ottawa: Royal Society of Canada.

Rowe, William D. 1977. An anatomy of risk. New York: Wiley.

Royal Society. 1981. Study Group on Risk. The assessment and perception of risk: A Royal Society discussion. London: The Royal Society.

Royal Society. 1983. Study Group on Risk. Risk assessment: A study group report. London: The Royal Society.

Salem, Steven L., Kenneth A. Solomon, and Michael S. Yesley. 1980. Issues and problems in inferring a level of acceptable risk. R-2561-DOE. Santa Monica, Calif.: Rand Corporation.

Schwing, Richard C., and Walter A. Albers, Jr., eds. 1980. Societal risk assessment: How safe is safe enough? New York: Plenum Press.

Siddall, E. 1981. Risk, fear and public safety. Report AECB-7404. Mississauga, Ontario: Atomic Energy of Canada Limited.

Sinclair, Craig, Pauline Marstrand, and Pamela Newick. 1972. Innovation and human risk: Human life and safety in relation to technical change. London: Centre for the Study of Industrial Innovation.

Slovic, Paul, Baruch Fischhoff, and Sarah Lichtenstein. 1979. Rating the risks. Environment 21 (3):14-20,36-39. Note: A slightly revised version of this article appears in Hohenemser and Kasperson (1982,141-166).

Starr, Chauncey. 1969. Social benefit versus technological risk. Science 165:1232-1238.

Starr, Chauncey, and Chris Whipple. 1982. Risks of risk decisions. In Risk in the technological society, ed. Christoph Hohenemser and Jeanne X. Kasperson, 217-239. AAAS Selected Symposium 65. Boulder, Colo.: Westview Press. Note: This article appeared originally in Science 208 (1980):1114-1119.

Viscusi, W. Kip. 1983. Risk by choice: Regulating health and safety in the workplace. Cambridge, MA: Harvard University Press.

Vlek, Charles A., and Pieter J. M. Stallen. 1981. Judging risks and benefits in the small and the large. Organizational Behavior and Human Performance 28:235-271.

Whyte, Anne V., and Burton, Ian, eds. 1980. Environmental risk assessment. SCOPE 15. New York: John Wiley for the Scientific Committee on Problems of the Environment (SCOPE).

Wilson, Richard. 1981. The role of health risk assessment in decision making. In Health Risk Analysis, ed. Chester K. Richmond, Phillip J. Walsh, and Emily D. Copenhaver, 73-83. Philadelphia: Franklin Institute Press.

Zeigler, Donald C., James R. Johnson, and Stanley D. Brunn. 1983. Technological hazards. Washington: Association of American Geographers.

Part One

Overview: Conceptualizing Hazards

The four chapters of Part 1 introduce concepts and empirical findings that contribute to a generic approach to technological hazards. One focus of Part 1 is **definitional** and establishes a language for thinking consistently and logically about technological hazards. A second focus is **descriptive** and portrays our conception of individual and societal response to technological hazards. A third focus is **comparative** and provides two distinct ways of ordering and classifying the large number of hazards that society faces.

Defining Hazards

Hazards are defined as "threats to humans and what they value." They are distinguished from **risks**, which chapter 4 defines as "quantitative measures of hazard consequences, usually expressed as conditional probabilities of experiencing harm." Usage of **risk** differs among authors; Slovic, Fischhoff, and Lichtenstein (chapter 5), for example, use a considerably broader risk concept.

In chapter 2, hazards are conceptualized as causal sequences of events that originate in human needs and wants and evolve in time to choice of technology, release of energy and materials, and possible human harm. Central to this idea of hazard is that technology, in its normal operation, may release energy or materials with sufficient intensity/concentration to threaten or cause harm. Chapter 2 introduces diagrammatic techniques for displaying the causal structure of technological hazards. The approach used is analogous to fault-tree diagrams, albeit in simplified or truncated form; it is also related to similar paradigms employed in the analysis of natural hazards.

Diagramming causal structure permits the mapping of hazard control interventions, negative and positive feedback processes, hazard management efforts, and a variety of analytic methods employed in hazard assessment. In chapter 3, causal structure provides a conceptual tool for evaluating societal hazard management; in chapter 4, it serves as a template for a causal taxonomy of hazards. For hazard management the most important attribute of the causal structure diagrams is that they provide a systematic way of asking questions about hazard control.

Describing Individual and Societal Response

Given that hazards are defined as threats to humans and what they value, how do individuals react, how does society respond, and what is the character of the actors, institutions, and processes that lead to societal decisions about hazards?

Chapter 3 offers one set of answers to these questions by detailing the cycle of societal hazard management, which runs from hazard assessment, to control analysis, to formulation of management strategy, to implementation and evaluation. The chapter is descriptive in seeking to portray what issues, values, and conflicts arise in reaching societal risk decisions; it is analytical in its attempts to be complete and logical in defining societal response; and it is evaluative in striving to judge the adequacy of the management process. The approach in chapter 3 emphasizes the management responsibilities of those entrusted with protecting health and safety and notes, too, that other theoretical perspectives will provide other "faces" of societal response to hazards. The chapter demonstrates clearly that hazard management is a strongly conflicted process in which values, bureaucratic behavior, and political imperatives interact with scientific and technical issues in shaping societal responses.

The emphasis on process and institutional issues in chapter 3 takes on credence in chapter 5, by a study of individual response to hazards. Of particular interest are psychometric estimates of such variables as **degree of desired risk reduction** and **desired degree of regulation.** Responses from a sampling of well-educated respondents indicate that most individuals desire substantial risk reduction for most hazards and are apparently willing to accept the degree of societal control and cost that this implies.

Chapter 5 also defines **perceived risk,** a psychometric variable that combines many risk characteristics into a single, global variable, which may prove useful in predicting people's attitudes toward a given hazard. Elicitations of this variable suggest that laypeople define risk broadly and are apt to take into account qualitative factors such as the catastrophic nature of the risk or the dread associated with its consequences. Attempts to characterize hazard, set safety standards, and make hazard-management decisions will founder in conflict if policy makers insist on narrow definitions (e.g., mortality-based) of risk that do not correspond to people's perceptions. Chapter 4 indicates that as much as 80 percent of the elicited variation in perceived risk can be explained by a set of scientifically derived descriptors of hazards. The door is thus open to predicting societal response via scientifically grounded characterization of hazards.

Comparative Hazard Analysis

A major challenge for hazard theory lies in finding an organizational principle for classifying and ordering the domain of hazards. The challenge is symbolized by the question: "What do saccharin, intrauterine devices, nuclear power plants, and the collapse of the Grand Teton Dam have in common, and how do they differ?" Two approaches are taken to this problem.

Chapter 4 describes a taxonomy of technological hazards in which 93 hazards are classified via 12 physical, biological, and social descriptors of causal structure. The descriptors are based on physical or categorical scales and are scored by detailed reference to the scientific literature on the hazard in question. Use of factor analysis collapses the 12 descriptors into five composite orthogonal dimensions that have substantial intuitive content.

One dimension, labeled **Biocidal**, describes the degree to which hazards produce nonhuman mortality and involve intentional killing; another, labeled **Delay**, describes the degree to which a hazard has long-term effects, either through persistence in the environment or through long exposure/consequence delays; a third dimension, called **Catastrophic**, specifies the degree to which a hazard combines episodic character and events involving a large number of simultaneous fatalities; a fourth dimension, labeled **Mortality**, simply measures the level of annual mortality; and a fifth, named **Global**, describes the magnitude of the population at risk and the degree of diffuseness of the hazard release. Taken together, these dimensions of technological hazards allow the classification of most or all hazards into distinctive classes, which appear to have quite different management patterns.

In a different but related approach to hazard comparison, chapter 5 orders hazards via a number of cognitive characteristics of associated risk. In contrast to chapter 4, scoring of these dimensions of risk depends on responses elicited from lay people. The measures of risk so defined are viewed as subjects' perceptions, which may or may not correspond to the scientific characterizations of the hazards. As in chapter 4, factor analysis allows collapse of 18 risk qualities into just two of three orthogonal composite dimensions.

A first composite dimension, called **Dread Risk,** is associated with lack of control, lethality, high catastrophic potential, a reaction of dread, inequitable distribution of risks and benefits, and the belief that the risks are increasing and not easily controlled. A second composite dimension, called **Unknown Risk,** is associated with hazards for which risks are unknown, unobservable, new, and delayed in their manifestation. A third composite dimension, labeled **Societal and Personal Exposure,** is primarily determined by the number of people exposed.

Chapter 5 shows that despite wide agreement among diverse groups about ratings assigned to risk qualities, attitudes on risk differ markedly between groups. This suggests that one may compare hazards via risk qualities elicited from almost any group, but one cannot predict, on this basis, the degree of desired regulation or more generally defined management responses that individuals prefer.

Taken together, the chapters of Part 1 are a beginning of a theory of hazard. Completing such a theory is a distant goal. Nevertheless, one can safely predict that it will involve integration of scientific analysis of hazard structure with a systematic description of societal and individual response.

2
Causal Structure

Christoph Hohenemser, Roger E. Kasperson,
and Robert W. Kates

If a tree falls in a forest far from human ears, is there a sound? This is a classic philosophical problem. For hazard control, the analogous question is: If a tree falls, is there a hazard? Trees fall from a variety of causes--disease, lightning, flood, fire, the sharp teeth of beavers, and the axes and chain saws of humans. Such occurrences may have no immediate human implications and we call them **events**. But trees may also crush people, maim livestock, and destroy buildings, dam streams, cause floods, and accelerate erosion. These and other impacts on humans we call **consequences**. As threats to humans and what they value, hazards consist at minimum of events and consequences, just as sound in the perceptual sense requires at minimum the physical excitation of sound waves and the receipt and perception of these by human ears and brains.

The division of hazards into events and consequences strongly implies three possible strategies for hazard control: (1) prevention of hazard events; (2) prevention of hazard consequences once events have taken place; and (3) mitigation of consequences once these have occurred. Prevention of events appears to be most fundamental, whereas mitigation is often regarded as unsatisfactory in the sense that "an ounce of prevention is worth a pound of cure." Yet, in any particular case, any one of the three strategies may be the most appropriate. Consider the following examples.

For catastrophic nuclear power accidents involving the release of massive amounts of radioactivity, neither consequence prevention nor consequence mitigation is especially feasible. Therefore, most control efforts concentrate on activities designed to prevent hazard events, that is, strategy (1) above.

In contrast, for intensive geophysical hazards (tornadoes, earthquakes), this strategy has little value since no one can prevent or significantly alter them. Even strategy (2), prevention of hazard consequences, is of only limited utility since geophysical hazards inevitably produce large losses, especially in developing countries (see chapter 6). Thus societies practice strategy (3), consequence mitigation, in the form of property-damage relief, medical attention to survivors, and reconstruction.

Most hazard management falls between the poles represented by catastrophic nuclear accidents and intensive geophysical hazards. For automobiles, for example, events (accidents) are preventable in principle, and society has expended much effort in this direction, as in eliminating curves in highways. Yet the high cost and relative ineffectiveness of accident prevention makes consequence prevention equally, if not more, relevant. Particularly in the last ten years, it has become clear that much can be done to block injury, via seatbelts or other restraints (see chapters 8 and 14), after an accident has occurred. Finally, coping with auto accidents involves a heavy dose of strategy (3), mitigation of consequences, usually in the form of insurance designed to distribute the burden of loss.

The Causal Anatomy of Hazards

How, specifically, do prevention and mitigation occur? To this end, it is useful to recognize that events and consequences are members of a causal sequence; that is, events **cause** consequences. As such, events and consequences are connected by causal pathways, the logical places for blocking the sequence for the purpose of hazard control. The causal sequence can be expanded arbitrarily to reflect the details of causal structure. For the purposes of this volume we generally use a seven-stage sequence, defined as follows.

We refer to two classes of events, **initiating events** and higher-order events, which we call **outcomes**. Initiating events include any number of occurrences that trigger hazardous failures of technological systems. Outcomes follow initiating events and are defined as releases of energy and materials that are direct threats to humans. The pathways that connect the two kinds of events vary in complexity. Often several nearly simultaneous initiating events lead to a given outcome, which in turn has several possible consequences. In the example illustrated in Figure 1, a loose, flammable garment, a strong wind, the wearer's distraction by conversation, and a nearby fire are all initiating events required to produce the outcome of garment ignition; and this, in turn, leads to three consequences—a destroyed garment, burned skin, and smoke inhalation.

In many cases, outcomes do not lead so directly to consequences, and it is useful to insert the stage **exposure** between outcomes and consequences. Grinding wheels may well release dust (outcome), but it does not follow that humans will be exposed or subsequently harmed. Exposure in this case occurs through dust inhalation, which may be blocked by appropriate respirators.

Upstream of initiating events (in the causal sense), we expand the sequence to include **choice of technology, human wants,** and **human needs.** Diagramming the full scope of hazard in this way is particularly important for cases in which downstream stages pose special control problems. In the case of certain pesticides, for example, there is at best only circumstantial evidence of carcinogenic potential, and specific downstream control interventions between events and consequences are poorly understood (Figure 2). In such a situation, prudent hazard control recognizes the ineffectiveness of downstream blocks and concentrates on upstream options such as choice of technology or modification of human wants. Examples of such strategies are the use of biodegradable, nonpersistent pesticides and the toleration of blemished fruit.

The seven stages of the causal sequence—human needs, human wants, choice of technology, initiating events, outcomes, exposure,

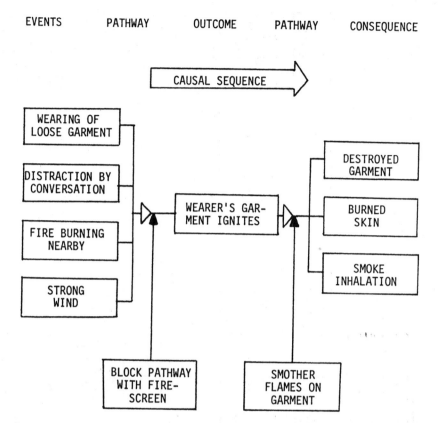

Figure 1. Events and consequences contributing to a fireplace accident. Note that several events contribute to the ignition of the wearer's garment and that this leads to several consequences. Coping with the hazard may be achieved by outcome prevention, illustrated here by a fireplace screen; and consequence prevention, illustrated by actions to smother the flames on the garment.

and consequences—are illustrated in Figure 3, which diagrams the full range of occurrences that lead from human needs to a burn injury. Also shown are appropriate control interventions that effectively block each step in the sequence. This simple example makes clear that there are many ways to control a hazard; and by implication, if one mode is ineffective or socially unacceptable, another may suffice.

Beyond the seven stages of Figure 3, one may usefully expand the sequence further if this yields additional meaningful opportunities for hazard control. Figure 4 provides two examples in which this is the case. In the first, several orders of consequences show how a burn may lead to eventual death. In the second, several orders of outcomes illustrate the process by which a corroded brake lining may result in an automobile crash.

28

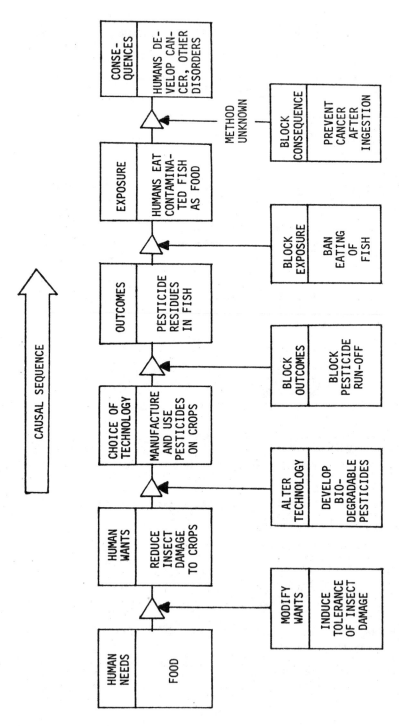

Figure 2. Hazard structure of pesticide use, illustrating the importance of upstream intervention when downstream intervention in hindered by insufficient knowledge.

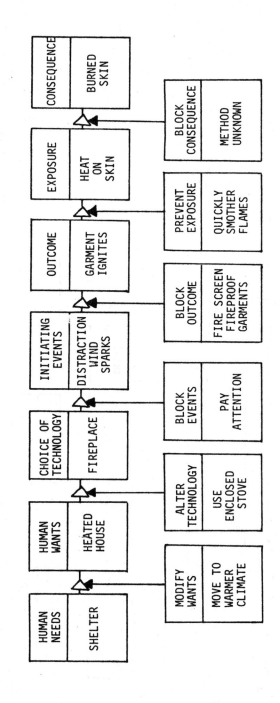

Figure 3. Seven-stage expansion of the hazard sequence, illustrated here for the case of the fireplace. Note the range of possible control interventions.

30

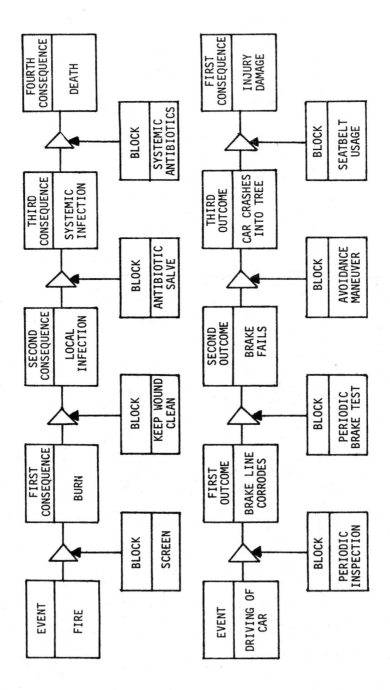

Figure 4. Expansion beyond the standard seven stages, illustrated here by the inclusion of several orders of consequences following a burn (top) and the inclusion of several orders of outcomes in an automobile accident (bottom). The variety of distinct control interventions available dictates the degree of expansion in each case.

Though the examples of Figures 1-4 involve a single causal chain, it happens quite often that a single technology generates **several** outcomes of significance. The entire cycle of coal-fuelled electric power, for example, involves release of air pollutants, coal dust, hot water, CO_2 and excessive kinetic energy which, respectively, may lead to respiratory disorders, black lung disease, damage to aquatic systems, climate change on a global scale, and a variety of injuries and fatalities. Each release involves a different causal sequence with a different set of consequences. As illustrated in Figure 5, the topology of the total hazard of "coal-fuelled electric power" resembles a pitchfork with a handle and several tines. More generally, the topology of complex hazard chains has a tree structure. In either case it is important to consider all branches and associated endpoints (consequences).

In its logic, the causal .chain of hazard that we have described is related to the partition of natural hazards into events and consequences (Burton, Kates, and White 1978) and to approaches widely used in risk assessment (Rowe 1977; Kates 1978). The causal chain may also be thought of as a simplified fault tree and as such is comparable to the methods used to analyze nuclear reactor safety (Nuclear Regulatory Commission 1975; Lewis 1978), to classify auto safety options (chapter 8), and to deal with a variety of consumer products (chapter 16).

Figures 1-5 show that the causal structure of hazards has several distinctive features:

- It focusses attention on outcomes and visualizes these as releases of energy and materials that exceed the level with which potential target organisms can cope.
- It is purposely simple, with a managerial focus designed to identify opportunities for blocking the evolution of hazard events.
- It is comprehensive and includes upstream options such as control of human wants and choice of technology.

In viewing the diagrams of Figures 2-5, it is essential to realize that for most stages of the model there are several causes similar in character to those noted in Figure 1. We have suppressed such multiple causes for two reasons. First, our main purpose--to describe where along the causal chain opportunities for control intervention reside--does not require detailed information on all possible necessary and sufficient conditions. Second, introducing all contributing multiple causes at each stage would complicate our diagrams beyond easy comprehension and is better left for such time as a full fault tree is needed. This is not to say that describing the full structure of multiple causation is unimportant. Indeed, such description will become central to the design of specific blocking actions or control interventions, which invariably involve the removal of one or more necessary conditions for a subsequent hazard stage.

The Dynamics of Hazard Control

Thus far we have indicated control points without considering the dynamics of the control process. In many cases of hazard

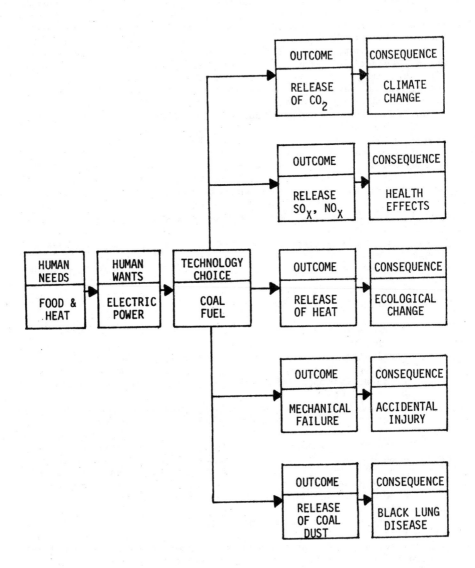

Figure 5. Illustration of the "pitchfork" topology of the hazards related to coal-fuelled electric power. There are at least five separate outcomes, each involving the release of different kind of energy or material and each leading to a distinct set of consequences.

control the simple sequence from "upstream" to "downstream" is an inappropriate description. Instead, a hazard is first recognized through an experienced release or consequence, and control action follows in time by inserting a block at appropriate upstream stages. In this sense, control intervention involves feedback: that is, information flows backward from downstream to upstream stages.

Feedback, in principle, may be either positive or negative. For reducing hazard, we desire negative feedback; that is, we seek upstream control intervention that blocks or reduces consequences. Unfortunately, hazard control has in many cases produced unintended positive feedback, or processes through which upstream control interventions increase the level of consequences.

In the field of hazard studies, one of the oldest and best-documented cases of unintended positive feedback involves the use of engineering technology to deal with the problem of flooding. Following the passage of the Flood Control Act of 1936 and subsequent amendments, the U.S. Army Corps of Engineers embarked on am ambitious program of flood dam, levee, and channel construction to protect flood plains. Not until 20 years later did research studies demonstrate that despite well engineered control interventions, flood damages were actually rising (Burton, Kates, and White 1968). In effect, the perceived safety of flood-plain location produced new settlements and overwhelmed the positive effects of less frequent flooding. The nature of the unintended feedback in the case of flood control is illustrated in Figure 6. Note that in addition to a negative or blocking action between initiating events and outcome, the control policy leads to a positive action—upstream of human wants—which intensifies wants.

In general, the unintended impacts of control actions are of two kinds: those that involve amplification of an existing hazard chain, and those that create new hazard chains. The case of engineered flood control illustrated in Figure 6, is an example of the first kind. The case of TRIS, a fire retardant that was later found to be carcinogenic, is an example of the second kind (see Figure 7).

The control loop structure may involve primary negative feedback at any of several points in the chain. Two examples taken from the field of auto safety illustrate the range. The first involves the increase in highway fatalities attributable to subsidized driver education (Figure 8), and stems from a control intervention that is largely focussed on blocking initiating events. The second involves the increase in highway fatalities due to nonuse of seatbelts (Figure 9) and is based on a control intervention that blocks consequences in the last 0.1 second before injuries are incurred. These cases are interesting from another point of view. Whereas the effects of subsidized driver education were quite unexpected, yet recently observed (Robertson and Zador 1978), the effect of seatbelt use was expected by some but found recently not to exist (Buseck 1980).

Though the principal benefit of diagramming the feedback loops of hazard control may lie in the potential for discovering **unexpected** positive feedback, our brief catalog of feedback diagrams would not be complete without noting a different but related case. There are obviously many instances where hazard control involves the **knowing acceptance** of new hazards as a price for control of an

34

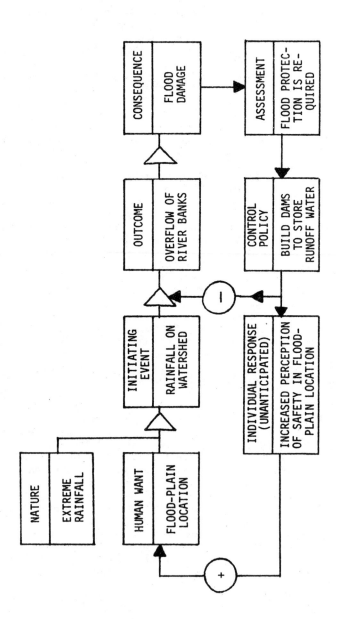

Figure 6. Feedback diagram for flood damage control. The unanticipated individual response of increased perceived safety in flood-plain location acts as a positive feedback and defeats the control policy.

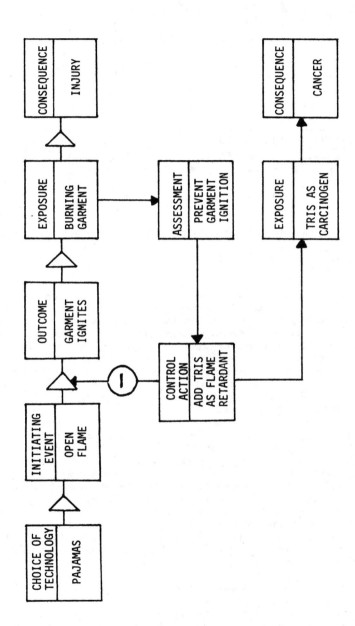

Figure 7. Feedback diagram for the use of flame retardants in children's pajamas. The control action leads to an unanticipated new hazard.

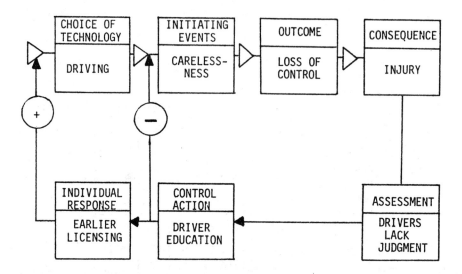

Figure 8. Feedback diagram associated with driver education. Driver education promotes earlier licensing, which increases the amount of driving by teenagers and defeats the initial control action.

existing hazard. A well-established example concerns the side effects of therapeutic drugs, illustrated in Figure 10 for the case of cancer chemotherapy. Other examples include the case of toxic waste dumps, where the risks of keeping large quantities in one place are traded for the higher risks of wide distribution; the case of non-persistent pesticides, in which short-term high-level exposure of workers is traded for long-term, low-level exposure of the general population; and the case of oral contraceptives (chapter 17) in which users trade the risks of unwanted births for a number of undesired medical side-effects.

Our analysis of the dynamics of hazard control suggests two conclusions. First, there are probably no "pure" control interventions that produce only their intended effects. Therefore, recognition of the amplification of an existing hazard or the creation of new hazards needs to be part of every control assessment. Second, it seems likely that diagramming a wide variety of control interventions will yield a small catalog of recurring types of interventions. Such a catalog could serve as a useful checklist for examining the efficacy of any proposed control intervention.

Beyond this, it is well to recognize that the present discussion abstracts a complex decision process and must therefore be applied with caution.

Applications

The causal structure model is applied at a number of points in this volume. As a guide to these applications we provide a brief description of the most important cases.

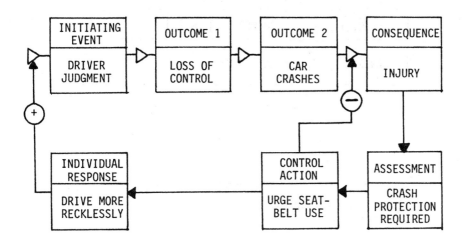

Figure 9. Hypothetical feedback diagram associated with the use of seatbelts. Research has shown that this type of feedback does not exist: i.e. belted drivers are less reckless than the average, not more reckless as indicated in the diagram.

Hazard Management and Its Limits

Chapter 3 shows that the idea of a feedback loop serves useful-ly in organizing the functions of hazard management. It also shows that the causal model may be used in the construction of "effort maps," measuring levels of regulatory actions by hazard stage. This helps raise fundamental questions about optimization of effort; for example, in considering nuclear power regulation, it prompts the question of why so much effort is spent on upstream control inter-ventions and almost none on consequence mitigation. A parallel application uses the model to conceptualize the timing of regulatory response.

The Causal Taxonomy of Hazards

In chapter 4 we utilize the causal structure model as a tem-plate, and through quantitatively expressed social, physical, and biological descriptors, applied to successive hazard stages, obtain a 12-descriptor profile for each of 93 hazards. Because they apply to all stages of hazard evolution, our descriptor profiles consider-ably extend the conventional consequence-centered definition of "risk." Through factor analysis, we show that five linearly inde-pendent composite dimensions underlie the descriptor profiles. A pilot comparison to lay perception shows that our 12 hazard descrip-tors or five factors capture a large fraction of lay people's con-cern with hazard. Perhaps the most surprising result of the work is that annual mortality, the measure most frequently used by scien-tists to quantify risk, explains only a small portion of the vari-ability of perceived risk.

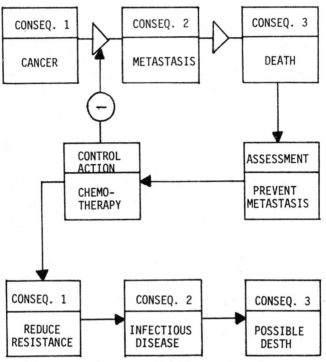

Figure 10. Feedback diagram for cancer chemotherapy. Initiating chemotherapy reduces the hazard of cancer but introduces new hazards of infectious disease. Unlike other cases where control action leads to new hazards, the hazard structure of chemotherapy is well-known to physicians, and the new hazard is accepted as the smaller of two risks.

Estimates of Consequences

By its divisions of hazards into events and consequences, the causal structure model lends itself naturally to an inventory of hazard consequences. Thus, chapters 6 and 7 give detailed estimates of mortality and economic losses, respectively. Though the separation of events and consequences, once recognized, is trivial, the tallies of mortality and economic loss obtained in chapters 6 and 7 are not and serve as our strong justification for society's continuing attention to the problem of technological hazards.

Management Options

The causal structure model provides useful descriptions of management options in the context of our case studies in Part 4. A particularly notable example is the discussion of highway and automobile safety (chapter 14), which shows that in recent years a major shift in regulatory effort has taken place. Other applications of the model occur in the discussion of PCBs (chapter 15) and the discussion of contraceptives (chapter 17).

Mapping the Full Scope of Analysis

Risk assessment, as described in chapter 11 and applied in Part 3 of this book, is only one of the family of methodologies related to hazard analysis. The causal structure model we have outlined lends itself well to mapping the full scope of hazard analysis. To diagram methods, we use double-line "forward jumpers," as shown in Figure 11. The loops created involve a forward flow of information and as such are distinct from the feedback loops used in describing control strategy. With this notation, the major methods of analysis may be summarized as follows.

Risk Assessment

This method links a specific technology design with subsequent stages of the model, including consequences. A good recent example is the **Reactor Safety Study** (Nuclear Regulatory Commission 1975), which proceeded from two specific reactor designs to event- and fault-tree analysis, and via explicit component failure proba- bilities to a range of outcomes. Each outcome involved specific radioactivity release and was assigned a specific probability. This was followed by a range of exposure models, each leading to a set of consequences.

Technology Assessment

This method is similar to risk assessment, but unlike the lat- ter focusses on several design alternatives, with associated conse- quence scenarios for each. It also deals with benefits. Typical of this genre is the assessment of the SST (U.S. Dept. of Transporta- tion 1976).

Environmental Impact Assessment

This method covers the same scope as technology assessment ex- cept that the consequence analysis is broadened to include environ- mental and social values to the fullest extent. A good example is the impact assessment of the breeder reactor (AEC 1974).

Comprehensive Assessment

Here, by movement of the origin further upstream, an effort is made to consider assumed or predicted human wants, expressed as "demand" in economic terms, or as "needs" in psychologic terms. Neither "demand" nor "need" is of course, value free. Thus, the use of "needs" for what are really wants purposefully blurs an important distinction; and the term "demand" implies an independence of other factors that is seldom found in industrial society. A good example of comprehensive assessment is the **CONAES** (Committee on Alterna- tive and Nuclear Energy Systems) **Report** (National Research Coun- cil 1980). Like the **Reactor Safety Study**, it is concerned with nuclear reactors; but the **CONAES Report** includes an assessment of undeveloped designs, other energy alternatives, and the possibil- ity of dampening human wants.

40

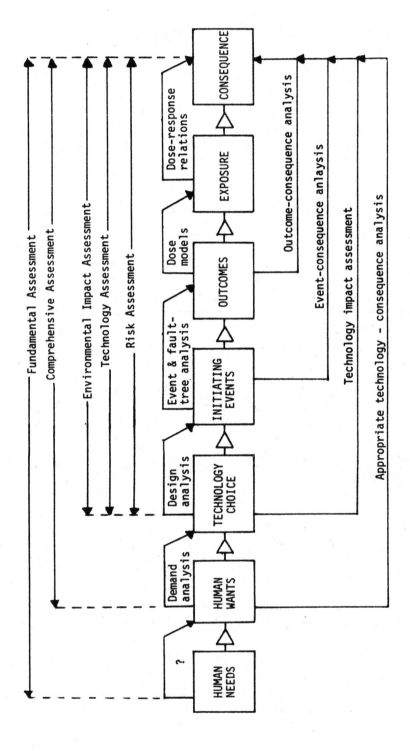

Figure 11. Modes of analysis is illustrated via the causal structure model of hazards.

Fundamental Assessment

Beyond human wants lie real, biologically determined needs. This ultimate origin of hazards must surely be recognized, though it is rarely, if ever, included in analysis. We therefore propose the term "fundamental assessment" to describe its inclusion. Interestingly, in the preparation of the CONAES Report, a primitive attempt was made to consider a "scenario" of life-style change involving a less "demanding" translation of needs into wants, but the Committee did not venture so far as to include it among the conventional "demand" scenarios. Thus, the discussion of the first link in fundamental assessment is left to the Maine Times, Mother Earth News, the Whole Earth Catalog, and the myriad of movements that argue for a simpler, less technological life. Meanwhile, a comprehensive assessment of hazards resulting from alternative life-styles remains to be made.

The lower portion of Figure 11 indicates a series of analytic methods that are largely implicit. In general, they seek only empirically derived correlations between earlier and later stages of hazard evolution. In later stages of the model, they are exemplified by actuarial statistics connecting, let us say, age- and sex-specific accident rates with particular geographical locations. In earlier stages, they fall within the purview of futurists, social critics, and philosophers who are concerned with technology evaluations that lack explicit treatment of the stages of hazard evolution. Both actuarial and implicit technology evaluation are useful in that they may alert society to potential problems and issues; but they do not suffice for design of hazard control.

Our discussion of methods of analysis in terms of the structure of hazards suggests two immediate conclusions:

- **Completeness.** The causal structure model easily accommodates all current practice of assessment and analysis, and does not require the use of new and unfamiliar terms; at the same time it provides a new sense of linkage that "puts into place" the current mélange of methods.
- **Potential for generalization.** It is likely that methods of analysis that work for one hazard may be generalized to other hazards and hazard groups with similar structure. Such generalization, in principle, will be a strong step toward a real discipline of hazard assessment.

Beyond this, it seems plausible that the more upstream the origin of analysis, the more fundamental are the derived results. For this reason, we prefer analysis that originates well upstream from the stage outcome. Our enthusiasm for such analysis is tempered only by the difficulties involved. Thus, the most reliable information is available in the area of empirically derived dose-response models, which incorporate only the last two stages of hazard; and analysis techniques that jump several stages but provide less adequate quantitative results are probably the least reliable.

Despite these reservations, we find powerful arguments for moving the origin of analysis upstream. It is becoming increasingly

clear that many traditional analyses, particularly for toxic chemicals, are fundamentally blocked by the trans-scientific nature of
the experiments required to establish reasonably explicit exposure-
consequence relations (Gori 1980).

Therefore, control analysis must move upstream or achieve nothing. In addition, from a strictly practical point of view, it is
clear that control analysis will lose its race with new hazards if
it persists in dealing with one hazard at a time (see chapter 4).

Summary and Conclusions

We have described the time development of hazards as a causal
sequence with well-defined stages and have indicated that the central stage in this sequence is the release of energy and materials.
The causal sequence, which indicates the nature of feedback control,
shows several characteristic patterns. We have shown how the causal
sequence lends itself to a classification of methods of analysis,
from narrowly focussed exposure-response relations to the most comprehensive methods.

In the next chapter we place causal structure analysis into the
larger context of hazard management.

REFERENCES

AEC (Atomic Energy Commission). 1974. Proposed final environmental
impact statement: Liquid metal fast breeder reactor program.
Washington: Atomic Energy Commission.

Burton, Ian, Robert W. Kates, and Gilbert F. White. 1968. The
human ecology of extreme geographical events. Toronto:
Department of Geography, University of Toronto.

Burton, Ian, Robert W. Kates, and Gilbert F. White. 1978. The
environment as hazard. New York: Oxford University Press.

Buseck, C. R. von, et al. 1980. Seat belt usage and risk taking in
driving behavior. SAE Paper No. 800388. Warrendale,
Pennsylvania: Society of Automotive Engineers.

Gori, Gio B. 1980. The regulation of carcinogenic hazards. Science 208:256-261.

Kates, Robert W. 1978. Risk assessment of environmental hazard.
Chichester: Wiley.

Lewis, Harold W. 1978. Risk Assessment Review Group report to the
Nuclear Regulatory Commission, NUREG/CR-0400. Washington:
Nuclear Regulatory Commission.

National Research Council. 1980. Committee on Nuclear and Alternative Energy Systems. Energy in transition. San Francisco:
Freeman.

Nuclear Regulatory Commission. 1975. Reactor safety study. WASH-
1400, NUREG 75/014. Washington: Nuclear Regulatory Commission.

Robertson, Leon S., and Paul L. Zador. 1978. Driver education and
fatal crash involvement of teenaged drivers. Washington:
Insurance Institute for Highway Safety.

Rowe, William D. 1977. An anatomy of risk. New York: Wiley.

U.S. Dept. of Transportation. 1976. The Secretary's decision on
the Concorde supersonic transport. Washington: Department of
Transportation.

3
Hazard Management

Roger E. Kasperson, Robert W. Kates,
and Christoph Hohenemser

Hazard management is the purposeful activity by which society informs itself about hazards, decides what to do about them, and implements measures to control them or mitigate their consequences. Management is not the only way society deals with hazards; people adapt to hazards biologically and culturally over the long term and hazard control and mitigation often occur as incidental byproducts of other activities.

In the United States today, society's management of technological hazards is a significant undertaking. Chapter 6 reveals that technological hazards in the United States are associated with 20-30 percent of male and 10-20 percent of female mortality. Tuller (chapter 7) estimates federal, state, and local expenditures on hazard management at $99-132 billion in 1979, with another $80-150 billion accounted for by damages and losses. Later in the volume (chapter 19), Branden Johnson shows that between 1957 and 1978 Congress passed 179 laws dealing with technological hazards. Coping with such hazards, it is evident, is a formidable managerial task.

In the discussion that follows, we focus on society's management of technological hazards. We describe the principal participants in the management process, discuss the structure of management activity, identify major problems, and indicate ways of utilizing these concepts in the analysis and praxis of hazard management.

Major Participants in Management

Who manages technological hazards? Although it is increasingly common to think of managers as regulators, regulators constitute only one of several classes of managers. In all likelihood, private individuals make the largest management effort in the United States, and industry, rather than government, undoubtedly carries the principal institutional management burden. In our view, there are five major types of hazard managers:

- **Individuals.** Historically, individuals have been the principal managers of hazards. Despite an increasing government and industry role, they are still the prime managers of hazards. And for many hazards, the individual is the most appropriate point of control in hazard management and some means of control (e.g. hazard labelling) specifically recognize this.

- **Technology sponsors.** These are either governmental agencies or private firms that develop or utilize technology to provide goods and services. Their management activity is based on the traditional assumption that technology sponsors should act with sufficient restraint to avoid endangering the public. Thus, most technologies incorporate in their designs purposeful measures to prevent or reduce hazard consequences.
- **Policy makers.** Included are not only legislators and their staffs at all levels of government, but executive branch members. The latter include standard-setting groups that may be quite autonomous.
- **Regulators.** These are officials formally charged by society with identifying and controlling hazards. Since they have customarily evolved in patchwork fashion, they differ widely in authority, resources, and legislative mandates.
- **Assessors.** Included are technical experts who increasingly support decision makers in the hazard-management process. Most prominent in the United States is the National Academy of Sciences/National Research Council, which has conducted over 250 risk assessments during the past five years. But also included are large consulting firms (e.g., Arthur D. Little, MITRE Corporation, Battelle Research Center, Stanford Research Institute), the national laboratories, and the universities.

Looking over the shoulders of the hazard managers are the self-appointed **hazard monitors.** They sound an early alert to the public, influence the agenda of policy makers and regulators, and provide a political counterbalance to the technology sponsor. Recently, as hazard monitors have increasingly become a professional lobby, the lines between officially designated hazard managers and these self-appointed monitors have become increasingly blurred. Involved are two major groups:

- **Adversarial groups.** The growth of "public-interest," environmental, and consumer groups has been one of the most remarkable changes in the American polity during this century. With increased scientific and legal capability and with a growing specialization in expertise, these groups provide a significant monitoring network of technological risks.
- **Mass media.** As principal risk communicators to policy makers and the public, the mass media play an essential role in shaping society's response to technological risk. Through selective attention, the mass media influence greatly what will be society's worry-beads, those issues that will be extensively aired and fretted over while other problems are neglected. They also constitute an early "alert" system for outbreaks of consequences or managerial failures.

Three Theoretical Perspectives

Three broad bodies of theory, each with its own strength and insight, are available for assessing the management process. These are: conflict analysis, self-preservation and expansion, and hazard control as part of society's management of technology.

Since various interests have stakes in the decisions that occur over technologies and their hazards, it is possible to conceive of the management process primarily as social conflict and to view decision making as conflict resolution. This perspective has a broad range of theory, from the materialist conceptions of Hobbes, Hegel, and Marx to modern theories of psychological aggression (Rapoport 1974). Political applications often define entities (such as industry, environmentalists), rather than individuals, as the conflicting parties and go on to assess the sources and objectives of conflict. One specific mode of analysis views conflicts as rational interactions, thereby permitting the application of bargaining and game theory to decision making; yet another sets forth theories of community decision making, elitist and pluralist.

Overlapping with the above is an approach that focusses not on the broad array of societal conflict but on those charged with management responsibility. Again, entities, or actors, must be defined, but this approach often disaggregates the entity into its component parts to understand why a manager, such as a government agency, takes a certain position. In such theories (e.g., see Blau and Meyer 1971;Crozier 1964), the management process is as much oriented to the political goals, systems of rules, organizational structure, and relationships to clients of the manager (as noted particularly by Weber and Mill) as to the responsibilities (e.g., controlling hazards) of the manager.

The third perspective, that adopted in the discussion to follow and emphasized throughout this volume, views management in a functional way, as a predominantly rational set of activities related to certain societal objectives. Regardless of the mix of motivations and sources of behavior, one can evaluate the manager in terms of performance on these objectives. This approach recognizes that the hazard control function occurs in the context of technology management as a whole, for hazard management is inextricably linked to the management of technological benefits a well as to broader societal goals.

All three perspectives must enter into a full understanding of how society responds to technological hazards. By adopting the third perspective for conceptualizing hazard management, we simply acknowledge that our results sketch one "face" of society's responses.

Managerial Activity

Hazard managers and monitors have two essential functions, **intelligence** and **control**. Intelligence provides the information needed to determine whether a problem exists, to make choices, and to assess whether success has been achieved. It is partly **prospective** in that the manager must identify and interpret hazards before the consequences are experienced and partly **retrospective** in that the effectiveness of control efforts must be evaluated. The **control** function consists of designing and implementing measures aimed to prevent, reduce, or redistribute the hazard, and/or to mitigate its consequences.

At any moment in time, the seven groups of managers and monitors are busily engaged in different aspects of intelligence and control. A large chemical company is testing the hazardousness of the thousands of chemicals it annually screens, the Consumer Product Safety Commission is monitoring accidents from consumer products as reported by 74 hospital emergency rooms, and the American Conference of Governmental Industrial Hygienists (ACGIH) is busily revising its threshold limit values (TLVs) for several hundred chemicals found in the air of factories and laboratories.

For any specific technological product or process, hazard management may be described, in simplified form, as a sequence of activity beginning with the identification of a hazard and assessment of its risk and concluding with efforts to control or mitigate the hazard (see chapter 2). To conceptualize this process, we begin with the causal structure of hazard (as outlined in chapter 2), extending from human needs and wants through choice of technology, to eventual human and biological consequences. In terms of this chain, management seeks to alter the flow of events in order to reduce or eliminate harmful consequences. The stimulus for such alteration may be multifaceted, but it generally originates in experienced or predicted events that lie downstream in the causal chain. Following a stimulus, management proceeds through a set of societal choices and actions that eventually produce a control intervention. Chapter 2 described this process as "negative feedback" and illustrated with several examples. In this chapter we go beyond the topology of feedback to inquire into the nature and content of managerial activity.

Figure 1 depicts the management process as a loop or cycle. In the center of the diagram, the structure of technological hazards is portrayed as a linked causal chain, through which the deployment of technology may cause harmful consequences for human beings and their environment, economy, and society. Four major managerial activities—**hazard assessment, control analysis, strategy selection, implementation and evaluation**—surround the chain. Each of these major activities characteristically involves normative judgements concerning scientific knowledge or social values, as illustrated, for example, by assigning priorities, judging tolerability, or allocating the risk.

This schematic diagram is, of course, an idealization and simplification of a process that in reality may not be linear. Each activity may occur in an order different than that diagrammed in Figure 1. It is not unusual, for example, for initial control actions to be instituted prior to a thorough assessment of the hazard. Nevertheless, Figure 1 provides a useful template for organizing an overview of management activity, to which we turn next.

Hazard Assessment

Hazard assessment is a least a four-step process involving hazard identification, assignment of priorities, risk estimation, and social evaluation.

Hazard Identification. Hazard managers do not like surprises. What is unacceptable, indeed downright dangerous, to the

47

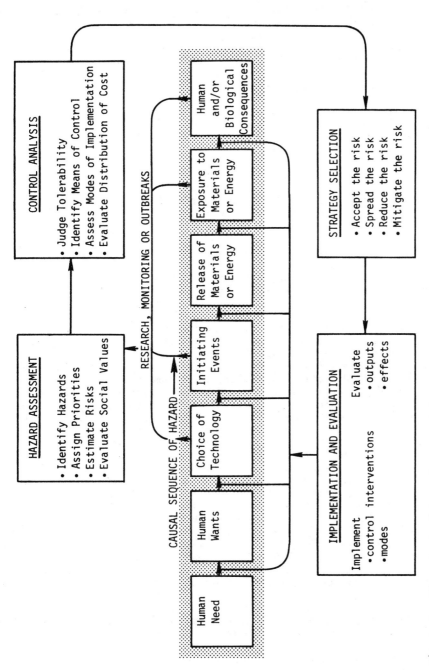

Figure 1. Flow chart of hazard management.

continued well-being of the manager is to miss the existence of a
hazard completely. Hazard managers have available to them a variety
of methods for identifying hazards, including research, engineering
analysis, screening, monitoring, and diagnosis (chapter 11). Some
of these sources of information are incidental to hazard identification: thus, new carcinogens may be found as byproducts of cancer
research, product failure mechanisms may be recognized in engineering analysis performed for other reasons, and surprise hazards
emerge as outbreaks or clusters. Other sources of information are
the product of purposeful efforts: thus, screening of chemicals
provides early warning of toxicity, and environmental monitoring of
pollutants affords estimates of potential health effects.

Despite the availability of this broad range of information and
continuous rapid improvement in scientific capability, the present
system of hazard identification is far from perfect. Entire classes
of hazards may go undetected because there is insufficient initial
suspicion to conduct the necessary analysis. On the other hand,
escalating monitoring capability is leading to information overload
in which society is confronted with more ambiguous warning than it
can comfortably **digest.** Perhaps most important, the current system of identification provides data largely on physical, health, and
ecological effects, while **neglecting** all but the most obvious
appraisals of mental health, social impacts, and political consequences.

At the same time, the scientific capability for hazard identification and measurement has improved remarkably. That progress is
Janus-faced, however, for it places ever larger demands on the whole
complicated intelligence apparatus that results. The growing capacity to identify potential hazards, in short, threatens to overwhelm
the more limited societal capability to respond.

Despite an enlarged capacity, some technological hazards escape
timely identification, of course, and become known through outbreaks
or experienced consequences. Thus, the threat of buried chemical
wastes at Love Canal was unrecognized until severe winter storms
raised the water table and injected noxious chemicals into the base-
ments of residences (Ember 1982). And the recognition that anxiety
was the most serious consequence of the Accident at Three Mile
Island came only months after the event that produced it (U.S.
President's Commission on the Accident at Three Mile Island 1979).

In such cases, vigilant monitors often act as hazard identifi-
ers, bringing such events to the attention of society and demanding action. Outbreaks are unwelcome news for both the technology
sponsor who has a substantial stake in the product and the manager
who "missed" the hazard. Despite the publicity that surrounds such
events, evidence exists that suggests that hazard identification is
becoming more rather than less effective over time and relatively
few hazards are escaping the various identification systems (Kasperson 1977;Lawless 1977).

Assigning Priorities. Hazard managers cannot, of course,
deal simultaneously with all identified hazards within their domains
of responsibility. Somehow the hazard domain must be ordered and
priorities attached to the many candidates competing for managerial
attention. There are choices to be made—choices between hazards
with better known acute consequences or poorly understood chronic
consequences, or choices between attending to serious hazards with

few available sources of control or lesser hazards with effective available means of control. The criteria for establishing priorities are laden with value considerations: Is it the aggregate risk or the distribution of risk that is more important? Should ecological risk receive lower priority than health risks? Should children enjoy a higher priority for protection than adults? Should present generations be valued higher than future generations? Inevitably, establishing priorities requires trading off some values to achieve others. Perhaps one of the most value-laden decisions a hazard manager, or monitor for that matter, makes is the initial one of what to work on.

Yet it is often political pressures rather than value conflicts that shape priorities. In its analysis of chemical regulation in the United States, a National Research Council (1975,33) study concluded rather pessimistically that "stories in the morning newspaper probably have more impact on what decisions come before the agency head than most internal agency processes of problem identification or priority-setting."

Crisis management, involving as it does case-by-case response, undermines hazard management because the domains involved are enormous in extent and heterogeneity. The Toxic Substances Control Act, for example, charges the Environmental Protection Agency (EPA) with the formidable task of screening the 70,000 or more chemical substances now in commerce and the thousand or so entering the market each year. Just keeping up would require (EPA) to rule on four new chemical applications every working day—clearly an impossible task (Culliton 1979). In fact, taken together, all federal agencies in the United States had, by the end of the decade, issued regulations to stop or reduce exposure to fewer than 30 carcinogenic substances, yet some evidence of carcinogenicity has been found for about another 400 chemicals (Carter 1979). And even many of these actions are quite incomplete. Although PCBs have recently been banned (see chapter 15), for example, there is still no comprehensive national program that deals with the 750 million pounds of PCBs already in existence.

The Consumer Product Safety Commission (CPSC), as discussed in chapter 16, provides a clear example of how things can go wrong in the absence of clear priorities. The Commission oversees annually a hazard domain that includes some 2.5 million firms, more than 10,000 products, and some 30,000 consumer deaths and 20 million consumer injuries. The CPSC compounded the problem of limited resources by failing in its early years to set clear priorities for action. By dispersing its efforts indiscriminately, the Commission produced regulations that fluctuated erratically between serious and trivial hazards. By 1981, the Commission was beginning to fashion more effective managerial approaches (especially in regard to chronic hazards), but it then encountered the antiregulatory efforts of the Reagan administration.

The message is clear. Effective hazard management requires a well-ordered risk domain. Taxonomic analysis, as described in chapter 4, can assist in that process of ordering. But since creating such a structure is intrinsically normative as well as scientific, it should be rationalized openly and in consultation with the various interested parties.

Estimation and Social Evaluation. Once a hazard has been identified, the next steps in assessment are: (1) to estimate and characterize scientifically the probabilities of specific events and related consequences and (2) to evaluate this characterization in social terms. In the view of many professional managers the steps are separate and distinct. These managers look to scientific experts to estimate the risk of death or injury, while searching their consciences or deferring to the political process for social valuation.

Unfortunately, there is no simple relation between scientifically estimated risks of death and injury and the social valuation of a given hazard. In fact, much to the chagrin of many scientific risk analysts, the public apparently does not respond equivalently to equal threats of mortality. A high mortality hazard such as an auto accident provokes relatively little fear, whereas some low mortality hazards (e.g., botulism or nuclear power) evoke great anxiety. A great deal of confusion and conflict in hazard management arises from this conundrum.

As shown by Slovic and colleagues in chapter 5, understanding the quixotic nature of the social valuation of hazards requires consideration of risk attributes other than mortality levels. Using cognitive data obtained from several lay groups, these authors have shown that the degree of dread (also termed the **perceived risk**) that people report vis-à-vis a given hazard includes such attributes as whether the hazard is voluntary or involuntary, new or old, prompt or delayed, or kills many or few at a time. Similarly, in a multivariate taxonomy of hazards based on characteristics of the causal sequence (see chapter 4), an appreciable fraction of the perceived risk measured by Slovic and collaborators can be explained only if a wide range of physical and categorical characteristics are included. In short, explaining the risk perceived by lay groups through mortality alone fails no matter how one approaches the problem.

The implications of these findings for management are far-reaching; unless scientific risk assessment embraces the full range of consequences that enter into the public response to hazards, the conflict between the scientific analysis of hazards and their social valuation will certainly continue.

Beyond the question of how best to characterize hazards, assessment presents other serious problems for the management process. There is the recurrent need to attend to secondary and tertiary effects, or, equivalently, to define the full range of possible consequences. A nuclear accident, for example, may produce fatalities, injuries, property damage, and a high level of anxiety. But it can also lead to regulatory change that results in subsequent shutdown of all similar plants, producing disruption, further anxiety, and possible power outages. Such secondary consequences are rarely predicted and infrequently analyzed, even though in many cases they are the most important consequences of a particular event (as at Three Mile Island).

Finally, hazard managers must deal with uncertainty which confounds nearly all estimates of hazard characteristics. Uncertainty arises because characterizing hazards involves extrapolation. Some hazard characteristics may be extrapolated from previous human experience; others require extrapolation from experience with animals;

others from analogous events or technologies, and still others may only be calculated theoretically, without direct basis in experience. In some cases, such as the nuclear reactor accident risk, the level of uncertainty is so great the even the best risk assessment, such as the **Reactor Safety Study** (Nuclear Regulatory Commission 1975), fails to provide an adequate basis for regulation.

Control Analysis

Following the risk assessment, control analysis judges the tolerability of the risk and rationalizes the effort that is made in preventing, reducing, and mitigating a hazard.

Judging Tolerability. The key link between hazard assessment and subsequent initiation of control actions is the judgment of whether a hazard is tolerable or not. One of the most perplexing issues facing hazard managers, it has been mislabeled the **acceptable risk** issue (Kasperson and Kasperson 1983). It is unlikely that any risk is "acceptable" if it is unaccompanied by benefits or is susceptible to easy reduction. Acceptability, in its strict dictionary meaning, bespeaks consent and this is seldom realized. Many risks are imposed upon individuals, often without warning or information. Such risks are better thought of as "tolerated;" they are suffered in practice, not accepted.

Fischhoff, Slovic, and Lichtenstein (chapter 12) discuss four methods of judging the tolerability of hazards: **cost/benefit analysis, revealed preferences, expressed preferences,** and **natural standards.** According to these methods a technology is deemed safe or tolerable if, respectively, its benefits outweigh its costs; if its risk are no greater than those of historically tolerated technologies of equivalent benefit; if people, when asked, say the risks are tolerable; and if its risks do not exceed those fixed by nature through the process of evolution. These methods of tolerable risk judgments are often in conflict, particularly the results of risk/benefit analysis and revealed preferences on the one hand and expressed preferences on the other.

In the hope of refocussing the debate on the level of risk rather than on unresolvable questions of values, a number of risk analysts (e.g., Okrent 1980; Starr and Whipple 1980; Deisler 1982) have called for quantitative risk standards. Yet clearly this problem of the hazard manager remains unsolved, for there is no adequate synthesis of the many ways of looking at the problem. If anything, a full accounting of the approaches to tolerable risk adds further complexity.

Table 1 provides a broader categorization of methods of determining tolerable risk. The three groups classify tolerable risk judgments according to whether they rely on historical experience, direct expression, or formal analysis. This division, includes **legal precedents** and **incremental decision making** as methods dependent upon historical experiences; **expressed judgments** of professionals, decisionmakers, and interest groups as approaches involving direct expression; and risk **comparisons** and **decision analysis** as methods involving formal analysis.

Whichever methods are used, the judging of risk tolerability employs basically four types of criteria. In the first, **risk aversion,** any level of risk is considered intolerable, either

TABLE 1
Approaches to determining tolerable risk

TYPE OF METHOD	DESCRIPTION
METHODS INVOLVING HISTORICAL EXPERIENCE	
Legal precedents	Judgments are guided by existing legislation and court decisions
Incremental decisions	Judgments are made in small increments following the pattern of earlier incremental decisions.
Revealed preferences	Risks are deemed tolerable if they are comparable to the risks of established technologies with comparable benefits.
METHODS INVOLVING DIRECT EXPRESSION	
By professionals	Judgments are made by professionals or groups of professionals with expert knowledge (e.g., doctors).
By decision makers	Judgments are made by public officials, technology managers, and others with responsibility (e.g., the commissioners of the Nuclear Regulatory Commission).
By interest groups	Judgments are expressed by groups representing a well-defined interest relative to the hazard (e.g., the National Rifle Association).
By lay persons	Judgments are expressed by laity through voting or survey instruments.
METHODS INVOLVING FORMAL ANALYSIS	
Risk comparisons	Tolerability is judged by comparing risks to standards, such as publicly agreed on quantitative risk standards, or "natural background."
Risk/benefit analysis	Tolerability is determined after comparing risks and benefits in commensurate units.
Decision analysis	Risk decisions are made after formal disaggregation into a sequence of choices.

because of the nature of the product, its use, or its consequences. Thus we ban biological weapons, the use of chlorofluorocarbon aerosols, thalidomide, DDT, food additives that exhibit carcinogenicity in animals, or a government-sponsored construction project that endangers an entire species. Risk aversion is drastic and is applied sparingly.

The other criteria all involve some type of **comparison**: of risks, of ways of reducing risks, of risks and benefits. By the second criterion, the projected risk level is **compared to other prevalent risks** (often with the assumption that risks should be balanced). Typically the comparisons are with natural background levels (or some fraction thereof), with similar technologies, with other risk stages of a fuel or production cycle, or with risks previously determined to be tolerable by a given risk manager. In the British chemical industry, for example, if a particular activity contributes more than 4 fatalities to the fatal accident frequency rate (FAFR)--the number of fatal accidents in a group of 1,000 men in a working lifetime (100 million hours)--risk reduction is undertaken (Kletz 1977). Several well-known, and oft-criticized, sets of comparisons are those of Wilson (1979), Cohen and Lee (1979), and the **Reactor Safety Study** (Nuclear Regulatory Commission 1975).

The third criterion focuses on the **cost-effectiveness of risk reduction**. The question at stake is how much society wishes to spend to avoid a particular consequence. It is well known that such expenditures vary widely. In Britain $2,000 was spent in 1972 to save an employee's life in agriculture, $200,000 in steel handling and $5 million in the pharmaceutical industry (Sinclair, Marstrand, and Newick 1972). Chapter 14 compares controls for reducing auto accidents in terms of average investment per fatality forestalled. Costs range from $500 for enforcing mandatory seat belt usage to $7.6 million for road alignment and gradient change.

Perhaps the most widely approved criterion is some mode of **comparison between risks and benefits**. This method recognizes as necessary some level of risk above zero and balances the benefits of the activity or technology against the risk to determine how much risk reduction should be undertaken. Benefit/risk analysis is essentially a subset of benefit/cost analysis, since risks are a component, often the principal one, of social costs. The quality of such analyses depends on such factors as the messiness of the problem, the skill of the analyst, the way in which the analytic question is posed, the existence of appropriate techniques, and the analyst's ability to fashion new ones (Fischhoff 1979).

None of the criteria treated above deals adequately with equity issues in hazard management. Characteristically those who enjoy the benefits of a technology or product are not the same as those who experience the risks. Risks are also seldom distributed evenly throughout society nor are they always confined to the present generation. Attempts to control risks often benefit groups different from those who pay for the controls.

Three major types of inequity, in our view, require analysis (Kasperson 1983). First is **inequity among social groups**. The adverse side-effects of technology are often concentrated in weak and powerless people. Nowhere is this more apparent than in the workplace where higher exposure standards are tolerated than those which protect the public generally (Derr et al. 1981). Second is the **inequity among regions.** This problem is apparent in the political controversy surrounding the location of noxious facilities, such as airports, prisons, chemical waste dumps, and dog tracks. Finally, there is **inequity among generations.** Increasingly there is concern over the risks that may be exported to future generations, particularly if the effects are irreversible.

Few hazards are judged tolerable. How much effort should be expended to control them is determined partly by the value judgment of relative tolerability and partly by the means of control available for preventing, reducing, or mitigating the hazard.

Identifying Means of Control. For identifying means of control, the causal structure of the hazard becomes the central concern. For many cases the use of simple causal chains (chapter 2) suffices. Complex cases require a full fault- or event-tree analysis. Whether simple or complex, an analysis of causal structure must be broadly based and include potential control actions that span the range from human needs and wants to exposure, consequences, and mitigation of consequences. To this end, it is useful to contemplate the control structure for the traditional technological hazard, the simple fireplace, discussed in chapter 2 and represented in Figure 2. Generalizing from Figure 2, within the seven-stage model of causal structure, the potential control actions include: (1) modify wants; (2) choose alternative technology; (3) prevent initiating events; (4) prevent releases; (5) restrict exposure; (6) block consequences; and (7) mitigate consequences.

In contemplating a given control intervention, it is important to recognize its potential to create new hazards. Chapter 2, under the general heading of positive feedback, describes a variety of examples, which demonstrate that ill-conceived hazard control can make things worse and surely accounts for some of the current skepticism toward government regulation of risk.

Modes of Implementation. In addition to identifying the technical means of control, each control action can be implemented in a number of different ways. There are, in our view, three major modes of inducing society to undertake the control action (Table 2). Society can: (1) **mandate** the action by law, administrative regulation, or court order and thereby ban or regulate the product or its use or distribution; (2) **encourage** the action through persuasion or by providing incentives, penalties, or insurance; or (3) **inform** those creating or suffering risk, allowing them voluntarily to reduce or tolerate the hazard. At any moment all of these modes may be utilized in connection with a specific hazard. Thus, as described in chapter 14, the hazard of driving is controllable by 37 different "highway safety countermeasures" and a comparable number of vehicle safety standards, each involving one or more stages of the causal structure of the hazard. For any given case, a lively ideological debate may erupt over which implementation modes are the most desirable.

Cost Analysis. An important aspect of control analysis is to inquire into the relative cost of control interventions (be they technical, behavioral, or informational) and modes of implementation. Known as **cost-effectiveness analysis,** this approach permits the hazard manager to select the most efficient actions available (Schwing 1979). Cost-effectiveness varies widely, both between actions employed for different hazards and for different actions

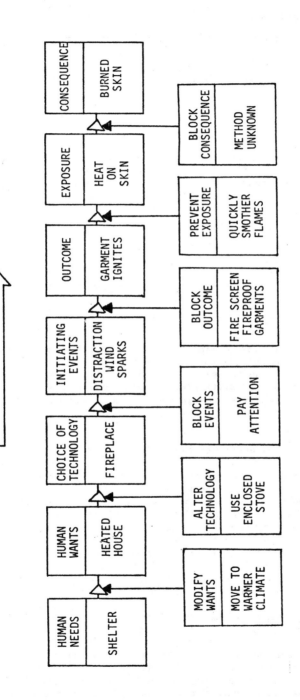

Figure 2. Seven-stage expansion of the hazard sequence, illustrated here for the case of the fireplace. Note the range of possible control interventions.

56

TABLE 2
Modes of implementation

MANDATE

　　Ban the product or process
　　Regulate the product or process (e.g., performance and design
　　standards; use and dissemination restrictions)

ENCOURAGE

　　Seek voluntary compliance
　　Provide incentives (e.g., credits or subsidies)
　　Penalize through indemnifying those harmed
　　• via the market (wages)
　　• via the courts (award damages)
　　• via transfer payments (taxes)
　　Provide insurance

INFORM

　　Inform hazard-makers (by monitoring and screening)
　　Inform those at risk (e.g., by labeling, advertising campaigns)

---

applied to the same hazard. Wilson (1975) for example, has estimated that the United States in 1975 expended $1,000 for avoiding a
death in the liquefied natural gas industry, as compared to $750,000
for nuclear power. Similarly, as noted earlier, alternative control
actions for highway safety in the United States had costs per fatality forestalled ranging from $500 to $7.6 million in 1976 (chapter
14).

Selecting a Management Strategy

　　Equipped with a hazard assessment and a control analysis, and
assuming the risk has been judged intolerable and thus requiring
action, the manager next must designate a hazard management strategy, consisting of an overall management goal and a package of control measures designed to achieve the goal. The control package
will specifically include both control interventions (oriented
toward the causal structure of hazards) and modes of implementation
(oriented toward alternative institutional means for control). Four
possible management goals—**risk acceptance, risk spreading, risk
reduction,** and **risk mitigation**—can drive management strategy.
　　Risk acceptance may be achieved by providing compensation,
as through higher wages for riskier work, or by seeking informed
consent, as in informational or warning labels on hazardous products. The purpose of the latter is to make the risk more voluntary
by enlarging information related to technology choice. It applies
to a broad range of hazards, including hazards with very large numbers of associated fatalities. Thus, risk acceptance is the basic

strategy (however imperfect) for society's effort to manage the 300,000 cigarette-related deaths each year.

Risk spreading seeks to transform a maldistribution of risk into a more equitable one, through redistribution of the risk over social groups, regions, or generations. The new distribution may also seek to equalize experienced risk, to make risk concordant with benefits or with the ability to bear risk. An interesting example of risk spreading is the introduction of tall stacks to transform a local pollution problem into a regional one. Here, by all appearances, the principal regional hazard, acid rain, was initially unsuspected, and only at a later time became recognized as a serious ecological and health threat.

Risk reduction, in contrast to spreading and acceptance of risk, involves decisive intervention in the causal sequence of hazards. It is therefore a step that may in some circumstances curtail the benefits of technology. An extreme case of reduction is aversion, as exemplified by the total ban of a technology. Whereas risk reduction is widely regarded the dominant mode of risk management, aversion has been practiced only in a few cases, as in the banning of the domestic uses of DDT and carcinogenic food additives.

Risk mitigation includes a variety of ways of modifying hazard consequences once they have occurred. Typical actions include disaster relief, medical intervention, family assistance, and compensation through insurance or other means. Risk mitigation is often an initial societal response when risks have not been anticipated or when the causal chain of the hazard is poorly understood, as exemplified by the thousands of court cases now pending against asbestos manufacturers.

Although presented here as an integrated managerial approach, strategies usually develop piecemeal, frequently lack internal logic, and may appear only through trial and error over time. However they develop, they must eventually strike a balance between reaping technological benefits and reducing unwanted risks. Control actions will range along the causal chains of hazards, reflecting optimal points of intervention. Thus, for control of cigarette smoking, an addictive activity that society regards as tolerable, upstream intervention (banning the manufacture and/or sale of cigarettes) would be effective, yet unacceptable. In contrast, for preventing the disruption of the ozone layer, a feared and intolerable consequence caused in part by a minor technology (aerosol cans), upstream intervention based on banning is both reasonable and acceptable. For most hazards, midstream and downstream strategies of intervention are appropriate. They interfere less with benefits and can be directed at specific targets. Typical of such interventions are the use of filters on cigarettes and the wearing of seatbelts in cars. Neither strongly affects the benefits, and each reduces hazard consequences. When the causal structure is poorly understood and unpredictable, society by necessity concentrates on mitigating consequences. An example of this is the case of environmentally caused cancer, where both agents and mechanisms are to a large extent unknown, if not unknowable.

A **mature** hazard management strategy is one which over time steers an optimal path between realizing technological benefits and reducing unwanted risks. It will normally employ a complex set of interventions along the hazard chain and utilize a variety of managerial modes of implementation. Such a system evolves partly through improved knowledge and partly through trial and error.

This concludes the sequence diagrammed in Figure 1. But effective hazard management follows up these control efforts with determined implementation and retrospective evaluation. Such evaluation includes monitoring of the control actions for their effectiveness and vigilance for unexpected surprise impacts that may occur.

Implementation and Evaluation

Implementation. The sequential flow of hazard management, in the idealized form presented herein, concludes with the implementation of management strategies, the "packages" of selected control actions and implementation modes designed to advance a designated management goal. As indicated in Figure 1, each of these is intended to block or modify one of the pathway links that govern the evolution of a hazard.

Implementation is a crucial and oft-neglected stage of hazard management. The lengthy review by the National Research Council (1977,36-41) of the performance of the U.S. Environmental Protection Agency indicates why control actions often fail at the implementation stage. First, administrative resources are often inadequate, particularly in a decentralized system where lower administrative levels face large enforcement burdens. Thus, states often lack the necessary technical and financial resources for monitoring and testing pollution or even issuing permits. Second, as suggested in our initial discussion of major theoretical perspectives, those charged with implementing health and safety control actions are often reluctant to do so because implementation conflicts with their own organizational and political interests. Third, hazard management strategies always contain implicit notions as to how hazard makers can be induced to take control measures. If the assumptions as to inducements are incorrect (as occurred in delay of water pollution control efforts), implementation fails. Finally, where managers lack monitoring and surveillance resources in their intelligence function, implementation becomes dependent on reports furnished by hazard makers and compliance becomes unreliable.

A number of these problems are evident in the control of PCBs, as discussed in chapter 15. Three years after the passage of PCB control regulations, the EPA inspection program may be missing as much as 80 percent of PCB facilities and the inspection priorities program may be missing major users because the program lacks resources and sufficient regional sensitivity (U.S. General Accounting Office 1981). Further, the penalties for noncompliance have lacked adequate deterrent value and have failed to produce widespread voluntary compliance. Finally, EPA oversight and informational systems have been inadequate to target enforcement priorities. Quite similar problems pervade the efforts of the Occupational Safety and Health Administration (OSHA) to implement its enabling legislation, specifically to effect compliance with occupational health and safety regulations (Mendeloff 1980,151-167).

Evaluation. Hazard management is not complete even when controls are implemented. Effective management requires continued monitoring of control effects reviewing the adequacy of control intervention in light of evolving knowledge and checking for the creation of new hazards.

Following Levy, Meltsner, and Wildavsky (1974), we recognize two classes of results of management actions: **outputs** and **effects.*** The theory of public policy has viewed outputs as the goods and services produced by government. **Outputs** in our usage refer to the concrete results of management efforts; thus they comprise the various interventions into the causal sequence of hazard and the associated modes of implementation. **Effects**, by contrast, are the "so-whats" of hazard management. They refer to the results, wrought by these actions, as determined by the application of social values. Put another way, effects are the consequences of outputs, the overall impacts upon society's experience of the hazard. Output analysis, then, is primarily descriptive and empirical, whereas effect analysis is primarily normative.

Output Analysis

Using the causal chain of hazard, it is possible to map the distribution of managerial effort and thereby to evaluate the breadth and timing of control actions. A level-of-effort map requires output indicators, such as work-force or budget allocations, the number of regulatory standards issued, or, as in chapter 19, the number of laws enacted. Effort maps illuminate the differences between theory and practice, the imbalance between upstream and downstream control interventions, and the change of effort by hazard stage over time. Effort maps lead naturally to a number of evaluative questions: Is the distribution of effort appropriate for the physical nature of the hazard, the perception of managers, the mandate of history, and the evolving understanding of the hazard?

To illustrate, we show in Figure 3 the distribution of regulatory guides, issued through 1975 by the Nuclear Regulatory Commission, on the question of reactor safety. Of the 95 guides issued, 63 focussed on initiating events, 29 on outcomes, 3 on consequences, and none on mitigation. Since Three Mile Island, consequence mitigation (e.g., emergency-response plans) has belatedly become a major priority. Via our level-of-effort map, we were able to recognize this need in 1976 (see chapter 10). A similar management-effort map, shown in Figure 4 (top), which categorizes highway safety standards issued by the U.S. Department of Transportation (see chapter 14), shows that 81 percent of the standards fall into the class of blocking initiating events, with little activity downstream. In contrast, a landmark highway safety report (U.S. Dept. of Transportation 1976) places 40 percent of potential activity downstream from initiating events (Figure 4, bottom).

It is also possible to construct effort maps as a function of time. In the case of auto-safety management (Figure 5), except for medical care administered to crash victims, the dominant early modes of management occurred far "upstream" in the causal sequence of

---

*Levy, Meltsner, and Wildavsky speak of **outputs** and **outcomes.** To avoid confusion with our usage of **outcomes** in the causal model (chapter 2), we substitute the term **effects.**

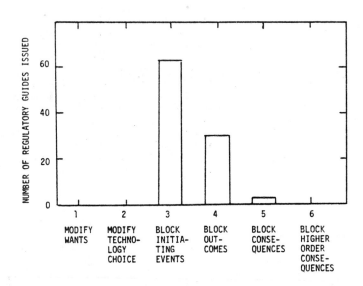

Figure 3. Number of Regulatory Guides by hazard stage, as issued by the Nuclear Regulatory Commission through 1975. Note the imbalance of regulation affecting upstream location and the near lack of regulation addressed to downstream stages.

hazard. The attempt to block injuries (that is, first-order consequences) once crashes have occurred is a rather recent development. In the case of Minamata disease (Figure 6) a diametrically opposite response pattern emerges. Control strategy begins downstream and in time moves steadily upward, leading finally to the elimination of the technology, or what is equivalent, its transfer to Thailand (chapter 9).

Recurring errors complicate the evaluation of hazard management. Characteristically, regulators overestimate the efficiency of their control actions, whereas technology sponsors often overstate the costs of those actions. As chapter 16 shows, safer products often become better products, and costs that are initially perceived as extremely high can be readily absorbed by design or engineering ingenuity. Control actions also produce **leveraged benefits,** those positive side-effects associated with industry's innovations to reduce the risks of products or the production process (Ashford 1980). Postaudits of the cost of control actions are equally needed.

Effects Analysis

Finally, there must be an overall social evaluation of what hazard management has accomplished, an assessment of the broad consequences of outputs. This involves the application of social

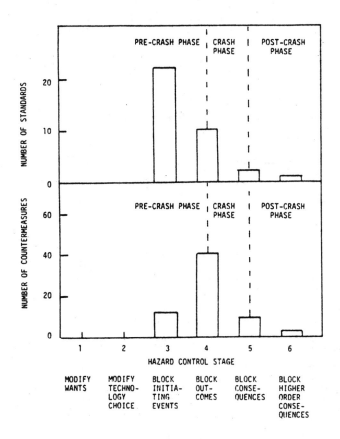

Figure 4. **Top:** Highway safety standards issued by the Department of Transportation, plotted by hazard control stage. **Bottom:** Highway safety countermeasures envisioned in the 1976 Highway Safety Needs Report. The distinctive shift toward more downstream intervention in the latter is noteworthy.

criteria to determine whether managerial "success" has been achieved. Broadly defined, success is the skillful steering between the enlargement of technological benefit and the minimization of technological harm. We propose four criteria for evaluating how skillful the steering has been.

First, the management actions must be **effective:** that is, if a product is deemed to be unsafe, how much risk reduction actually occurs? Such effectiveness, of course, requires sound performance in the various stages of both intelligence and control functions. In particular, a full analysis of the causal chain and prospective feedback is required as well as the determined implementation of control actions chosen in the face of social conflict.

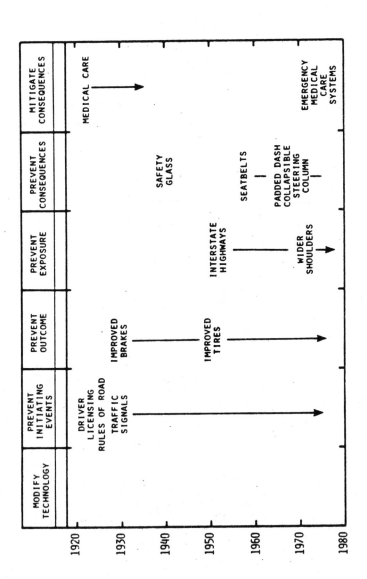

Figure 5. Chronological distribution of hazard control intervention by hazard stage for the case of automobile safety.

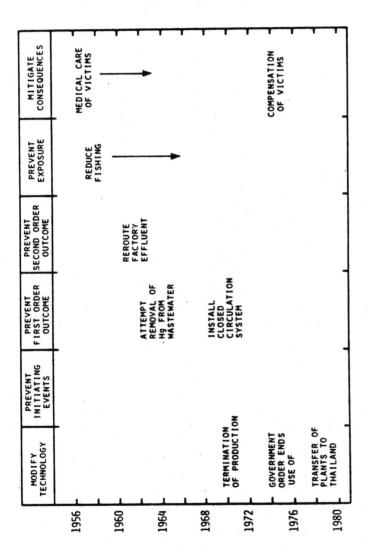

Figure 6. Chronological distribution of hazard control intervention by hazard stage for the case of Minamata disease.

Such risk reduction, at least for certain types of risk, should be amenable to quantitative statement.

Second, management must be **efficient**. There are two ways of viewing such efficiency, and both should be employed. One measure treats the simultaneous juggling of technological benefit and hazard. Clearly society does not seek risk reduction at any price and, as the energy/environment conflict makes clear, reasonable actions are needed that consider benefit and risk in tandem. The other measure is cost-effectiveness in risk reduction opportunity. Given a commitment to reduce risk, the most efficient measures, as measured in sociopolitical as well as economic terms, should be employed.

Third, management must be **timely.** Clearly, society's expectation, however unreasonable, is that those charged with responsibility for identifying and responding to hazards will do so promptly. Indeed, it is evident that managers, however large their domains of responsibility, are expected to identify prospective hazards and to take action before the hazard grows. As noted above, both previous research and our ongoing studies suggest that, despite public alarm over the hazard-of-the-week syndrome, society's hazard management appears to be improving over time on this criterion.

Finally, hazard management must be **equitable.** Unfortunately, although this criterion finds wide appeal, its application presents a formidable challenge often difficult to ascertain, because distributions of impacts are difficult to determine, and principles of social justice frequently conflict. Some forms of inequity—such as the more permissive standard for risk in the workplace—either remain quite concealed from the public view or are apparently tolerated. Only where equity issues attract social attention, as in compensation for black lung disease or exposure to atomic weapons testing, does action to reduce inequity usually occur. Management strategies that decrease inequities, however, are often deemed preferable.

Summary and Conclusion

The foregoing discussion provides a framework for analysis of society's management of technological hazards. Hazard management engages at least seven classes of major participants who make, influence, or match decisions. Our model of hazard management began with the causal chain of hazard and defined steps in the sequence of management, running from hazard assessment to control analysis, selection of management strategy, and implementation and evaluation. Throughout this management process (Figure 1), value-laden considerations, such as setting priorities, judging tolerable risk, and framing management goals, were important ingredients.

Of the participants responsible for hazard management, we know least about technology sponsors, little about individual citizens as hazard managers, and a great deal comparatively about public managers who are regulators and scientific risk assessors. Within the sequence of management activities we know most about hazard assessment, less about control analysis, little about how management strategies are formulated, and least about implementation. Meanwhile, comparative evaluation of managerial outputs and effects is just beginning.

REFERENCES

Ashford, Nicholas A. 1980. The limits of cost-benefit analysis in regulatory decisions. Technology Review 82 (May):70-72.

Blau, Peter M., and Marshall W. Meyer. 1971. Bureaucracy in modern society. 2d ed. New York: Random House.

Carter, Luther J. 1979. Yearly report on carcinogens could be a potent weapon in the war on cancer. Science 203:525-528.

Cohen, Bernard L., and I-Sing Lee. 1978. A catalog of risks. Health Physics 36 (June):707-722.

Crozier, Michel. 1964. The bureaucratic phenomenon. Chicago: University of Chicago Press.

Culliton, Barbara. 1979. Toxic substances legislation: How well are laws being implemented? Science 201:1198-1199.

Deisler, Paul F. 1982. Dealing with industrial health risks: A step-wise, goal-oriented concept. In Risk in the technological society, ed. Christoph Hohenemser and Jeanne X. Kasperson, 241-258. AAAS Selected Symposium, 65. Boulder, Colo.: Westview Press.

Derr, Patrick, Robert Goble, Roger E. Kasperson, and Robert W. Kates. 1981. Worker/public protection: The double standard. Environment 23 no. 7 (September):6-15, 31-36.

Ember, Lois R. 1982. Uncertain science, politics, and law. In Risk in the Technological Society, ed. Christoph Hohenemser and Jeanne X. Kasperson, 77-102. AAAS Selected Symposium, 65. Boulder, Colo.: Westview Press.

Fischhoff, Baruch. 1979. Behavioral aspects of cost-benefit analysis. In Energy risk management, ed. Gordon T. Goodman and William D. Rowe, 269-283. London: Academic Press.

Kasperson, Roger E. 1977. Societal management of technological hazards. In Managing technological hazards: Research needs and opportunities, ed. Robert W. Kates, 49-80. Program on Technology, Environment, and Man, Monograph 25; Boulder: Institute of Behavioral Science, University of Colorado.

Kasperson, Roger E., ed. 1983. Equity issues in radioactive waste management. Cambridge, Mass.: Oelgeschlager, Gunn and Hain.

Kasperson, Roger E., and Jeanne X. Kasperson. 1983. Determining the acceptability of risks: Ethical and policy issues. In Risk: A symposium on the assessment and perception of risk to human health in Canada, October 18 and 19, 1982, Proceedings, ed. J. T. Rogers and D. V. Bates, 135-155. Ottawa: Royal Society of Canada.

Kletz, Trevor. 1977. The risk equation: What risks should we run? New Scientist 74 (12 May):320-322.

Lawless, Edward W. 1977. Technology and social shock. New Brunswick, N.J.: Rutgers University Press.

Levy, Frank S., Arnold J. Meltsner, and Aaron Wildavsky. 1974. Urban outcomes: Schools, streets, and libraries. Berkeley and Los Angeles: University of California Press.

Mendeloff, John. 1980. Regulating safety: An economic and political analysis of occupational safety and health policy. Cambridge, Mass.: MIT Press.

National Research Council. 1975. Committee on Principles of Decision Making for Regulating Chemicals in the Environment.

Decision making for regulating chemicals in the environment. Washington: National Academy of Sciences.

National Research Council. 1977. Committee on Environmental Decision Making. Decision making in the Environmental Protection Agency. Analytical Studies for the U.S. Environmental Protection Agency, vol. 2. Washington: National Academy of Sciences.

Nuclear Regulatory Commission. 1975. Reactor safety study. WASH 1400, NUREG 75/014. Washington: The Commission.

Okrent, David. 1980. Comment on societal risk. Science 208:372-375. Note: A slightly revised version of this article appears in Risk in the technological society, ed. Christoph Hohenemser and Jeanne X. Kasperson, 203-215. AAAS Selected Symposium, 65. Boulder, Colo.: Westview Press, 1982.

Rapoport, Anatol. 1974. Conflict in man-made environment. Harmondsworth, England: Penguin Books.

Schwing, Richard C. 1979. Longevity benefits and costs of reducing various risks. Technological Forecasting and Social Change 13:333-345. Note: A slightly revised version of this article appears in Risk in the technological society, ed. Christoph Hohenemser and Jeanne X. Kasperson, 259-280. AAAS Selected Symposium, 65. Boulder, Colo.: Westview Press, 1982.

Sinclair, Craig, Pauline Marstrand, and Pamela Newick. 1972. Innovation and human risk: The evaluation of human life and safety in relation to technical change. London: Centre for the Study of Industrial Innovation.

Starr, Chauncey, and Chris Whipple. 1980. Risk of risk decisions. Science 208:1114-1119. Note: A slightly revised version of this article appears in Risk in the technological society, ed. Christoph Hohenemser and Jeanne X. Kasperson, 217-239. AAAS Selected Symposium, 65. Boulder, Colo.: Westview Press, 1982.

U.S. Dept. of Transportation. 1976. The national highway safety needs report. Washington: The Department.

U.S. General Accounting Office. 1981. EPA slow in controlling PCBs. CED-82-21. Washington: GAO.

U.S. President's Commission on the Accident at Three Mile Island. 1979. The need for change: The legacy of TMI. Washington: Government Printing Office.

Wilson, Richard. 1975. The costs of safety. New Scientist 68: 274-275.

Wilson, Richard. 1979. Analyzing the daily risks of life. Technology Review 81 (February):41-46.

4
A Causal Taxonomy[1]

Christoph Hohenemser,
Robert W. Kates, and Paul Slovic

Despite the burden imposed by technological hazards and the broad regulatory effort devoted to their control, few studies have compared the nature of technological hazards in terms of generic characteristics. Existing studies are limited to case studies (Lawless 1977), comparative risk assessments of alternative technologies (Inhaber 1979;National Research Council 1980), lists of comparable hazards (Wilson 1979;Cohen and Lee 1979), and comparative costs of reducing loss (U.S. Dept. of Transportation 1976;Schwing 1979;Lave 1981).

A first step in ordering the domain of hazards should be classification. Today technological hazards are classified by the technology source (automotive emissions), use (medical x-rays), potentially harmful ·events (explosions), exposed populations (asbestos workers), environmental pathways (air pollution), or varied consequences (cancer, property damage). A single scheme is chosen, often as a function of historical or professional choice and regulatory organizations, although a given technology usually falls into several categories. For example, a specific chemical may be a toxic substance, a consumer product, an air or land pollutant, a threat to worker health, or a prescription drug. Indeed, a major recent achievement has been the cross-listing of several of these domains of hazardous substances by their environmental pathways (Greenwood, Kingsbury, and Cleland 1979).

In this chapter, we identify common differentiating characteristics of the domain of technological hazards in order to simplify hazard analysis and management. We conceptualize technological hazards as involving potentially harmful releases of energy or materials; characterize the stages of hazard causation via 12 physical biological, and social descriptors expressed on quantitative scales; score 93 technological hazards on these scales and analyze their correlative structure; and consider the implications of hazard structure for understanding hazards, their perception, and their management. The following is a highly condensed account of our detailed analysis (Hohenemser et al. 1983).

Measures of Hazardousness

We distinguish between **hazard** and **risk.** We define **hazards** as threats to humans and what they value and we define **risks** as quantitative measures of hazard consequences expressed as conditional probabilities of experiencing harm. Thus, we think of automobile driving as a hazard but say that the average American's lifetime risk of dying in an automobile crash is 2-3 percent of all ways of dying.

As already explained in chapter 2, we describe hazards as a sequence of causally connected events that lead from human needs and wants, to choice of technology, to initiating events, to possible release of materials and energy, to human exposure, and eventually to harmful consequences. To differentiate among types of hazards, we define 12 appropriate measures for describing individual hazards at each stage of this causal chain. We chose descriptors, which are identified in Figure 1 and explained in Table 1, that would be universally applicable to all technological hazards, comprehensible to ordinary people, and capable of being expressed by common units and distinctions.

As Figure 1 indicates, one variable describes the degree to which hazards are intentional, four characterize the release of energy and materials, two deal with exposure, and five apply to consequences. Only one descriptor, human mortality (annual), is closely related to the traditional concept of risk as the probability of dying; the others considerably expand and delineate the idea of hazardousness.

As Table 1 indicates, four of 12 scales involve categorical distinctions, and eight are logarithmic. The latter are practical where successive occurrences range over a factor of 10 or more in magnitude and where estimated errors easily differ by the same amount. Compared to linear scales, logarithmic scales may also better match human perception, as seen by the success of the decibel scale for sound intensity and the Richter scale for earthquake intensity.

Hazards were selected from a variety of sources (Lawless 1977; Slovic, Fischhoff, and Lichtenstein 1980;chapter 5) and, after scoring, were found to be well distributed on the 12 scales (Figure 2). Where appropriate, hazards were scored by reference to the scientific literature. Many cases were discussed by two or more individuals or referred to specialists for clarification of available information. After completion of scoring, a series of consistency checks led to alteration of 8 percent of the scores by 1-2 sclae points and less than 1 percent by 3 or more scale points. We therefore believe replicability to be within ± 1 scale point in most cases.

Hazard Classification

Many authors have developed descriptive classifications of technological hazards. These include distinctions between voluntary and involuntary exposure (Starr 1969) and between natural and technological hazards (Burton, Kates, and White 1968). They also include lists of "considerations" (Lowrance 1976), risk factors (Rowe 1977;Litai, Lanning, and Rasmussen 1983) and psychometric qualities (chapter 5). Though mindful of this work, we based our classification on the causal structure descriptors defined in Table 1.

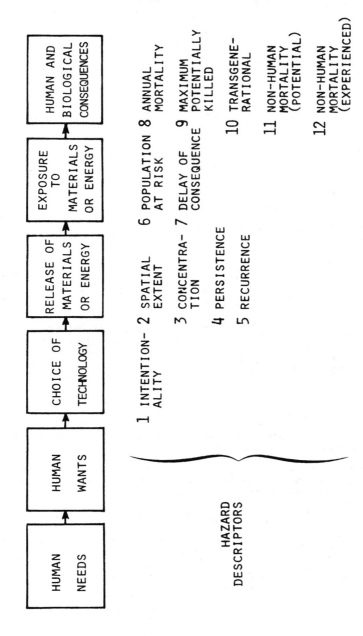

Figure 1. Causal structure of technological hazards illustrated via a simplified causal sequence. Hazard descriptors used in our classification of hazards are shown below the stage of causal evolution to which they apply.

TABLE 1
Hazard descriptor scales

TECHNOLOGY DESCRIPTOR

1. Intentionality

Measures the degree to which technology is intended to harm.

Score	Categorical Definition
3	Not intended to harm living organisms.
6	Intended to harm nonhuman living organisms.
9	Intended to harm humans.

RELEASE DESCRIPTORS

2. Spatial Extent

Measures the maximum spatial extent over which a single release exerts a significant impact. The quantitative scale is based on lineal dimensions, the categorical scale on common geographical units.

Score	Distance Scale	Categorical Definition
1	1 m	Individual
2	1–10 m	Small Group
3	10–100 m	Large Group
4	100–1000 m	Neighborhood
5	1–10 m	Small Region
6	10–100 km	Region
7	100–1000 km	Subcontinental
8	10^3–10^4 km	Continental
9	$>10^4$ km	Global

3. Concentration

Measures the degree to which concentration of released energy or materials is above natural background.

Materials and nonthermal radiation: the scale is based on the ratio, R, defined as the concentration averaged over the release scale divided by the natural background.

Score	Concentration Scale
1	$R < 1$
2	$R \simeq 1$
3	$1 < R < 10$
4	$10 < R < 100$
5	$100 < R < 1000$
6	$10^3 < R < 10^4$
7	$10^4 < R < 10^5$
8	$10^5 < R$

TABLE 1
Hazard descriptor scales (continued)

R E L E A S E   D E S C R I P T O R S

Mechanical energy: the quantitative scale is based on the acceleration, a, to which humans are subjected, expressed in units of the acceleration of gravity, $g = 9.8$ m/s^2.

Score	Acceleration Scale	Categorical Equivalent
1	$a < 1$ g	Protected ordinary life
2	$a \simeq 1$ g	Ordinary life, small falls
3	$2 < a < 5$ g	Very few fatalities
4	$5 < a < 10$ g	A few unlucky fatalities
5	$10 < a < 20$ g	Significant fatalities
6	$20 < a < 40$ g	Protected individuals survive
7	$40 < a < 80$ g	Some protected individuals survive
8	$80 < a$ g	Rare survivors

Thermal energy: the quantitative scale is based on the thermal flux, f, to which a human is subjected, expressed in units of the solar flux, $s = 2$ cal/cm^2/min.

Score	Thermal Flux Scale	Categorical Equivalent
1	$f < 1$ s	Protected ordinary life
2	$f \simeq 1$ s	Ordinary life: 1st degree burn possible
3	$2 < f < 5$ s	1st-degree burn in minutes
4	$5 < f < 10$ s	2nd-degree burn possible; few deaths
5	$10 < f < 20$ s	2nd-degree burn in minutes; some deaths
6	$20 < f < 40$ s	3rd-degree burns possible
7	$40 < f < 80$ s	3rd-degree burns in minutes; many deaths
8	$80 < f$ s	Rare survivors

4. Persistence

Measures the time period over which the release remains a significant threat to humans.

Score	Time Scale
1	1 min.
2	1-10 min.
3	10-100 min.
....	
8	10^6-10^7 min.
9	$>10^7$ min.

5. Recurrence

Measures the time period over which the minimum significant release recurs within the U.S. Use the scale for Persistence.

TABLE 1
Hazard descriptor scales (continued)

EXPOSURE DESCRIPTORS

6. Population at risk

Measures the number of people in the U.S. exposed or potentially exposed to the hazard.

Score	Number of People
1	0-10
2	10-100
...	...
8	10^8
9	$> 10^8$

7. Delay

Measures the delay time between exposure to the hazard release and the occurrence of consequences. Use the scale for Persistence.

CONSEQUENCE DESCRIPTORS

8. Human mortality (annual)

Measures the average annual number of deaths in the U.S. due to the hazard in question. Use the scale for population at risk.

9. Human mortality (maximum)

Measures the maximum credible number of people that could be killed in a single event. Use the scale for population at risk.

10. Transgenerational

Measures the number of future generations that are at risk for the hazard in question.

Score	Categorical Definition
3	Hazard affects the exposed generation only.
6	Hazard affects children of the exposed, no others.
9	Hazard affects more than one future generation.

11. Nonhuman mortality (potential)

Score	Categorical Definition
3	No potential nonhuman mortality.
6	Significant potential nonhuman mortality.
9	Potential or experienced species extinction.

12. Nonhuman mortality (experienced)

Measures nonhuman mortality that has actually been experienced.

Score	Categorical Definition
3	No experienced nonhuman mortality.
6	Significant experienced nonhuman mortality.
9	Experienced species extinction.

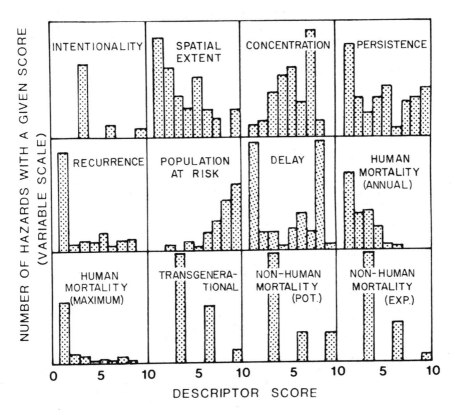

Figure 2. Descriptor frequency distribution for 93 hazards. The vertical scales are chosen to fit the space available and are different for the various descriptors.

Energy versus Materials Hazards

One of the simplest, yet significant, distinctions is the division of hazards into those that involve energy and materials releases, respectively. As illustrated in Figure 3, comparison of 33 energy hazards and 60 materials hazards leads to four striking differences: (1) Energy releases have short persistence times, averaging less than a minute; materials releases have long persistence times, averaging a week or more. (2) Energy hazards have immediate consequences, with exposure-consequence delays of less than a minute; materials hazards have exposure-consequence delays averaging one month. (3) Energy hazards have only minor transgenerational effects; materials hazards affect on the average one future generation. (4) Energy hazards have little potential nonhuman mortality; materials hazards have significant potential effects on nonhuman mortality.

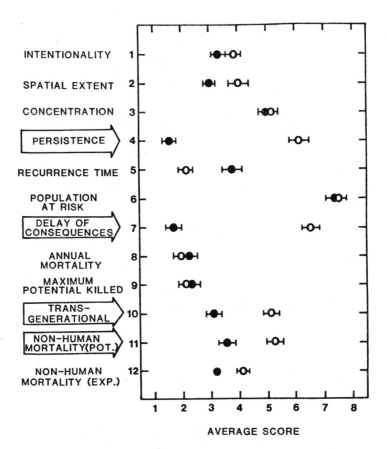

Figure 3. Average scores for energy hazards (solid circles) and materials hazards (open circles) on 12 descriptor scales. Significant differences (more than three standard deviations) are indicated by arrows on the vertical axis. Error bars indicate the standard deviation of the mean.

Reducing the Number of Dimensions

Beyond simple division of hazards by release class, we explored the extent to which hazards may be grouped according to causal structure. To this end we used principal component factor analysis[2] to derive five orthogonal composite dimensions (factors) that "explain" 81 percent of the variance of the sample. This means that the causal structure of each of the 93 hazards, and probably others to be scored in the future, can be described by five variables, rather than by twelve.

The relation of the derived factors to the original set of descriptors is summarized in Table 2. The names given to the factors—**biocidal, delay, catastrophic, mortality,** and **global**—are intended to aid the intuition and are related to the descriptors

TABLE 2
Factor structure

No.	Name	Variance Explained[a] (%)	Name	Factor Loading[b]
	FACTOR		HAZARD DESCRIPTOR	
1.	BIOCIDAL	33	nonhuman mortality (experienced)	0.87
			nonhuman mortality (potential)	0.79
			intentionality	0.81
2.	DELAY	19	persistence	0.81
			delay	0.85
			transgenerational effects	0.84
3.	CATASTROPHIC	11	recurrence	0.91
			human mortality (maximum)	0.89
4.	MORTALITY	11	human mortality (annual)	0.85
5.	GLOBAL	9	population at risk	0.73
			concentration	−0.73
	RESIDUAL		spatial extent	

[a]The percentages given for "variance explained" differ somewhat from those in previous work (Hohenemser, Kates, and Slovic 1983, 380), which was subject to erroneous reading of the computer output.
[b]Factor loadings are the result of varimax rotation.

that define each factor. It is noteworthy that whereas the first four factors involve descriptors whose scores increase as the factor increases (positive factor loadings), the factor **global** is different. Here, because of negative loading of **concentration**, hazards scoring highest on **global** are high in **population at risk** and low in **concentration** (i.e., diffuse). The factor **global** thus picks out a special combination of hazardousness involving widespread exposure and a concentration of release that is modest with respect to background.

Several tests indicate that the factor structure does not change significantly when hazards are added and deleted from the sample, or when scoring changes comparable to the estimated scoring errors are made. Thus an initially chosen set of 66 hazards yielded the same factor structure as the final 93; changing 20 percent of the scores by 1-2 scale points had no significant effect; and removing 24 hazards with the most extreme factor scores produced only minor changes in factor structure. Particularly the last finding is remarkable and quite unexpected, since extreme scores often dominate a factor solution.

TABLE 3
Descriptor and factor codes for 93 hazards

The descriptor code for each hazard consists of a digit for each
descriptor, and represents scores on the scales defined in Table 1.
To help visualize the factor structure, the descriptors have been
grouped by factor in the order defined in Table 2. The factor code
consists of a single digit for each factor, identifies extreme
scores by "1" and nonextreme scores by "0", and also follows
the order defined in Table 2. Hazards with two or more extreme
factors are identified with *.

	HAZARD	DESCRIPTOR CODE	FACTOR CODE
	ENERGY HAZARDS		
1.	Appliances – fire	333-333-42-3-95-2	00000
2.	Appliances – shock	333-113-21-3-95-1	00000
3.	Auto – crashes	333-113-11-5-96-2	00010
4.	Aviation – commercial – crashes	333-113-63-3-97-4	00100
5.	Aviation – commercial – noise	333-213-11-1-85-5	00000
6.	Aviation – private – crashes	333-113-32-4-97-4	00010
7.	Aviation – SST noise	333-313-41-1-76-5	00000
8.	Bicycles – crashes	333-113-11-3-84-2	00000
9.	Bridges – collapse	333-113-53-1-95-3	00000
10.	Chainsaws – accidents	666-113-11-1-74-2	10000
11.	Coal mining – accidents	333-233-53-3-64-3	00000
12.	Dams – failure	693-423-74-2-85-5	10100*
13.	Downhill skiing – falls	333-113-21-2-63-1	00000
14.	Dynamite blasts – accidents	333-113-32-2-65-3	00000
15.	Elevators – falls	333-113-52-2-96-2	00000
16.	Fireworks – accidents	333-113-31-1-83-2	00000
17.	Handguns – shootings	369-113-41-4-96-1	10010*
18.	High construction – falls	333-113-71-1-28-2	00000
19.	High voltage wires – electric fields	333-173-11-1-74-3	00000
20.	LNG – explosions	363-213-85-1-86-5	00100
21.	Medical x-rays – radiation	333-189-11-4-92-2	00011*
22.	Microwave ovens – radiation	333-173-11-1-84-2	00000
23.	Motorcycles – accidents	333-113-11-4-76-2	00010
24.	Motor vehicles – noise	333-213-11-1-83-3	00000
25.	Motor vehicles – racing crashes	333-113-52-2-67-2	00000
26.	Nuclear war – blast	699-213-87-4-98-6	10110*
27.	Power mowers – accidents	333-113-21-2-73-2	00000
28.	Skateboards – falls	333-113-11-3-73-1	00000
29.	Skydiving – accidents	333-113-51-2-48-1	00000
30.	Skyscrapers – fire	333-113-53-3-85-4	00000
31.	Smoking – fires	333-433-32-3-85-1	00000
32.	Snowmobiles – collisions	333-113-41-2-73-2	00000
33.	Space vehicles – crashes	333-313-84-1-98-5	00100
34.	Tractors – accidents	333-113-41-2-74-2	00000
35.	Trains – crashes	333-213-53-3-84-3	00000
36.	Trampolines – falls	333-113-51-1-74-2	00000
	MATERIALS HAZARDS		
37.	Alcohol – accidents	333-313-11-4-95-2	00010
38.	Alcohol – chronic effects	333-486-11-5-85-1	00010
39.	Antibiotics – bacterial resistance	666-563-11-3-97-1	10000
40.	Asbestos insulation – toxic effects	333-583-11-3-56-3	00000

41.	Asbestos spray - toxic effects	333-583-11-1-83-3	00000
42.	Aspirin - overdose	333-456-11-3-97-1	00000
43.	Auto - CO pollution	333-346-11-2-94-4	00000
44.	Auto - lead pollution	663-976-11-2-95-5	01000
45.	Cadmium - toxic effects	663-986-11-2-74-6	01000
46.	Caffeine - chronic effects	333-566-11-1-95-1	00000
47.	Coal burning - NO_x pollution	693-566-11-3-95-7	10000
48.	Coal burning - SO_2 pollution	693-563-11-4-94-7	10010*
49.	Coal mining - black lung	333-483-11-4-64-3	00010
50.	Contraceptive IUDs - side effects	333-763-11-2-67-1	00000
51.	Contraceptive pills - side effects	333-586-11-3-74-1	00000
52.	Darvon - overdose	333-556-11-4-77-1	00010
53.	DDT - toxic effects	996-886-32-1-87-5	11000*
54.	Deforestation - CO_2 release	696-993-11-1-91-9	10001*
55.	DES - animal feed - human toxicity	333-586-11-1-93-1	00001
56.	Fertilizer - NO_x pollution	393-686-11-1-93-9	00001
57.	Fluorocarbons - ozone depletion	393-883-11-1-97-9	00000
58.	Fossil fuels - CO_2 release	393-993-11-1-92-9	00001
59.	Hair dyes - coal tar exposure	333-286-11-1-87-1	00000
60.	Hexachlorophene - toxic effects	666-363-11-2-87-1	10000
61.	Home pools - drowning	333-223-41-3-83-1	00000
62.	Laetrile - toxic effects	333-553-11-1-55-1	00000
63.	Lead paint - human toxicity	333-773-11-3-75-2	00000
64.	Mercury - toxic effects	663-986-13-2-85-5	01000
65.	Mirex pesticide - toxic effects	696-886-22-1-67-5	11000*
66.	Nerve gas - accidents	669-836-73-1-77-5	10100*
67.	Nerve gas - war use	699-836-87-3-97-7	10100*
68.	Nitrite preservative - toxic effects	336-786-11-1-91-1	00001
69.	Nuclear reactor - radiation release	363-969-86-1-96-7	01100*
70.	Nuclear tests - fallout	663-989-73-3-91-9	01101*
71.	Nuclear war - radiation effects	699-989-88-4-97-9	11110*
72.	Nuclear waste - radiation effects	363-989-15-1-82-6	01001*
73.	Oil tankers - spills	663-763-61-1-15-6	00000
74.	PCBs - Toxic effects	663-976-13-1-97-6	01000
75.	Pesticides - human toxicity	996-886-12-2-97-5	11000*
76.	PVC - human toxicity	333-486-11-2-77-4	00000
77.	Recombinant DNA - harmful release	393-869-97-1-97-9	01100*
78.	Recreational boating - drowning	333-223-51-4-83-2	00010
79.	Rubber manufacture - toxic exposure	333-986-11-3-57-4	01000
80.	Saccharin - cancer	333-486-11-1-87-1	00000
81.	Smoking - chronic effects	333-486-11-6-85-1	00010
82.	SST - ozone depletion	393-893-11-1-93-9	00001
83.	Taconite mining - water pollution	663-983-11-1-67-6	00000
84.	Thalidomide - side effects	333-456-51-1-17-1	00000
85.	Trichloroethylene - toxic effects	333-983-11-1-87-4	00000
86.	Two, 4,5-T herbicide - toxic effects	696-886-22-1-77-5	11000*
87.	Underwater construction - accidents	333-223-61-1-44-3	00000
88.	Uranium mining - radiation	333-989-12-2-64-5	01000
89.	Vaccines - side effects	696-556-11-2-84-1	10000
90.	Valium - misuse	333-566-11-3-87-1	00000
91.	Warfarin - human toxicity	666-653-11-1-87-1	10000
92.	Water chlorination - toxic effects	666-583-11-1-97-5	10000
93.	Water fluoridation - toxic effects	333-786-11-1-82-5	00001

Hazard scoring and derived factor structure are summarized in Table 3. Individual descriptor scores have been grouped by factor into a 12-digit **descriptor code**, and extreme scores on each factor have been identified through a five-digit **factor code**, using truncated factor scores.[3] The code sequence is defined in Table 3.

Inspection of Table 3 permits quick identification of dimensions that dominate hazardousness in specific cases. For example, commercial aviation (crashes) is high in the **catastrophic** factor, and nondistinctive in the other four; power mower accidents are extreme in none of the five factors; nuclear war (radiation effects) is extreme in four.

The distinctions offered in Table 3 led naturally to a seven-class taxonomy with three major groupings (Table 4). The first group, **multiple extreme hazards**, includes cases with extreme scores in two or more factors; the second, **extreme hazards** comprises cases with extreme scores on one factor; and the third, **hazards**, contains all other hazards. The group into which a hazard falls depends, of course, on the cutoff for the designation **extreme**. Although the location of the cutoff is ultimately a policy question, our preliminary definition is arbitrary.[4]

How appropriate and useful is our approach to hazard classification? To succeed it must approximate the essential elements that make specific hazards threatening to humans and what they value, reflect the concerns of society, and offer new tools for managing hazards. On the first point, we invite the review and evaluation of specialists; on the second and third points, we have additional evidence that we discuss next.

Comparing Perceptions

The scores for 93 hazards are products of judgments relying on explicit methods, a scientific framework, and deliberate efforts to control bias. None of these are necessarily attributes of public perception. Indeed, many scientists believe that lay judgments of hazards vary widely from scientifically derived judgments (Kasper 1980). Since hazard policy in our society is determined to a large extent by people inside and outside government who are **not** scientists or hazard assessment experts, it is important to know whether lay people are able to understand and judge our hazard descriptors and whether these descriptors capture their concerns. Although we cannot offer a definitive answer to these questions, we can report on the results of a pilot study of a group of 34 college-educated people (24 men, 10 women, mean age 24) living in Eugene, Oregon. (The study employed methods similar to those described further in chapter 5.)

To test the perception of these subjects we created nontechnical definitions and simple scoring instructions for the causal descriptors of hazards and asked our subjects to score our sample of 93 hazards.[5] After an initial trial, **concentration** was judged too difficult for our respondents to score. For similar reasons, 12 of the less familiar hazards were omitted. The subjects, using only our instructions and their general knowledge, reasoning, and intuition, then scored 81 hazards on 11 measures.

TABLE 4
A seven-class taxonomy

CLASSES	EXAMPLES
1. MULTIPLE EXTREME HAZARDS (extreme in more than one factor)	nuclear war – radiation, recombinant DNA, pesticides, nerve gas – war use, dam failure.
2. EXTREME HAZARDS (extreme in one factor)	
a. intentional biocides	chain saws, antibiotics, vaccines.
b. persistent teratogens	uranium mining, rubber manufacture.
c. rare catastrophes	LNG explosions, commercial aviation crashes.
d. common killers	auto crashes, coal mining – black lung.
e. diffuse global threats	fossil fuel – CO_2, SST – ozone depletion.
3. HAZARDS (extreme in no factor)	saccharin, appliances, aspirin, skateboards, power mowers, bicycles.

The results indicate reasonably high correlations between the scores derived from the scientific literature and the mean judgments of our lay sample, with correlation coefficients ranging from 0.65 to 0.96 (Table 5). As illustrated in three sample scatter plots (Figure 4), despite high correlation coefficients, deviations of a factor of 1000 between scientific and lay estimates were encountered. This suggests that there were significant biases in lay perceptions for some descriptors and some hazards. Also, the subjects tended to compress the scale of their judgment; in effect, lay judgments exhibited systematic overvaluation of low scoring hazards and systematic undervaluation of high scoring hazards. This effect was not an artifact of regression toward the mean, for it appears in the scores of individual subjects as well. Similar effects were found by Lichtenstein et al. (1978) in comparisons of **perceived risk** with scientific estimates of annual mortality.

To test whether our causal structure descriptors capture our subjects' overall concern with risk, we collected judgments of **perceived risk**, a global risk measure whose determinants have been explored in psychometric studies (chapter 5). Subjects were asked to consider "the risk of dying across all of U. S. society," as a consequence of the hazard in question, and to express their judgment on a relative scale of 1 to 100. Modest positive correlations between perceived risk and our descriptor scores were obtained in 9 of 12 cases (Table 6, top). Each hazard descriptor thus explains only a small portion of the variance in perceived risk.

In Table 6, bottom, we show the modest positive correlations of the five factors with perceived risk. Because the factors are

TABLE 5

Correlation of lay and scientific judgments of hazard descriptors

HAZARD DESCRIPTOR	CORRELATION COEFFICIENTS		
	ENERGY HAZARDS	MATERIALS HAZARDS	ALL HAZARDS
T E C H N O L O G Y D E S C R I P T O R			
1. Intentionality	0.95	0.84	0.89
R E L E A S E D E S C R I P T O R S			
2. Spatial Extent	0.83	0.89	0.87
3. Concentration	N/A	N/A	N/A
4. Persistence	0.33	0.62	0.79
5. Recurrence	0.85	0.73	0.80
E X P O S U R E D E S C R I P T O R			
6. Population at risk	0.77	0.73	0.74
7. Delay	0.88	0.92	0.96
C O N S E Q U E N C E D E S C R I P T O R S			
8. Human mortality (annual)	0.79	0.77	0.76
9. Human mortality (maximum)	0.89	0.75	0.79
10. Transgenerational	0.34	0.56	0.65
11. Nonhuman mortality (potential)	0.82	0.75	0.78
12. Nonhuman mortality (experienced)	0.63	0.73	0.71

linearly independent, the summed variance of the factors may be used to determine the total variance explained. From the sample of 34 young Oregonians we find that our descriptors account for about 50 percent of the variance in perceived risk.

Perhaps the most striking aspect of these results is that perceived risk shows no significant correlation with the factor **mortality**. Thus, the variable most frequently chosen by scientists to represent risk appears not to be a strong factor in the judgment of our subjects.

When the analysis of Table 6 is carried out using not our descriptor scores but average ratings obtained from our 34 subjects, correlations with perceived risk increase substantially, and factor scores derived from the subjects' descriptor ratings explain 85 percent (not 50 percent) of the variance in perceived risk. We conclude, therefore, that our hazard descriptors were well understood

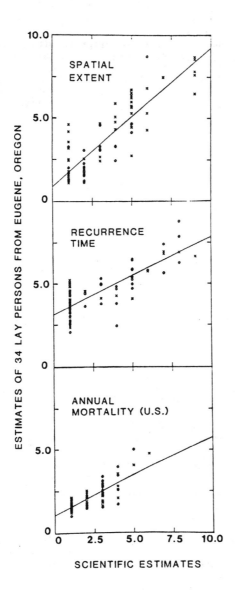

Figure 4. Scatter plots with linear regression lines indicating correlations between mean lay judgments and our estimates of hazard descriptors. The three cases are illustrative of the principal features of these correlations: (1) a generally high degree of correspondence between the two types of judgments; (2) some deviations corresponding to a factor of as high as 1000 (three scale points) on quantitatively defined logarithmic scales; (3) except for the case of spatial extent (top graph), a significant compression of scale for lay judgments, indicated by a slope of less than unity.

TABLE 6
Correlation of causal structure descriptors with psychometrically
determined values of "perceived risk" across 81 hazards

HAZARD DESCRIPTOR	CORRELATION COEFFICIENTS (only r-values at greater than 0.95 confidence level are given.)
T E C H N O L O G Y D E S C R I P T O R	
1. Intentionality	0.28
R E L E A S E D E S C R I P T O R S	
2. Spatial Extent	0.57
3. Concentration	—
4. Persistence	0.42
5. Recurrence	—
E X P O S U R E D E S C R I P T O R S	
6. Population at risk	0.42
7. Delay	0.30
C O N S E Q U E N C E D E S C R I P T O R S	
8. Human mortality (annual)	—
9. Human mortality (maximum)	0.53
10. Transgenerational	0.43
11. Nonhuman mortality (potential)	0.53
12. Nonhuman mortality (experienced)	0.30
FACTORS	
1. Biocidal	0.32
2. Delay	0.41
3. Catastrophic	0.32
4. Mortality	—
5. Global	0.30
Variance explained $= \Sigma r^2$	0.50

by our pilot sample of nonexperts and that they captured most of the
global concern with risk that is expressed in the variable perceived
risk. Nonetheless, before these conclusions can be cast in a more
general form, much additional work is needed with larger, more
representative samples.

Applications to Hazard Management

In addition to improving our understanding of hazards, our conceptualization of hazardousness may assist in selecting the social
and technical controls that society employs to ease the burden of
hazards. Though detailed discussion of hazard management is not
intended here (See chapter 3), we envision three ways of improving
the management process.

Comparing Technologies

Basic to hazard management are comparisons and choices among
competing technologies. For example, in debates on electricity
generation, comparisons between coal and nuclear power are common.
Insofar as such comparisons involve hazards, they are invariably
couched in terms of mortality estimates. A controversial example is
the estimate by Inhaber (1979) that coal has a mortality rate 50
times that of nuclear power (Figure 5, top). Quite aside from the
validity of Inhaber's methods, which have been questioned (Holdren
et al. 1979; Herbert, Swanson, and Reddy 1979), such one-dimensional
comparisons create considerable controversy and dissatisfaction
because they ignore other important differences, including other
apsects of hazardousness, between the two technologies (Holdren
1982).

Our broader conceptualization of hazardousness offers a partial
solution. To illustrate, we apply our multidimensional hazard profile for coal and nuclear power (Figure 5, bottom). This profile
was obtained by combining descriptor scores for each of several
hazard chains that make up the total hazard of coal and nuclear
power.[6] Coal exceeds nuclear in human mortality, as would be
expected from Inhaber's analysis, and it also exceeds nuclear in
nonhuman mortality (that is, environmental effects). Nuclear power,
on the other hand, dominates in transgenerational effects and the
catastrophic factor. The two technologies show little difference in
persistence, delay, population at risk, and diffuseness.

We believe that our 12-descriptor profile captures the complexity of choice in energy risk assessment and management better than
the common mortality index. At the same time it in no way settles
the problem of choice but raises an interesting new and largely
normative question: how should society weight the different dimensions of hazardousness?

Dealing with the Hazard of the Week

Analysis of national news media shows that 40 to 50 hazards
receive widespread attention each year (Kates 1977). Each new
hazard goes through a sequence of problem recognition, hazard
assessment, and managerial action. Often there is need for early
managerial response of some kind. To this end, our descriptors of
hazardousness provide a quick profile that allows new hazards to be
grouped with other hazards that have similar profiles.

To illustrate this possibility we used available information to
score the new hazard **tampons—toxic shock syndrome.** Profile
comparisons enabled us to determine that this hazard was most similar in structure to the previously scored hazards **contraceptive**

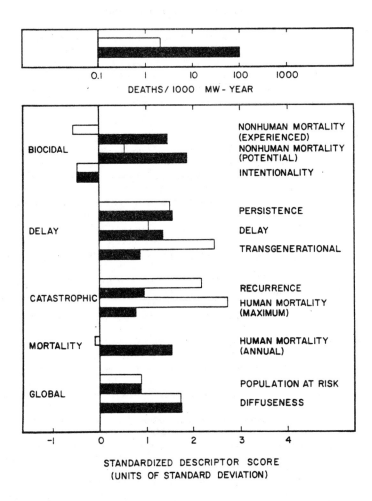

Figure 5. Comparison of nuclear and coal-fired electric power, shown by light and dark shading, respectively, using Inhaber's analysis (top) and our hazardousness concept (bottom). Labels on the left are factor names, labels on the right are names of descriptors belonging to each factor. For the method of computing the combined descriptor scores plotted here, see note 5.

IUDs—side effects, aspirin—overdose, Valium—misuse, and Darvon—overdose. Such comparisons may provide industrial or governmental hazard managers immediate access to relevant, albeit incomplete, precedents, as well as warning of unexpected problems, a range of suggested managerial options, and, at the very least, a measure of consistency in public policy. Indeed, subsequent societal response on tampons has paralleled that of IUDs, the hazard in our inventory closest in structure to tampons.[7]

A Case for Triage?

As a society we cannot make extraordinary efforts on each of the 100,000 chemicals or 20,000 consumer products in commerce. If our causal structure and its descriptors reflect key aspects of hazards--threats to humans and what they value--then our taxonomy provides a way of identifying those hazards worthy of special attention. Cases with extreme scores in each of the five composite dimensions of hazard have already been identified in Table 3, and these lend themselves naturally to a proposal for **triage**: extraordinary attention for multiple extreme hazards, distinctive effort for each of the groups of extreme hazards, and an ordered, routine response for the remainder.

Although we regard the suggestion of triage an important outcome of our analysis, it is well to remember that many of the extreme hazards, such as nuclear weapons, are among a group that has defied solution for a long time and that special efforts expended on them may produce few concrete results. This leads some to argue that society should focus its effort on cases of proven cost-effectiveness--cases with the maximum reduction in hazardousness per unit expenditure.

We regard neither triage nor adherence to cost-effectiveness criteria as adequate foundations for managing hazards; rather, we see them as the horns of a familiar dilemma: whether to work on the "big questions" where success is limited, or to work on the normal, where success is expected.

Summary and Conclusions

All taxonomies are based on explicit or implicit assumptions, and ours is no different. We assume that technological hazards form a single domain, that they are defined by causal sequences, and that these sequences are usefully measured by a few physical, biological, and social descriptors. Our picture leads us to distinguish between energy and materials releases and provides a method for constructing profiles of hazardousness that considerably extend the conventional concept of **risk** as annual human mortality.

A pilot study has shown that our profiles of hazardousness appear comprehensible to lay people and that they capture a significant fraction of our subjects' concern with hazardousness. This suggests that some conflict between experts and lay people may be resolved by clarifying the definition of hazardousness.

We expect that our approach can improve the quality and effectiveness of hazard management. In particular, it may help in comparing the hazards of competing technologies, provide a quicker, more orderly response to new hazards, and offer society a rational approach to triage.

ACKNOWLEDGMENTS

This chapter was prepared by the authors for the Clark University Hazard Assessment Group and Decision Research, a Branch of Perceptronics. Whereas we take full responsibility for the contents, we acknowledge the participation of a number of people at both institutions. In addition to ourselves, participating members of the Clark group were R. Goble, A. Goldman, J. X. Kasperson, R. E. Kasperson, M. P. Lavine, M. Morrison, and B. Rubin. Participating members of the Decision Research group were B. Fischhoff, M. Layman and S. Lichtenstein. In addition we thank R. C. Harriss and T. C. Hollocher for help in conceptualizing the causal structure of hazard; B. Johnson and N. Winter for assistance in hazard scoring; D. McGregor for assistance in collecting risk perception data; and P. Collins for computer analysis. Support was received from the National Science Foundation under grants ENV 77-15334, PRA79-11934, and PRA81-16925.

NOTES

1. Except for editorial changes appropriate for this volume, the contents of this chapter appeared previously under the title "The nature of technological hazard," _Science_ 220 (1983):378-384. Copyright 1983 by the American Association for the Advancement of Science. Reproduced by permission.

2. Factor analysis was done using the package Biomedical Computer Program, Program BMDP:P4M, developed by the Health Sciences Computing Facility, U.C.L.A., available in _BMDP, P-series, 1979_, ed. W. J. Dixon and M. B. Brown (Los Angeles: University of California Press, 1979). Orthogonal rotation was performed according to the varimax criterion, which maximizes the variance of the squared factor loadings.

3. In identifying extreme scores on each factor, we might have used exact factor scores generated by the factor analysis. These, however, include significant off-diagonal contributions, so that two hazards with identical descriptors may have significantly different factor scores. Because we believe that the significance of factor analysis lies in descriptor grouping, and not in the mathematical abstraction called a "factor," we have used truncated factor scores (consisting of sums of descriptors belonging to a given factor) to generate the extreme scoring hazards designated in Table 3. Our truncation procedure and its validity may be described as follows. By using the raw descriptor scores D_{ik} for the ith descriptor and the kth hazard, we obtained truncated factor scores

$$F_{jk}^t = \sum_i D_{ik},$$

where i runs over just the salient descriptors belonging to the jth factor. This suppresses contributions from descriptors that load weakly on the jth factor. In contrast, the factor analysis program obtains exact standardized factor scores through the 12-term sum

$$F_{jk} = \sum_i d_{ik} f_{ij},$$

where the d_{ik} are standardized descriptor scores belonging to the ith descriptor and the kth hazard, and f_{ij} is the 12x5 factor score coefficient matrix, given in the following table. In a statistical sense, there is little difference between the two methods: the correlation coefficients between F_{jk}^t and F_{jk} are (0.94, 0.96, 0.97, 0.85, and 0.96) for $j = (1,2,3,4,5)$, respectively.

VARIABLE			FACTOR SCORE COEFFICIENTS FOR FACTORS				
No.	Mean	Stdev.	1	2	3	4	5
12	3.9	1.5	.42	-.07	-.15	-.03	-.08
11	4.6	2.4	.31	-.05	-.00	-.14	.15
1	3.7	1.6	.41	-1.2	-.04	.18	.09
4	4.4	3.0	.01	.32	-.02	-.10	-.08
7	4.8	3.1	-.03	.38	-.18	.01	-.07
10	4.4	1.9	-.21	.47	.15	.25	-.10
5	2.7	2.4	-.09	-.08	.47	-.10	-.08
9	1.8	1.6	-.06	.09	.44	.10	.03
8	2.1	1.2	.02	.08	-.00	.67	.03
6	7.6	1.7	.11	-.06	.02	.30	.57
3	5.1	1.8	.14	.07	.05	.17	-.59
2	3.5	2.5	.02	.06	.19	-.29	.27

4. We define **extreme** hazards as those with truncated factor scores of 1.2-1.5 standard deviations above the mean.

5. Instructions to subjects sought to follow as closely as possible the scale definitions described in Table 1. For example lay instructions for the descriptor **persistence** are as follows:

> Rate the persistence over time of the damage-producing activity or substance. For example, collisions or explosions usually last one minute or less. For environmental pollutants, persistence time is the length of time they remain active in the environment. For prescription drugs, rate the time they remain active in the body. Use the following scale.

1:	Less than 1 minute	6:	1 week-2 1/2 months
2:	1-10 minutes	7:	2 1/2 months-2 years
3:	10-100 minutes	8:	2 years-20 years
4:	2 hours-17 hours	9:	More than 20 years
5:	17 hours-1 week		

6. To obtain "combined" hazard profiles, the hazards of coal-fired electric power were taken to be numbers 11, 47, 48, 49, and 58 in Table 3, and those of nuclear electric power numbers 69, 72, and 88. Consistent with the logarithmic character of most of the descriptor scales, corresponding descriptor scores from different hazard chains were combined through the addition

algorithm: score (a + b + c...) = maximum (a,b,c, ...). In effect, combined hazardousness on a given descriptor is determined by the highest scoring component hazard. Because of the negative loading of **concentration** on the **global** factor, "minimum" was substituted for "maximum" in applying the above algorithm to the descriptor **concentration**.

7. Management for both IUDs and tampons included three responses: (1) removal of specific products most associated with health effects; (2) stricter classification and scrutiny by the regulatory agency; and (3) warnings and recommendations for use packaged with all other products in the generic class. For details on IUDs, see chapter 17.

REFERENCES

Burton, Ian, Robert W. Kates, and Gilbert F. White. 1968. The human ecology of extreme geographical events. Natural Hazard Working Paper no. 2. Toronto: Department of Geography, University of Toronto.

Cohen, Bernard L., and I-Sing Lee. 1979. A catalog of risks. Health Physics 36:707-722.

Greenwood, D. R., G. L. Kingsbury, and J. G. Cleland. 1979. A handbook of key federal regulations and criteria for multimedia environmental control. EPA-600/7-79-175. Washington: Environmental Protection Agency.

Herbert, John H., Christina Swanson, and Patrick Reddy. 1979. A risky business: Energy production and the Inhaber report. Environment 21 no. 6 (July/August):28-33.

Hohenemser, Christoph, et al. 1983. Methods for analyzing and comparing technological hazards: Definitions and factor structures. CENTED Research Report no. 3. Worcester, Mass.: Center for Technology, Environment, and Development (CENTED), Clark University.

Holdren, John P. 1982. Energy hazards: What to measure, what to compare. Technology Review 85 no. 3:32-38, 74-75.

Holdren, John P., Kent Anderson, Peter H. Gleick, Irving Mintzer, Gregory Morris, and Kirk W. Smith. 1979. Risk of renewable energy sources: A critique of the Inhaber report. ERG 79-3. Berkeley: Energy and Resources Group, University of California, June.

Inhaber, Herbert. 1979. Risk of energy production, Report AECB-1119, rev. 3, 4th ed. Ottawa: Atomic Energy Control Board.

Kasper, Raphael G. 1980. Perceptions of risk and their effects on decision making. In Societal risk assessment: How safe is safe enough?, ed. R. C. Schwing and W. A. Albers, Jr., 71-84. New York: Plenum.

Kates, Robert W. 1977. Summary report. In Managing technological hazard: Research needs and opportunities, ed. R. W. Kates, 1-48. Boulder: Institute for Behavioral Science, University of Colorado.

Lave, Lester B. 1981. Conflicting objectives in regulating the automobile. Science 212:893-899.

Lawless, Edward W. 1977. Technology and social shock. New Brunswick, N.J.: Rutgers University Press.

Lichtenstein, Sarah, Paul Slovic, Baruch Fischhoff, Mark Layman, and Barbara Combs. 1978. Judged frequency of lethal events. Journal of Experimental Psychology: Human Learning and Memory 4:551-558.

Litai, D., D. D. Lanning, and N. C. Rasmussen. 1983. The public perception of risk. In The analysis of actual vs. perceived risks, ed. V. T. Covello, W. G. Flamm, J. V. Rodricks, and R. G. Tardiff, 213-224. New York: Plenum Press.

Lowrance, William W. 1976. Of acceptable risk: Science and the determination of safety. Los Altos, Calif.: William Kaufmann.

National Research Council. 1980. Committee on Nuclear and Alternative Energy Systems. Energy in transition, 1985-2010. San Francisco: W. H. Freeman.

Rowe, William D. 1977. An anatomy of risk. New York: Wiley.

Schwing, Richard C. 1979. Longevity benefits and costs of reducing various risks. Technological Forecasting and Social Change 13:333-345.

Slovic, Paul, Baruch Fischhoff, and Sarah Lichtenstein. 1980. Facts and fears: Understanding perceived risk. In Societal risk assessment: How safe is safe enough?, ed. R. C. Schwing and W. A. Albers, Jr., 181-214. New York: Plenum Press.

Starr, Chauncey. 1969. Social benefit versus technological risk. Science 165:1232-1238.

U. S. Department of Transportation. 1976. National highway safety needs report. Washington: Department of Transportation.

Wilson, Richard. 1979. Analyzing the daily risks of life. Technology Review 81 no. 4 (February):40-46.

5
Characterizing Perceived Risk

Paul Slovic, Baruch Fischhoff,
and Sarah Lichtenstein

> How extraordinary! The richest, longest-lived, best-protected, most resourceful civilization, with the highest degree of insight into its own technology, is on its way to becoming the most frightened.
> Is it our environment or ourselves that have changed? Would people like us have had this sort of concern in the past?...today, there are risks from numerous small dams far exceeding those from nuclear reactors. Why is the one feared and not the other? Is it just that we are used to the old or are some of us looking differently at essentially the same sorts of experience?
>
> (Wildavsky 1979, 32)

Wildavsky's concerns are likely to reverberate throughout the 1980s as society continues to grapple with the question: "How safe is safe enough?" Over the past few years, we have been attempting to address such concerns by asking people to characterize and evaluate hazardous activities and technologies in a variety of ways. As do our colleagues at Clark University, we define **hazards** as threats to humans and what they value. When we speak of **risk,** however, we include a wide range of cognitive dimensions that extend well beyond the idea of risk as "quantitative measures of hazard consequences expressed as conditional probabilities of experiencing harm" (see chapter 4).

Our research is descriptive. It aims to discover what people mean when they say that something is risky; to develop a psychological taxonomy of risk that can be used to understand people's perceptions and predict societal response; and to develop methods for assessing public opinion about risk in a way that is useful for informing policy decisions.

In this chapter, we report on three psychometric scaling studies, summarized in Table 1. In each study, participants rated a given set of hazards on a range of risk characteristics and indicated the degree of risk reduction and regulation they desired. Based on this data, we explore the relationships among risk characteristics and a smaller number of dimensions (factors) derived from them. We also relate risk characteristics to people's perception of risk and their desire for risk reduction and regulation. Our work builds on and extends earlier studies with a smaller number of hazards and risk characteristics (Fischhoff et al. 1978).

TABLE 1
Studies of perceived risk: Overview[a]

STUDY	DATE		SUBJECTS	TASKS
1	Spring	1976	76 League members	
	July	1977	69 college students	Rate perceived risk, adjusted risk, and
	October	1977	47 Active Club members	nine risk character- istics for each of 30 hazards
	January	1978	15 professional risk assessors	
	February	1978	38 college students	Rate the 30 items on a scale of desired stringency of regula- tion
2	March	1979	175 college students	Rate perceived risk, adjusted risk, de- sired regulation, and 18 risk charac- teristics for each of 90 hazards
3	August	1980	34 college students	Rate perceived risk, adjusted risk, de- sired regulation, and 18 risk characteris- for each of 81 hazards

[a]Data from League members in Study 1 were reported in detail by Fischhoff et al. (1978). Other data from Study 1 were reported in Slovic, Fischhoff, and Lichtenstein (1979). Parts of Study 2 were reported in Slovic, Fischhoff, and Lichtenstein (1980a).

We find that psychometric scaling can quantify similarities and differences among groups with regard to risk perception and atti-tudes; that expert judgments differ substantially from nonexpert judgments; and that different groups agree generally on risk qualities but not on attitudes toward risk. We believe our results can aid technology managers to understand public response to haz-ards. Like the analysis of causal structure in chapter 2, our results appear to provide a useful perspective on risk comparisons. A significant conclusion is that a broad definition of risk, incor-porating many of the qualities we have studied, is necessary for understanding people's concerns. Attempts to characterize or com-pare risk, set safety standards, and make risk decisions will founder in conflict if policy makers insist, as they often have, on the narrow definition of risk as a conditional probability of dying.

We turn next to a brief description of methods and follow this with a presentation and interpretation of the results from the three studies.

Methods

Four different groups participated in Study 1. Three of these—members of the League of Women Voters and their spouses, college students, and members of the Active Club (a community service organization—were from Eugene, Oregon. The fourth consisted of a nationwide sample of experts on risk assessment. Members of all four groups rated each of 30 hazardous activities, substances, and technologies on the basis of (1) its perceived risk of death, (2) the acceptability of its current level of risk, and (3) its position on each of nine scales defining characteristics of risk.

In Study 2, the number of hazards was increased from 30 to 90, the number of risk characteristics from 9 to 18, and participants were restricted to University of Oregon students. As in Study 1, participants rated perceived risk, the acceptability of current levels of risk, and the risk characteristics. Because of the large number of judgments involved, each person performed only a subset of the whole task.

Study 3 extended Study 2, exploring the sensitivity of results to the composition of the set of hazards by examining 81 activities, substances, and technologies considered in the causal-structure taxonomy discussed in chapter 4. Thirty of the hazards studied overlapped with those used in Study 2; the remaining 51 consisted of various chemicals (not included in the first two studies because they were thought hard for laypersons to judge) and numerous activities and technologies that were labeled more specifically than in the earlier studies. Thus, for example, automobile risks were partitioned into four separate categories treating (a) accidents to vehicle occupants, pedestrians, and cyclists, (b) automobile racing, (c) carbon monoxide exhaust, and (d) airborne lead. Smoking risks were treated in two categories, one dealing with accidental fires and the other dealing with health effects. Thirty-four college-educated individuals participated in Study 3, and every participant rated all 81 hazards on 18 risk characteristics.

Lists of hazardous activities, data on participants, and the instructions and procedures for all three studies are discussed in more detail in the Appendix.

Results of Study 1

Perceived Risk

To determine perceived risk, participants were asked for each of the 30 hazards "to consider the risk of dying (across all U. S. society as a whole) as a consequence of this activity or technology." Table 2 shows the results for each of four groups. Because arithmetic means tend to be influenced unduly by occasional extreme values, geometric means are used in Table 2.

TABLE 2
Geometric means of perceived risks in study 1

ACTIVITY OR TECHNOLOGY	LEAGUE	STUDENTS	ACTIVE CLUB	EXPERTS
1. Nuclear power	250	449	74	138
2. Motor vehicles	247	169	104	8,332
3. Handguns	220	193	137	1,516
4. Smoking	189	192	94	6,766
5. Motorcycles	176	155	121	532
6. Alcoholic beverages	161	150	92	3,793
7. General (private) aviation	114	75	60	250
8. Police work	111	116	84	178
9. Pesticides	105	188	39	426
10. Surgery	104	94	70	1,264
11. Fire fighting	92	98	86	159
12. Large construction (dams, bridges, etc.)	91	82	54	238
13. Hunting	82	61	66	124
14. Spray cans	73	85	25	68
15. Mountain climbing	68	43	55	40
16. Bicycles	65	38	40	190
17. Commercial aviation	52	71	30	181
18. Electric power	52	59	29	421
19. Swimming	52	20	31	311
20. Contraceptives	50	102	26	258
21. X rays	45	66	25	450
22. Skiing	45	31	31	40
23. Railroads	37	41	29	152
24. High school and college football	37	30	27	55
25. Food preservatives	36	90	18	221
26. Food coloring	31	54	15	137
27. Prescription antibiotics	30	48	21	103
28. Power mowers	29	22	23	44
29. Home appliances	25	24	19	132
30. Vaccinations	17	22	16	79
All Responses	69	71	42	887

NOTE: The correlations between log Perceived Risk for pairs of groups were as folows:

	1	2	3	4
1. League	–			
2. Students	.85	–		
3. Active Club	.92	.68	–	
4. Experts	.66	.60	.59	–

The data clearly indicate that experts employ a much greater range of values to discriminate among the various hazards. Thus, the median ratio of the most risky to least risky hazard was 45 for nonexperts and 7,900 for experts. We believe this is because most experts equate **risk** with something akin to **yearly fatalities,** whereas laypeople do not.

Intergroup correlations are given at the foot of Table 2.[1] They show that the League of Women Voters group was similar to both the students and the Active Club members; but the Active Club data correlated less well with the student data, and the correlations between the experts and the three lay groups were quite modest.

The observed intergroup correlations can be understood by looking at the data for specific hazards. Note that all four groups judged handguns, smoking, motor vehicles, and alcohol as relatively high in risk; and power mowers, antibiotics, home appliances, and vaccinations as relatively low in risk. Experts, however, saw nuclear power as much less risky, and X rays, surgery, and non-nuclear electric power as much more risky than did the three lay groups.

The November, 1976 nuclear power referendum in Oregon provided a check on our finding that the League group rated nuclear power much more risky than did the Active Club group. We found that 95 percent of the League group voted against nuclear power, whereas 85 percent of the Active Club respondents voted for nuclear power.

Adjusted Risk

After rating the risks, each group judged the acceptability of the risk by specifying a risk-adjustment factor. For this variable, a value of n larger than unity implies the risk should be reduced by a factor n; a value n less than unity implies the risk may tolerably be increased by $1/n$, and a value of 1.0 implies the risk needs no adjustment (see the Appendix for the specific instructions).

Table 3 presents the geometric mean risk-adjustment ratings for each of the four groups. As indicated by the preponderance of hazards with mean risk-adjustment factors larger than unity, people thought that most of these activities should be made safer, even though this calls for serious societal action. At the same time, respondents wanted **much** improved safety for only a few activities and technologies. For the League members and students, these included handguns, nuclear power, pesticides, smoking, and spray cans; for the Active Club, smoking headed the list. Experts judged handguns as most in need of risk reduction.

Because mean risk-adjustment scores for most hazards were close to unity, intergroup correlations, shown at the bottom of Table 3, are determined primarily by the few hazards judged as needing large adjustments. Hence, the agreement between the League and student groups in this regard leads to a high correlation (r = .93) as does the agreement between the Active Club and experts (r = .80). However, adjustments desired by League members and students are not closely related to those desired by the Active Club and the experts. The most striking discrepancy in the opinions of these two pairs of groups concerns nuclear power.

The higher the perceived risk of an activity or technology, the higher was the desired reduction of risk. Correlations[2] between

TABLE 3
Geometric means of risk adjustment for subjects in study 1[a]

ACTIVITY OR TECHNOLOGY	LEAGUE	STUDENTS	ACTIVE CLUB	EXPERTS
1. Nuclear power	29.0	32.3	2.4	2.3
2. Motor vehicles	6.1	5.0	1.6	10.5
3. Handguns	17.3	9.0	5.1	62.9
4. Smoking	15.2	11.2	10.2	15.5
5. Motorcycles	5.3	4.0	3.9	3.2
6. Alcoholic beverages	4.4	3.4	2.6	3.0
7. General (private) aviation	2.1	2.0	1.9	1.7
8. Police work	1.8	1.8	1.4	1.4
9. Pesticides	9.5	12.4	2.0	8.3
10. Surgery	1.9	2.8	1.4	3.5
11. Fire fighting	1.1	1.2	1.5	.8
12. Large construction (dams, bridges, etc.)	1.7	2.1	1.2	2.4
13. Hunting	2.5	2.6	2.7	3.1
14. Spray cans	7.8	10.6	1.8	9.5
15. Mountain climbing	1.0	.8	1.4	.6
16. Bicycles	1.5	1.0	1.2	1.3
17. Commercial aviation	1.3	2.2	1.0	1.1
18. Electric power	1.0	1.3	.7	1.8
19. Swimming	1.0	.8	1.2	.8
20. Contraceptives	2.0	5.5	1.0	3.9
21. X rays	1.7	3.6	.8	3.1
22. Skiing	1.0	1.0	1.0	.7
23. Railroads	1.2	1.2	1.0	1.1
24. High school and college football	1.7	1.5	1.2	2.2
25. Food preservatives	2.7	5.6	.9	9.7
26. Food coloring	3.0	4.8	.7	13.9
27. Prescription antibiotics	1.3	2.2	1.0	1.0
28. Power mowers	1.5	1.2	.8	2.1
29. Home appliances	1.1	1.2	.8	1.8
30. Vaccinations	.8	1.2	.8	1.2
All Responses	2.4	2.7	1.4	2.7

[a]Values greater than one mean that the item should be safer; values less than one mean that the item could be riskier.

NOTE: Between-group correlations for log risk adjustments were as follows:

		1	2	3	4
1.	League	−			
2.	Students	.93	−		
3.	Active Club	.57	.36	−	
4.	Experts	.54	.27	.80	−

these two variables were .74, .83, and .78 for the League, student, and Active Club groups, respectively; for experts, the correlation was a weaker .55. The principal exceptions to these positive correlations were several hazards (spray cans, food preservatives, food coloring) that had low perceived risk but high risk-reduction factors and several hazards (fire fighting, police work, general aviation) with the opposite configuration.

Risk Regulation

As a way of further elaborating the meaning of risk adjustment, we asked students to judge the current and desired degree of hazard regulation, using a scale 0-5, ranging from **do nothing** to **ban the product or activity.** The results, shown in Table 4, indicate that nuclear power is viewed as the most strictly regulated activity at present and is also seen as one of the activities most in need of increased regulation. Other hazards judged to need much more regulation were food coloring, food preservatives, handguns, pesticides, smoking, and spray cans. Respondents wanted minor to modest increases in regulation for most hazards and identified only one activity--alcoholic beverages--as over-regulated. By and large, the respondents shunned the extremes. Only 5 percent of the responses--concentrated among football, mountain climbing, skiing, and swimming--fell into the **do nothing** category; similarly, less than 6 percent of responses called for an outright ban, and most of these were distributed among nuclear power, handguns, and food coloring.

Perceived risk and risk adjustment showed moderate positive correlations with the desired level of regulation, with values of .65 and .71, respectively. Exceptions to these positive relationships occurred for motor vehicles, alcoholic beverages, police work, and fire fighting, all of which had high perceived risk and low desired regulation; and food coloring, food preservatives, and prescription antibiotics, which had low perceived risk and relatively high levels of desired regulation.

Risk Characteristics

Participants were asked to rate each of the 30 hazards on nine seven-point scales, each of which represents a characteristic that has been hypothesized to influence risk perception (Lowrance 1976; Rowe 1977; Starr 1969). These scales are described in Table 5. All 30 hazards were rated on one characteristic before the next characteristic was considered.

Means. In contrast to the judgments of perceived risk, judgments of risk characteristics were quite similar for the three groups of laypersons. When means for a particular characteristic were compared between any two lay groups, differences were less than .5 units on the seven-point scale, and only 1 percent of the differences exceeded 1.0. Therefore we concluded that the differences among lay groups were immaterial.

Coefficients of concordance, indicating the average agreement between pairs of individuals, were computed for each group. The results showed that experts agreed with one another in their ratings of risk characteristics more than did members of the other groups.

TABLE 4
Mean judgments of present and desired levels of regulation for 30 activities and technologies[a]

ACTIVITY OR TECHNOLOGY	REGULATION		DIFFERENCE
	PRESENT	DESIRED	
1. Nuclear power	3.11	4.11	1.00
2. Motor vehicles	2.55	2.95	.40
3. Handguns	2.47	3.68	1.21
4. Smoking	1.79	2.84	1.05
5. Motorcycles	2.37	2.68	.31
6. Alcoholic beverages	2.95	2.32	-.63
7. General (private) aviation	3.03	3.18	.15
8. Police work	2.50	3.16	.66
9. Pesticides	2.79	3.79	1.00
10. Surgery	2.92	3.34	.42
11. Fire fighting	2.16	2.29	.13
12. Large construction (dams, bridges, etc.)	2.95	3.53	.58
13. Hunting	2.68	3.60	.92
14. Spray cans	2.74	4.03	1.29
15. Mountain climbing	1.05	1.42	.37
16. Bicycles	1.97	1.97	.00
17. Commercial aviation	3.05	3.05	.00
18. Electric power	1.89	2.61	.72
19. Swimming	1.32	1.55	.23
20. Contraceptives	1.82	1.79	-.03
21. X rays	2.37	2.87	.50
22. Skiing	.87	1.03	.16
23. Railroads	2.53	2.55	.02
24. High school and college football	1.53	1.66	.13
25. Food preservatives	1.87	3.11	1.24
26. Food coloring	1.89	3.11	1.22
27. Prescription antibiotics	2.66	3.18	.52
28. Power mowers	1.11	1.76	.65
29. Home appliances	1.53	1.97	.44
30. Vaccinations	2.18	2.87	.50
All Responses	2.18	2.87	.50

[a]Student subjects; n = 38;

NOTE: Regulation scale labels were:

    0 = do nothing
    1 = monitor and inform
    2 = mild restriction
    3 = moderate restriction
    4 = severe restriction
    5 = ban

TABLE 5
Risk characteristics rated by League members, students, Active Club, and experts

Voluntariness of risk
Do people face this risk voluntarily? If some of the risks are voluntarily undertaken and some are not, mark an appropriate spot towards the center of the scale.

risk assumed
voluntarily   1   2   3   4   5   6   7   risk assumed
involuntarily

Immediacy of effect
To what extent is the risk of death immediate—or is death likely to occur at some later time?

effect
immediate   1   2   3   4   5   6   7   effect
delayed

Knowledge about risk
To what extent are the risks known precisely by the persons who are exposed to those risks?

risk level
known
precisely   1   2   3   4   5   6   7   risk level
not
known

To what extent are the risks known to science?

risk level
known
precisely   1   2   3   4   5   6   7   risk level
not
known

Control over risk
If you are exposed to the risk, to what extent can you, by personal skill or diligence, avoid death?

personal risk
can't be
controlled   1   2   3   4   5   6   7   personal risk
can be
controlled

Newness
Is this risk new and novel or old and familiar?

new   1   2   3   4   5   6   7   old

Chronic-catastrophic
Is this a risk that kills people one at a time (chronic risk) or a risk that kills large numbers of people at once (catastrophic risk)?

chronic   1   2   3   4   5   6   7   catastrophic

Common-dread
Is this a risk that people have learned to live with and can think about reasonably calmly, or is it one that people have great dread for—on the level of a gut reaction?

common   1   2   3   4   5   6   7   dread

Severity of consequences
When the risk from the activity is realized in the form of a mishap or illness, how likely is it that the consequence will be fatal?

certain
not to be
fatal   1   2   3   4   5   6   7   certain
to be
fatal

This greater homogeneity in the expert group caused their mean judgments to be more extreme. As a result, means of experts differed from those of laypersons more than means of laypersons differed from each other. Nevertheless, only about 10 percent of the experts-versus-laypersons comparisons reached any statistical significance (p < .05; two-tailed), and even those discrepancies were small (the largest being only 1.97 units).

Interestingly, both laypersons and experts judged most hazards as **better known to science** than **known to those exposed**. The principal exceptions to this tendency were mountain climbing, police work, and fire fighting--activities about which those exposed to the risk were judged to know more than scientists.

Factor Analysis. The intercorrelations among the mean risk characteristics were sufficiently large to suggest that they might be explained by a few basic dimensions of risk underlying the nine characteristics. In order to identify such underlying dimensions, we used a procedure known as **principal components factor analysis** (Harman 1967). In effect, factor analysis combines highly correlated characteristics into a single composite dimension, called a factor. We obtained for each group a two-factor solution, with the first and second factors' accounting respectively for 56-58 percent and 21-26 percent of the variance among risk characteristics. No other potential factor accounted for more than 8 percent of the variance.

Factor 1 is primarily determined by the characteristics **unknown to those exposed** and **unknown to science,** and to a lesser extent by **newness, involuntariness,** and **delay of effect.** Factor 2 tends to be defined most strongly by **severity of consequences** (certainty of being fatal), **dread,** and **catastrophic potential. Controllability** contributes to both factors.

Factor loadings,[3] shown in Table 6, indicate the degree to which each risk characteristic correlated with the two factors. The commonality index, h^2, reflects the extent to which the two factors accounted for the ratings of each risk characteristic. The fact that commonalities were high indicates that the two-factor solution did a good job of representing the data. As one would expect from the high level of intergroup agreement on the nine risk characteristics (see above), the factor loadings in Table 6 are remarkably similar across groups.

The effective rating of hazards on each factor was determined by computing factor scores.[4] To explore similarities and differences between hazards, we show in Figure 1 a plot of the 30 hazards in a space defined by the factor scores. The location of the hazards within this space conforms to the characteristics defining each factor. The upper extreme of Factor 1 (plotted vertically) is associated with new, unknown, and involuntary hazards whose consequences tend to be delayed. Hazards low on this first factor tend to be familiar, voluntary activities whose risks are well known to science and to those exposed. High scores on Factor 2 (to the right in Figure 1) were associated with events whose consequences are seen as certain to be fatal, often for large numbers of people, in the event of a mishap. Hazards low on Factor 2 are seen as causing injury rather than death to one person at a time. On the basis of these relationships, we have labeled Factor 1 **Unknown Risk** and Factor 2 **Dread Risk.**

TABLE 6
Rotated factor loadings for nine risk characteristics

CHARACTERISTIC	FACTOR 1	FACTOR 2	h^2
L E A G U E			
Not known to exposed	.97	.06	94
Not known to science	.92	.16	87
New/Unfamiliar	.84	.27	78
Effect delayed	.84	-.08	71
Involuntary	.80	.41	80
Not controllable	.64	.60	76
Certainly fatal	-.32	.86	85
Dread	.32	.85	82
Catastrophic	.29	.78	69
% variance explained	50	30	(80)
S T U D E N T S			
Not known to exposed	.98	-.02	96
Not known to science	.92	.12	87
Involuntary	.84	.39	86
New/Unfamiliar	.84	.32	80
Effect delayed	.80	-.17	67
Not controllable	.73	.60	89
Dread	.20	.91	87
Certainly fatal	-.30	.90	91
Catastrophic	.34	.80	76
% variance explained	51	32	(84)
A C T I V E C L U B			
Not known to exposed	.97	.14	95
Not known to science	.88	.16	81
Effect delayed	.84	-.02	70
New/Unfamiliar	.82	.42	84
Involuntary	.75	.47	78
Dread	.16	.88	80
Catastrophic	.16	.83	72
Not controllable	.55	.74	84
Certainly fatal	-.65	.70	91
% variance explained	49	33	(82)
E X P E R T S			
Not known to science	.94	-.12	91
Not known to exposed	.93	-.19	91
Effect delayed	.86	-.10	75
Involuntary	.85	.17	76
New/Unfamiliar	.80	.02	63
Not controllable	.79	.37	76
Dread	.69	.45	68
Certainly fatal	-.20	.90	86
Catastrophic	.14	.83	71
% variance explained	56	21	(77)

Figure 1 also shows the high degree of similarity in factor scores across groups. Although the differences between experts and lay groups were occasionally greater than the differences among lay groups, the hazards were located in essentially the same sector of space for each group. One of the striking features about Figure 1 is the unique and isolated position of nuclear power. Clearly, the individuals in each of our groups judged the characteristics of the risks from nuclear power to be different from those of the other hazardous activities. In particular, risks from nuclear power were judged to be extremely involuntary, catastrophic, dread, fatal, unknown to those exposed, uncontrollable, and unfamiliar.

Relationships

How do the ratings on the nine risk characteristics (and the two factors that summarize them) relate to the judgments of current and desired levels of risk? The answer to this question lies in the correlations among these variables, shown in Table 7. Both perceived risk and desired magnitude of adjustment correlated fairly highly with **dread** for League and student subjects and with **certain to be fatal** for all three nonexpert groups. The risk judgments made by the Active Club showed moderate correlation with delay of effect and lack of knowledge, characteristics that did not strongly influence the judgments of the other three groups. This difference appears again in the analysis of factor scores, which shows that the perceived risk judgments of League and students were related to Factor 2 (**Dread Risk**), whereas the judgments of the Active Club correlated most highly with Factor 1 (**Unknown Risk**). Weighted combinations of three to four risk characteristics led to highly accurate predictions of perceived risk (multiple correlations between .90 and .95) for each of the three groups of laypersons.

Of special interest is the finding that the experts' risk judgments did not correlate highly with any of the characteristics or factors. Their risk judgments were, however, closely related ($r = .92$) to technical estimates of the average annual fatalities from each activity (Slovic, Fischhoff, and Lichtenstein 1979). In other words, the experts seemed to equate riskiness with annual fatalities, which may account for the great range of values they used to discriminate among hazards. Risk judgments of laypersons were less closely related to fatality estimates (the correlations were between .50 and .62).

No clear pattern emerges from the correlations between desired risk adjustment and the various characteristics and factor scores. Dread and certain to be fatal, both Factor 2 characteristics, correlated most highly with desired adjustments, especially for the League and student groups. Judgments of desired regulatory stringency, made only by students, were related to dread, catastrophic potential, voluntariness, control, and certainty of being fatal. As a result, desired regulation correlated substantially with Factor 2, but very little with Factor 1.

Results of Studies 2 and 3

Studies 2 and 3 measured perceived risk, desired regulation, and an expanded set of risk characteristics. The set of hazards was

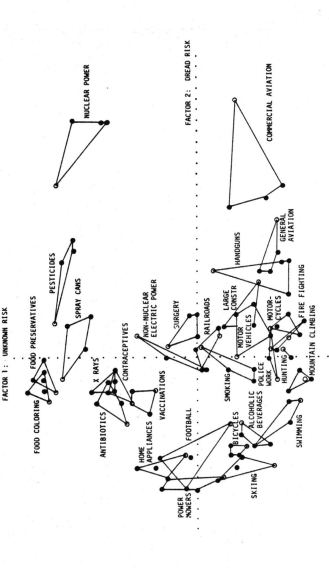

Figure 1. Location of 30 hazards within the two-factor space obtained from League, student, Active Club, and expert groups. Connected lines join or enclose the loci of four group points for each hazard. Open circles represent data from the expert group. Unattached points represent groups that fall within the triangle created by the other three groups.

TABLE 7
Correlations between risk characteristics and measures of perceived
and desired regulation

RISK CHARACTERISTIC		LOG PERCEIVED RISK	LOG RISK ADJUSTMENT FACTOR	DESIRED REGULATION
Involuntary	League	-.04	.36	
	Students	.27	.54	.62
	Active Club	-.39	-.32	
	Experts	.13	.50	
Effect delayed		-.13	.38	
		.10	.48	.16
		-.48	-.10	
		.18	.48	
Unknown to those exposed		-.34	.23	
		-.06	.37	.28
		-.67	-.37	
		-.08	.31	
Unknown to science		-.24	.29	
		.00	.37	.29
		-.58	-.35	
		-.13	.40	
Controllable		.06	-.24	
		-.38	-.57	-.64
		.24	.18	
		-.07	-.38	
Familiar		.14	-.36	
		-.21	-.53	-.39
		.44	.23	
		.23	-.26	
Catastrophic		.23	.38	
		.44	.47	.63
		-.05	.00	
		-.04	-.11	
Dread		.58	.75	
		.81	.79	.70
		.34	.41	
		-.14	.57	
Certain to be fatal		.75	.53	
		.74	.51	.62
		.67	.63	
		.31	.11	
Factor 1 (Unknown Risk)		-.36	.22	
		-.01	.40	.28
		-.72	-.44	
		-.02	.47	
Factor 2 (Dread Risk)		.61	.56	
		.73	.60	.72
		.31	.28	
		.19	.10	

NOTE: Desired Regulation was judged only by the student group.

also expanded, to 90 in Study 2 and to 81 in Study 3. College students made the judgments in both studies. Because of the similarity in design for the two studies, their results are reported together in this section.

Perceived Risk and Desired Regulation

In Study 2, nuclear weapons, warfare, DDT, handguns, crime, and nuclear power were judged the most risky hazards. Solar electric power, jogging, sunbathing, cosmetics, and roller coasters were rated least risky. In Study 3, the highest risk ratings were assigned to nuclear weapons (warfare and test fallout), nuclear reactor accidents, automobile accidents, and radioactive wastes. Tractors, snowmobiles, power mowers, and swimming pools were judged to pose the least risk.

Most of the mean risk adjustments were greater than 1.0, again demonstrating that people wanted a reduction in current risk levels for most hazards. Very great reductions in risk were desired for nuclear weapons, warfare, handguns, crime, nuclear power, terrorism, and nerve gas. Over all the hazards, about half of the means for desired regulation were above 3.0, indicating a desire for moderate to strong restrictions. As in Study 1, mean risk adjustment and perceived risk were highly correlated (r = .91 and r = .89 in Studies 2 and 3, respectively),[5] indicating that people wanted stricter regulation of the hazards they viewed as most risky. This is also implied by the correlation between perceived risk and desired regulation (.78 and .73 in Studies 2 and 3, respectively.

Risk Characteristics

In Study 2, as in Study 1, most risks were thought to be known better to science than to those who were exposed. The only risks for which those exposed were thought to be more knowledgeable than scientists were those from police work, marijuana, contraceptives (judged relatively unknown both to science and to those exposed), boxing, skiing, hunting, and several other sporting activities. Only 25 of the 90 hazards were judged to be decreasing in riskiness; two of them (surgery and pregnancy/childbirth) were thought to be decreasing greatly. Risks from 62 hazards were judged to be increasing, 13 of them markedly so. The risks from crime, warfare, nuclear weapons, terrorism, national defense, herbicides, and nuclear power were judged to be increasing most. None of the risks was judged to be easily reduced, with the lowest (home appliances and roller coasters) scoring 3.2 on a seven-point scale. Judgments in Study 3 followed similar patterns.

Factor Analysis. Principal components factor analysis followed by varimax rotation indicated that, in both studies, the pattern of correlations among risk characteristics could be represented by three factors, accounting for 79 percent and 85 percent of the total variance in Studies 2 and 3, respectively. No other potential factor accounted for more than 5 percent of the variance. Factor loadings and commonalities, h^2, are shown in Table 8. The high values of h^2 for all but a few characteristics indicate that the three factors did a good job of accounting for the variance in the individual risk characteristics.

TABLE 8
Rotated factor loadings for 18 risk characteristics

S T U D Y 2

CHARACTERISTIC	FACTOR 1	FACTOR 2	FACTOR 2	h^2
Severity not controllable	.92	.07	-.14	.87
Globally catastrophic	.91	.09	.24	.89
Dread	.91	-.02	.02	.83
Certainly fatal	.88	-.16	-.15	.83
Little preventive control	.88	-.06	-.26	.84
Inequitable	.85	.11	.33	.85
Catastrophic	.84	.06	.16	.73
Threatens future generations	.78	.23	.45	.86
Not easily reduced	.75	-.21	.00	.60
Risk increasing	.71	-.11	.40	.68
Involuntary	.68	.43	.25	.72
Much personal exposure	.67	.08	.58	.86
Not observable	-.10	.89	.22	.85
Not known to exposed	.09	.88	.32	.87
Effect delayed	-.20	.80	.45	.88
New/Unfamiliar	.31	.76	-.16	.70
Not known to science	-.08	.66	-.12	.45
Many people exposed	.05	.20	.90	.86
% variance accounted for	46	20	13	79

S T U D Y 3

CHARACTERISTIC	FACTOR 1	FACTOR 2	FACTOR 2	h^2
Severity not controllable	.93	-.01	.02	.86
Dread	.88	.19	.19	.85
Certainly fatal	.88	-.30	-.07	.87
Not easily reduced	.83	.24	.26	.81
Inequitable	.81	.30	.26	.81
Little preventive control	.79	.40	.28	.87
Catastrophic	.76	.27	.41	.82
Globally catastrophic	.73	.40	.48	.92
Threatens future generations	.67	.56	.43	.95
Risk increasing	.66	.41	.28	.68
Involuntary	.63	.42	.48	.81
Not observable	.17	.87	.28	.87
Effect delayed	-.19	.87	.24	.85
Not known to science	.35	.85	.13	.86
New/Unfamiliar	.34	.80	-.17	.78
Not known to exposed	.16	.79	.41	.82
Many people exposed	.14	.21	.93	.92
Much personal exposure	.47	.16	.81	.91
% variance accounted for	41	27	17	85

Further examination of Table 8 shows that the factors from both studies were extremely similar; moreover, the first two factors in Studies 2 and 3 seem much the same as the two factors that emerged in Study 1.[6] Factor 1 is associated with lack of control, lethality, high catastrophic potential, reactions of dread, inequitable distribution of risks and benefits (including transfer to future generations), and the belief that risks are increasing and not easily controlled; it thus corresponds closely to the factor **Dread Risk** of Study 1. Factor 2 is associated with risks that are unknown, unobservable, new, and delayed in their manifestation; it therefore corresponds closely to the factor **Unknown Risk** of Study 1. Factor 3 is primarily determined by the number of people exposed. One's own personal exposure contributes clearly to it in Study 3 and secondarily to it in Study 2. The label **Societal and Personal Exposure** seems appropriate for this third factor.

Factor scores for Factors 1 and 2 are plotted in Figures 2 and 3. The hazards located at the extremes in this factor space give further support for our choice of factor labels. At the high (right) end of Factor 1 are found such dreaded hazards as nerve gas, nuclear power accidents, radioactive waste, nuclear weapons, terrorism, warfare, and crime (Study 2); and radioactive wastes and fallout, reactor accidents, and nuclear weapons (in Study 3). In contrast, hazards at the opposite end of Factor 1 pose risks that elicit little concern (e.g., home appliances, bicycles, caffeine). Extreme hazards on the positive vertical axis in both figures fit the theme of unknown risk, whereas the negative vertical axis is populated with hazards that are known, familiar, and observable.

Not shown in Figures 2 and 3 is Factor 3. In the two studies, hazards high on Factor 3 included automobiles, caffeine, alcoholic beverages, smoking, and food preservatives, all known for their high level of societal and personal exposure. In contrast, hazards low in Factor 3 included lasers, sport parachutes, laetrile, auto racing, solar electricity, scuba diving, steeplejacking, and uranium mining and milling.

In general, the subset of hazards common to Studies 2 and 3 occupies roughly similar positions in Figures 2 and 3. Risks of nuclear energy retained their unique locations in the upper right quadrant of the space defined by the first two factors, although each nuclear power component (Study 3) was relatively less unknown than nuclear power in the aggregate (Study 2). Activities such as crime, warfare, terrorism, and use of nerve gas were as much dreaded (Factor 1) as nuclear power, but not as unknown (Factor 2).

One noteworthy shift occurred with recombinant DNA hazards, which rose from 30th highest on **Dread Risk** in Study 2 to seventh in Study 3. DNA technology was also judged much higher in perceived risk in Study 3. Whether these changes were due to sampling fluctuation, to the slight change in wording (from **DNA research** to **DNA technology**), or to a shift in students' attitudes during the period March 1979 to August 1980 is not known. If the shift is real, the change could portend trouble for this technology, since the combination of dread with lack of familiarity and knowledge appears to be very threatening to society and difficult to manage, as exemplified by the problems associated with nuclear power.

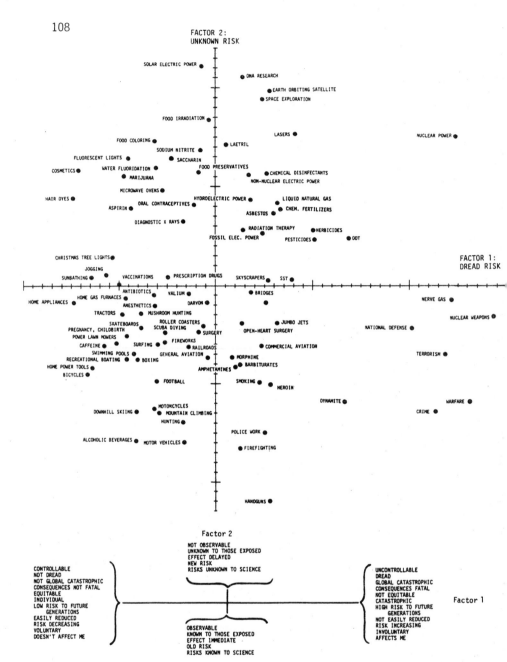

Figure 2. Hazard locations on Factors 1 and 2 of the three-dimensional structure derived from the interrelationships among 18 risk characteristics in Study 2. Factor 3 (not shown) reflects the number of people exposed to the hazard and the degree of one's personal exposure. The diagram beneath the figure illustrates the characteristics that comprise the two factors.

Figure 3. Factors 1 and 2 of the three-dimensional structure derived from the interrelationships among 18 risk characteristics in Study 3. Factor 3 (not shown) reflects the number of people exposed to the hazard and the degree of one's personal exposure.

Relationships

Table 9 presents the correlations between measures of perceived and desired regulation, the risk characteristics, and the factors underlying these characteristics. In both studies, perceived risk, adjusted risk, and desired regulation were closely related to the factor **Dread Risk** (see Figure 4). The individual characteristics **dread, threat to future generations,** potential for global **catastrophe,** and **perceived inequity** were especially highly correlated with the three dependent measures. Factor 3, **societal and personal exposure,** correlated modestly with perceived risk but related less well to desired risk adjustment and regulation. In Study 2, the factor **Unknown Risk** and several of its main components (unknown to those exposed to science, not observable) had weak negative relationships with the three risk and regulation measures. In Study 3, the factor **Unknown Risk** displayed weak to moderate positive relationships with those measures. This small but reliable difference seems to be attributable to the large number of chemical hazards examined in Study 3, most of which were judged as relatively unknown, unobservable, and high in risk.

TABLE 9
Correlations between perceived risk, risk adjustment, risk regulation, and risk characteristics

CHARACTERISTIC	PERCEIVED RISK		LOG RISK ADJUSTMENT		DESIRED REGULATION	
	Study 2	Study 3	Study 2	Study 3	Study 2	Study 3
Dread	.83	.82	.87	.93	.84	.85
Threatens future generations	.80	.82	.77	.85	.76	.80
Globally catastrophic	.78	.87	.82	.87	.74	.79
Certain to be fatal	.74	.52	.74	.59	.80	.75
Risk increasing	.73	.80	.76	.80	.56	.62
Affects me personally	.70	.84	.65	.68	.61	.59
Risks & benefits inequitable	.68	.68	.73	.80	.74	.84
Not easily reduced	.63	.77	.69	.77	.64	.76
Severity not controllable	.63	.53	.65	.63	.73	.78
Little preventive control	.51	.64	.57	.68	.66	.80
Catastrophic	.50	.71	.54	.71	.62	.80
Involuntary	.39	.63	.42	.63	.54	.75
Many people exposed	.25	.67	.14	.43	.12	.39
New/Unfamiliar	.17	.27	.16	.39	.33	.38
Effects immediate	.11	.17	.17	.22	.12	.09
Not known to those exposed	-.06	.35	-.09	.33	.06	.40
Not observable	-.19	.42	-.23	.47	-.08	.44
Not known to science	-.27	.42	-.22	.49	-.23	.46
Factor 1 (Dread Risk)	.74	.66	.79	.77	.83	.84
Factor 2 (Unknown Risk)	-.22	.15	-.22	.25	-.03	.20
Factor 3 (Societal and Personal Exposure)	.41	.58	.29	.30	.11	.24

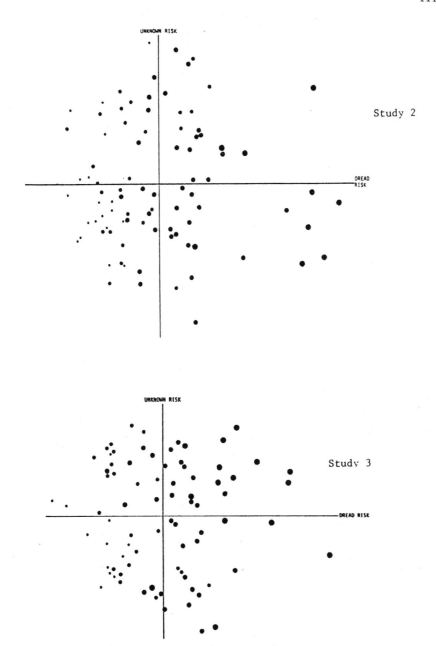

Figure 4. Attitudes towards regulation of the hazards shown in Figures 2 and 3. The larger the dot, the greater the desire for strict
regulation to reduce risk. The data reflect correlations greater
than .83 with the horizontal factor (Dread Risk) and minimal correlation with the vertical factor.

Table 9 shows that certain individual risk characteristics and factors can do a good job by themselves in predicting the measures of perceived and desired risk. Further analysis indicated that it is possible to develop regression equations that predict these measures quite accurately on the basis of combinations of risk characteristics and factors. These equations produced multiple correlations in the range .89 to .95.

Annual mortality rates for the 81 hazards in Study 3 have been estimated by our colleagues at Clark University (see chapter 4). The correlation between these rates and perceived risk was only .17, lending further support to the conclusion that laypersons' perceptions of risk are not much influenced by annual mortality. A plot of this relationship is presented in Figure 5.

Summary of Key Results

The present research extended the design employed by Fischhoff et al. (1978) to examine multiple groups, greater numbers of risk characteristics, and a broader domain of hazards. Several new results were obtained. The examination of diverse groups showed that psychometric scaling can identify and quantify similarities and differences among groups in their risk perceptions and attitudes. These data promise to be politically relevant, as is illustrated by the judgments about nuclear power made by League and Active Club members and their corresponding voting behavior in a subsequent nuclear power referendum.

Experts' judgments of risk differed systematically from those of nonexperts. As we have noted in a previous report (Slovic, Fischhoff, and Lichtenstein 1979), experts' risk perceptions correlated quite highly with technical estimates of annual number of fatalities; their perceptions also reflected the vast range, from high to low risk, inherent in the statistical measures. In contrast, nonexperts' perceptions of risk were compressed into a small range and were not as highly correlated with annual mortality statistics. When asked about perceived risk, it appears as though experts translated the task into one of judging well-defined technical statistics, whereas laypeople gave a judgment of risk that was influenced by a variety of factors and used a psychophysical scale whose units have no intrinsic meaning.

Despite the sizable group differences in risk perceptions and attitudes toward specific hazards, there was remarkable agreement across groups of laypeople and experts displayed remarkable agreement in characterizing hazards on various aspects or riskiness such as knowledge, controllability, dread, etc. Subsets of these characteristics correlated quite highly with laypersons' risk perceptions, but not with experts' perceptions. Several trends in these characteristics were observed in more than one study. In particular, the risks from most hazardous activities were judged to be increasing, not easily reduced, and better known to science than to the people exposed to the risks.

The risk characteristics exhibit a high correlation with one another. Factor analysis of the nine characteristics examined in Study 1 showed that these were reducible to two factors that were virtually identical for each of the four groups in that study. One factor discriminated between known and unknown risks, with hazards

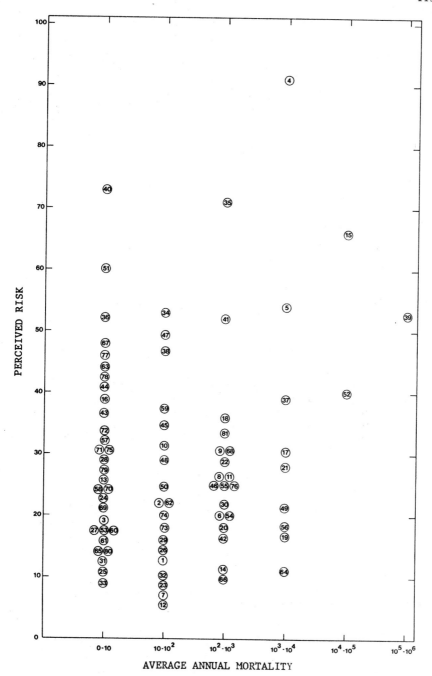

Figure 5.   Perceived risk vs. annual mortality for 81 hazards in
Study 3.   The hazards associated with each circled number are listed
in Table 12.

scoring high in the unknown direction also tending to be seen as new, unfamiliar, involuntary, and having delayed effects. The second factor, labeled **Dread Risk**, discriminated between hazards whose adverse consequences were seen to be fatal, often to large numbers of people, and hazards causing injuries, rather than fatalities, to single individuals. Within this two-factor space, the positions of hazards were similar for each of the four groups studied. Nuclear power consistently appears alone in the highly unknown, highly dreaded sector of the factor space.

Although the hazards in Study 3 were more precisely specified and included more chemicals than the hazards examined in Study 2, analysis of the 18 risk characteristics found them reducible to the same three underlying factors in both studies. Factors 1 and 2 corresponded closely to the two factors that emerged from the much smaller hazard set in Study 1 and were thus assigned the same labels, **Dread Risk** and **Unknown Risk.** Factor 3 was primarily determined by the number of people exposed to the risk and secondarily determined by whether one was personally exposed to the risk. Despite the addition of a number of particularly noxious hazards (e.g., nerve gas, crime, warfare), the risks from nuclear energy retained their unique position in the highly unknown and highly dreaded sector of the factor space.

In each of the three studies, the factor labeled **Dread Risk** correlated most highly with the laypersons' judgments of perceived and desired risk. This factor's correlations with desired stringency of regulation were especially high. The factor labeled **Unknown Risk** tended not to correlate very highly with perceived risk and desire for reduced risk or regulation. Factor 3, **societal and personal exposure,** tended to correlate moderately with perceived risk but only slightly with risk adjustment and desired regulation.

The regulation questionnaire developed here appears to be a useful first step for scaling the present and desired strictness of social control for any activity or technology. Desired stringency of regulation was correlated highly with perceived risk and the degree of adjustment needed to make risk levels acceptable, but it was not synonymous with these considerations (e.g., certain activities, such as drinking alcoholic beverages, driving automobiles, and smoking, were judged as highly risky but not in need of strict regulation). Regulatory stringency also correlated very highly with characteristics loading on the factor labeled **Dread Risk.**

Policy Implications

The present study tells an incomplete story. Despite our attempt to study diverse groups of people, we have by no means obtained a representative sample of opinions. Our respondents were all highly educated persons, most of them from one geographic locale. National surveys and surveys of diverse interest groups, regulators, and legislators are in order, although it will take substantial effort to redesign the questionnaires and sets of items to meet the special demands of such studies.

Until such studies are conducted, we believe that the present study can help technology managers understand and forecast public response and future acceptance and rejection of specific technologies, albeit in a preliminary way. A case in point is nuclear power, whose isolated position in the factor space reflects our respondents' view that its risks are exceptionally dreaded and unknown. Once such concerns are identified, one can ask how likely they are to change over time in response to information and experience, such as a good safety record or an accident. We have argued that people's strong fears of nuclear power and their political opposition to it are not irrational but can be understood as a logical consequence of their concerns about considerations such as equity, catastrophic potential, and the safety of future generations (Slovic, Lichtenstein, and Fischhoff 1979).

We also contend that accidents occurring with unknown and potentially catastrophic technologies will be seen as signals that portend loss of control over the technology and indicate further, and possibly more severe, losses (Slovic, Fischhoff, and Lichtenstein 1980a; Slovic, Lichtenstein, and Fischhoff 1984). Hence, a small accident's occurring in a technology that occupies the unknown and dread quadrant of the risk space is likely to have much greater impact than an accident, causing similar immediate damage, occurring in a technology located elsewhere in the factor space. That even a "small" accident in the nuclear industry will likely have immense impact carries direct implications for the setting of safety standards (Slovic, Fischhoff, and Lichtenstein 1980b).

Based on our student respondents, it appears DNA technology is located in the unknown and dread portion of the factor space. If similar views are held by the general population or its politically active members, DNA technologists have cause for concern; for they may have to face some of the same problems and opposition now confronting the nuclear industry.

Beyond forecasting public response, we believe our work is relevant for comparing risks. One frequently advocated approach to broadening people's perspectives is to present quantified risk estimates for a variety of hazards. These presentations typically involve elaborate tables and even "catalogs of risks" in which some unidimensional index of death or disability is displayed for a broad spectrum of life's hazards. These indices include risks per hour of exposure (Sowby 1965), annual probabilities of death (Wilson 1979), and reductions in life expectancy (Cohen and Lee 1979; Reissland and Harries 1979). Compilers of such information typically assume that the data will be useful for decision making even though such comparisons have no logically necessary implications (Fischhoff et al. 1981).

The present studies suggest that mortality comparisons will not, by themselves, be adequate guides to personal or public decisions. As we have shown, laypeople's risk perceptions and attitudes are determined not only by annual mortality rates or mean loss of life expectancy but also by numerous other characteristics of hazards.

It is possible that the broad characterization of risk given here can be applied directly to comparing hazards. In its multivariate character it is analogous to the taxonomy of hazards described in chapter 4, where hazards are classified via biological, physical, and social dimensions of causal structure. Unfortunately we do not yet know how to incorporate such broadened conceptualizations into the processes of policy analysis and decision making. Cost/benefit and decision analysts have made some progress towards developing procedures for treating such cases (Fischhoff et al. 1981), but there are still many problems to be overcome before practical and logically defensible techniques will be available for decision makers.

In whatever way these issues are resolved, it seems clear from our work that attempts to characterize hazards, set safety standards, and make decisions about hazards will founder in conflict if policy makers insist on a narrow definition of risk that does not correspond to people's concerns.

APPENDIX

Study 1 Methodology

Hazards. The 30 hazards (recall Table 2) included eight examined in Starr's (1969) pioneering study of societal risk preferences along with 22 additional hazards chosen to vary widely in the nature and magnitude of their risks and benefits. We selected some hazards whose risks are undertaken voluntarily and others that are involuntary, some whose effects are immediate and others with delayed effects, some whose risks are well known and others whose risks are rather unknown, and so forth.

Subjects. Four groups of subjects participated in Study 1. One group consisted of members of the Eugene, Oregon, League of Women Voters and their spouses, who took part in return for a contribution to the organization's treasury. In all, 76 individuals (52 women and 24 men) returned completed, anonymous questionnaires. Student subjects were recruited by means of an advertisement placed in the University of Oregon student newspaper. One sample of students, consisting of 33 women and 36 men, performed the same tasks as the League sample. Another student sample (27 women and 11 men) completed only one task, in which they rated the 30 items according to existing and desired regulatory stringency. Group 3 consisted of 47 members (3 women and 44 men) of the Eugene, Oregon, "Active Club," an organization of business and professional people devoted to community service activities. The fourth group included 15 persons (3 women and 12 men), selected nationwide for their professional interest and expertise in risk assessment. This expert group ranged in age from 29 to 68 (median 42) and included a geographer, an environmental policy analyst, an economist, a lawyer, a biologist, and a government regulator of hazardous materials.

Instructions: Perceived Risk. Participants were asked, for each of the 30 hazards, "to consider the risk of dying (across all U. S. society as a whole) as a consequence of this activity or technology." To simplify the evaluation task, we listed each hazard on a separate index card. Respondents were told first to study the hazards individually, thinking of all the possible ways someone might die from each (e.g., fatalities from non-nuclear electricity were to include deaths resulting from the mining of coal and other energy-production activities as well as electrocution; motor vehicle fatalities were to include collisions with bicycles and pedestrians). Next, the respondents ordered the cards from least to most risky and, finally, they assigned numerical risk values by giving a rating of 10 to the least risky item and making the other ratings accordingly. They also received additional suggestions, clarifications, and encouragement to do as accurate a job as possible.

Instructions: Risk Adjustment. After rating the risks, each group of participants judged the acceptability of the level of risk currently associated with each hazard. The instructions were as follows:

> The acceptable level of risk is not the ideal risk. Ideally, the risks should be zero. The acceptable level is a level that is "good enough," where "good enough" means you think that the advantages of increased safety are not worth the costs of reducing risk by restricting or otherwise altering the activity. For example, we can make drugs safer by restricting their potency; cars can be made safer at a cost, by improving their construction or requiring regular safety inspection. We may, or may not, believe such restrictions are necessary.
>
> If an activity's present level of risk is acceptable, no special action need be taken to increase its safety. If its riskiness is unacceptably high, **serious action**, such as legislation to restrict its practice, should be taken. On the other hand, there may be some activities or technologies that you believe are currently safer than the acceptable level of risk. For these activities, the risk of death could be higher than it is now before society would have to take serious action.

Participants were asked to record on their answer sheets one of the following response options: (a) **Could be riskier: It would be acceptable if it were _____ times riskier;** (b) **It is presently acceptable;** and (c) **Too risky: To be acceptable, it would have to be _____ times safer.**

Instructions: Risk Characteristics. Participants rated each of the 30 hazards on nine seven-point scales (Table 5) each of which represented a characteristic that had been hypothesized to influence risk perceptions (Lowrance 1976;Rowe 1977;Starr 1969). All 30 hazards were rated on one characteristic before the next characteristic was considered.

Instructions: Risk Regulation. The risk adjustment task, described above, provides one index of the acceptability of risks, but its relevance to behavior and policy making is unclear. As Fischhoff et al. (1978,150-151) noted:

> If people assert that motor vehicles should be five times safer, does this mean that they would accept an immediate Draconian step designed to attain that goal? Does it mean that a five-fold reduction in risk is a long-term goal for society and that meaningful (but not necessarily drastic) steps should be taken until that goal is reached, or does it mean that the adjustment ratios...only measure relative concerns about the risk levels of various activities? A more behaviorally relevant scale of acceptability should be developed, with clearer implications for regulatory actions.

As a first step in this direction, we designed a questionnaire in which respondents judged the current and the desired strictness of regulatory action for each hazard. In Study 1, this regulation questionnaire was administered to a new group of 38 students. The instructions began with information about the nature and stringency of various regulatory actions. Next, a six-category scale of increasing regulatory strictness was defined as follows:

0: Do nothing
1: Monitor the risk and/or inform those exposed
2: Place **mild** restriction on (a) who can do the activity, use the product, or operate the technology and (b) how, when, and where something can be done or used
3: Place **moderate** restriction on who, how, when, and where
4: Place **severe** restriction on who, how, when, and where
5: **Ban** the product or activity

The instructions stressed that as one moves from 0 to 5 on the scale, several things tend to happen:

(a) freedom to use the product or do the activity becomes increasingly limited,
(b) risk decreases,
(c) benefits also decrease, and
(d) the economic costs of regulation increase

For each of the basic 30 hazards, subjects first indicated, to the best of their knowledge, which category of regulation presently exists. They then indicated the category of regulation they believed **should** exist. If they indicated Categories 2, 3, 4, or 5, they were also allowed to indicate Category 1 if they wished to do so. If they did mark more than one category, only the higher-numbered (more restrictive) category was used in the data analysis.

Study 2 Methodology

Study 2 was designed to examine a much broader set of hazards (90 instead of 30, as shown in Table 10), and risk characteristics (18 instead of 9). The participants were 175 University of Oregon students (90 women and 85 men) who were asked to judge the perceived risk, desired risk adjustment, and the need for regulation of these hazards, and to rate them on various risk characteristics. Insofar as Study 2 sought to examine three times as many hazards and twice as many risk characteristics as Study 1, no participant completed the entire task. Instead, two groups (42 individuals in one and 45 in the second) shared the assignment of judging the perceived risk and the need for risk adjustment. Each group judged a different set of 45 hazards. Judgments of the 18 risk characteristics were made by four groups of 22 persons; each individual rated the 90 hazards on either four or five characteristics. As a further step to keeping the judgment task manageable, we replaced the cumbersome

TABLE 10

Hazards rated in studies 1 and 2

STUDY 1

1. Alcoholic beverages
2. Bicycles
3. Commercial aviation
4. Contraceptives
5. Non-nuclear electric power
6. Fire fighting
7. Food preservatives
8. Food coloring
9. General aviation
10. Handguns
11. High school and college football
12. Home appliances
13. Hunting
14. Large construction
15. Motorcycles
16. Motor vehicles
17. Mountain climbing
18. Nuclear power
19. Pesticides
20. Power mowers
21. Police work
22. Prescription antibiotics
23. Railroads
24. Skiing
25. Smoking
26. Spray cans
27. Surgery
28. Swimming
29. Vaccinations
30. X rays

STUDY 2

1. Home gas furnaces
2. Home appliances
3. Home power tools
4. Microwave ovens
5. Power lawn mowers
6. Handguns
7. Terrorism
8. Crime
9. Nerve gas
10. Nuclear weapons
11. National defense
12. Warfare
13. Bicycles
14. Motorcycles
15. Motor vehicles
16. Railroads
17. General aviation
18. SST
19. Jumbo jets
20. Commercial aviation
21. Anesthetics
22. Vaccinations
23. Pregnancy, childbirth
24. Open-heart surgery
25. Surgery
26. Radiation therapy
27. Diagnostic X rays
28. Alcoholic beverages
29. Caffeine
30. Water fluoridation
31. Food coloring
32. Saccharin
33. Sodium nitrite
34. Food preservatives
35. Food irradiation
36. Earth orbit satellite
37. Space exploration
38. Lasers
39. Asbestos
40. Police work
41. Firefighting
42. Christmas tree lights
43. Cosmetics
44. Fluorescent lights
45. Hair dyes
46. Chemical disinfectants
47. DNA research
48. Liquid natural gas
49. Smoking
50. Tractors
51. Chemical fertilizers
52. Herbicides
53. DDT
54. Pesticides
55. Aspirin
56. Marijuana
57. Heroin
58. Laetrile
59. Amphetamines
60. Barbiturates
61. Darvon
62. Morphine
63. Oral contraceptives
64. Valium
65. Antibiotics
66. Prescription drugs
67. Boxing
68. Downhill skiing
69. Fireworks
70. Football
71. Hunting
72. Jogging
73. Mountain climbing
74. Mushroom hunting
75. Recreational boating
76. Roller coasters
77. Scuba diving
78. Skateboards
79. Sunbathing
80. Surfing
81. Swimming pools
82. Fossil electric power
83. Hydroelectric power
84. Solar electric power
85. Non-nuclear electric power
86. Nuclear power
87. Dynamite
88. Skyscrapers
89. Bridges
90. Dams

TABLE 11
Additional risk characteristics rated in studies 2 and 3

9. Control over risk
Risks can be controlled either by preventing mishaps or by reducing the severity of mishaps after they occur. To what extent can people, by personal skill or diligence, prevent mishaps or illnesses from occurring?

Much preventive
control   1   2   3   4   5   6   7   Little preventive control

10. Control over risk
Risks can be controlled either by preventing mishaps or by reducing the severity of mishaps after they occur. After a mishap or illness does occur, to what extent can proper action reduce the likelihood or number of fatalities (i.e., the severity)?

Severity can't
be controlled   1   2   3   4   5   6   7   Severity can be controlled

11. Exposure
How many people are exposed to this risk in the United States?

Few   1   2   3   4   5   6   7   Many

12. Equity
To what extent are those who are exposed to the risks the same people as those who receive the benefits?

Risk and benefits
matched   1   2   3   4   5   6   7   Risks and benefits mismatched

13. Future generations
To what extent does present pursuit of this activity or technology pose risks to future generations?

Very little threat 1   2   3   4   5   6   7   Very great threat

14. Personal exposure
To what extent do you believe that you are personally at risk from this activity, substance or technology?

I am not at risk   1   2   3   4   5   6   7   I am very much at risk

15. Global catastrophe
To what extent does pursuit of this activity, substance, or technology have the potential to cause catastrophic death and destruction across the whole world?

Very low cata-
strophic potential 1   2   3   4   5   6   7   Very high cata-strophic potential

16. Observability
When something bad is in the process of happening because of this activity, substance, or technology, to what extent is the damage observable?

Observable   1   2   3   4   5   6   7   Not observable

17. Changes in risk
Are the risks from this activity, substance, or technology changing?

Increaing greatly 1   2   3   4   5   6   7   Decreasing greatly

18. Ease of reduction
How easily can risks from this activity or technology be reduced?

Easily reduced   1   2   3   4   5   6   7   Not easily reduced

TABLE 12
Hazards rated in study 3

1.	Downhill skiing	45.	Cadmium usage
2.	Dynamite	46.	Lead paint
3.	Fireworks	47.	Automobiles (airborne
4.	Nuclear weapons (use in war)		lead)
5.	Handguns	48.	Mercury
6.	Coal-mining accidents	49.	Coal mining (occupational
7.	Snowmobiles		exposure)
8.	Electric wiring & appliances (electric shock)	50.	Uranium mining & milling (occupational exposure)
9.	Smoking (accidental fires)	51.	Radioactive waste (expo-
10.	Large dams		sure of workers and
11.	Railroad trains (collision)		general public)
12.	Farm tractors	52.	Alcohol (liver damage and
13.	Microwave ovens		other disorders)
14.	Skateboards	53.	Saccharin
15.	Automobile accidents (includ-	54.	Valium
	ing pedestrians, cyclists)	55.	Oral contraceptives
16.	LNG storage & transport	56.	Darvon
17.	Diagnostic medical X rays	57.	Mirex (insecticide: expo-
18.	Electrical wiring & appliances (fires)		sure of workers and general public)
19.	General (private) aviation	58.	Water fluoridation
20.	Bicycles	59.	Polyvinyl chloride (expo-
21.	Motorcycles		sure of workers and
22.	Commercial aviation		general public)
23.	Power mowers	60.	Caffeine
24.	Bridges (construction & use)	61.	Coal-tar hair dyes
25.	Chainsaws	62.	Vaccines
26.	Elevators	63.	DDT
27.	High construction & repair (by steeplejacks)	64.	Recreational boating
		65.	Laetrile
28.	High tension electrical wires (exposure to ionizing radia-	66.	Home swimming pools
	tion)	67.	Fossil fuels (atmospheric CO_2 pollution)
29.	Automobile racing	68.	Antibiotics (effects of
30.	Skyscrapers (fires)		resistant bacteria)
31.	Orbiting space satellites (crashes to earth)	69.	Chlorination of drinking water
32.	Sport parachutes	70.	D-con (rat poison)
33.	Trampolines	71.	DES
34.	Automobiles (CO exhaust)	72.	Nitrogen fertilizers (ni-
35.	Nuclear weapons testing (fallout)		trogen oxides pollution)
		73.	Hexachlorophene
36.	Recombinant DNA technology	74.	IUDs
37.	Alcohol (accidents involving machinery)	75.	Nitrites
		76.	Rubber manufacture
38.	Pesticides	77.	SST (ozone depletion)
39.	Smoking (heart disease, cancer, and other disease)	78.	2,4,5-T
		79.	Trichloroethylene
40.	Nuclear reactors (accidents)	80.	Underwater construction
41.	Coal burning (atmospheric pollution)		and repair
		81.	Asbestos insulation
42.	Aspirin		(exposure of workers and
43.	Nerve gas (accidents)		general public)
44.	PCBs		

card-sorting method with an instruction to rate individual items on a 0-100 scale from **not risky** to **extremely risky.**

The eighteen risk characteristics included eight (excluding controllability) used in Study 1 and 10 new scales, as shown in Table 11. Several of the old scales were changed as follows: (a) the labels on the **chronic-catastrophic** scale were changed to read **individual** (1) versus **catastrophic** (7); (b) the labels on the **common-dread** scale were changed to read **not dread** (1) versus **dread** (7); (c) the **controllability** scale was split into two separate characteristics representing control over the occurrence of a mishap (preventive control) and control over the consequences given that something did go wrong (control of severity). The remaining characteristics were selected to represent additional concerns thought to be important by risk assessment researchers. As in Study 1, each characteristic was rated on a seven-point bipolar scale representing the extent to which the characteristics described the hazard.

Study 3 Methodology

Participants in Study 3 were 10 women and 24 men who responded to an advertisement in the student newspaper at the University of Oregon. They judged each of the 81 hazards (Table 12) in terms of perceived risk, need for risk adjustment, current and desired regulation, and standing on each of 18 risk characteristics. With exception of one procedural difference—each participant spent six hours over a two-day period rating all 81 hazards on every task and all characteristics—tasks and instructions were identical to those of Study 2.

Appended to the instructions was a glossary providing brief descriptions of 16 hazards (13 chemicals, recombinant DNA technology, high-tension electric wires, and high construction and repair by steeplejacks) found in a pretest to need some elaboration. The descriptions primarily described the nature or use of the item and did not indicate the quality degree of hazardousness (for example, "Mirex, an insecticide for control of fire ants and pineapple insects"). Respondents were told not to judge hazards about which they felt inadequately informed. As a result, only 11 responses (of 34 possible responses) were obtained for cadmium, 16 for trichloroethylene, 20 for DES and Mirex, 21 for Hexachlorophene, 22 for PCBs and polyvinyl chloride, and 23 for coal-tar hair dyes.

ACKNOWLEDGEMENTS

This research was supported by the Technology Assessment and Risk Analysis Program of the National Science Foundation under Grants PRA 7911934 and PRA 8116925. Any opinions, findings, and conclusions or recommendations expressed in this study are those of the authors and do not necessarily reflect the views of the National Science Foundation. Our thanks to Robert Kates, Chris Hohenemser, and Roger Kasperson for their encouragement and advice during the various phases of this research.

NOTES

1. Numerous other psychometric studies of risk perception have been conducted in recent years. Readers interested in this work should consult Brown and Green (1980), Gardner et al. (1982), Green (1980), Otway (1980), Renn (1981), Tversky and Johnson (1983), Vlek and Stallen (1981), and von Winterfeldt, John, and Borcherding (1981).

2. Logarithms of perceived risk and risk adjustment were used in order to reduce the effects on the correlations of one or two hazards with extreme values.

3. Fischhoff et al. (1978) have described results from the unrotated factor analysis for the League group. To facilitate interpretation of the factors, varimax rotation to simple structure was performed in the analyses in the present study. Subsequently, various oblique rotations were performed, in which the factors were allowed to become correlated. The oblique rotations did not change or clarify the picture provided by the orthogonal rotations. Hence, our interpretations of the factors will be based on the results of the varimax (orthogonal) rotations.

4. Factor scores were computed for each of the hazards by weighting the ratings on each of the risk characteristics proportionately to the importance of the characteristic for the factor and summing over all characteristics. This weighted summation gives individual hazards high or low scores if their ratings are high or low on the variables most closely involved with the factor.

5. Correlations in Studies 2 and 3 are based on the logarithm of mean **risk adjustment.** They utilize perceived risk ratings without logarithmic conversion, however, because the scale for this variable was bounded by 0 and 100 and produced a symmetric distribution of means.

6. Slovic, Fischhoff, and Lichtenstein (1980a) concluded that the factors obtained in Study 2 were not the same as those obtained in Study 1. That statement differs from the conclusion expressed in the present study primarily because Slovic et al. were comparing **unrotated** factors from Study 1 with the **rotated** factors from Study 2. The present study shows that rotation of factors in Study 1 brings them into closer agreement with the factors obtained in Study 2 (and Study 3).

REFERENCES

Brown, Richard A., and Colin H. Green. 1980. Precepts of safety assessment. Journal of the Operational Research Society 31:563–571.

Cohen, Bernard, and I-Sing Lee. 1979. A catalog of risks. Health Physics 36 (June):707-722.

Fischhoff, Baruch, Sarah Lichtenstein, Paul Slovic, Stephen L. Derby, and Ralph L. Keeney. 1981. Acceptable risk. New York: Cambridge University Press.

Fischhoff, Baruch, Paul Slovic, Sarah Lichtenstein, Stephen Read, and Barbara Combs. 1978. How safe is safe enough? A psychometric study of attitudes towards technological risks and benefits. Policy Sciences 8:127-152.

Gardner, Gerald T., Adrian R. Tiemann, Leroy C. Gould, Donald R. DeLuca, Leonard W. Doob, and Jan A. J. Stolwijk. 1982. Risk and benefit perceptions, acceptability judgments, and self-reported actions toward nuclear power. Journal of Social Psychology 116:179-197.

Green, Colin H. 1980. Risk: Attitudes and beliefs. In Behaviour in fires, ed. D. V. Canter, 277-291. Chichester: Wiley.

Harman, H. 1967. Modern factor analysis. 2nd ed. Chicago: University of Chicago Press.

Lowrance, William W. 1976. Of acceptable risk: Science and the determination of safety. Los Altos, California: William Kaufmann.

Otway, Harry. 1980. Risk perception: A psychological perspective. In Technological risk: Its perception and handling in the European Community, ed. M. Dierkes, S. Edwards, and R. Coppock, Cambridge, Mass.: Oelgeschlager, Gunn and Hain.

Reissland, J., and V. Harries. 1979. A scale for measuring risks. New Scientist 83:809-811.

Renn, Ortwin. 1981. Man, technology, and risk: A study on intuitive risk assessment and attitudes towards nuclear power. Report Jül-Spez 115. Jülich, West Germany: Kernforschungsanlage Jülich.

Rowe, William D. 1977. An anatomy of risk. New York: Wiley.

Slovic, Paul, Baruch Fischhoff, and Sarah Lichtenstein. 1979. Rating the risks. Environment 21 no. 3:14-20, 36-39.

Slovic, Paul, Baruch Fischhoff, and Sarah Lichtenstein. 1980a. Facts and fears: Understanding perceived risk. In Societal risk assessment: How safe is safe enough?, ed. R. C. Schwing and W. A. Albers, Jr., 181-214. New York: Plenum Press.

Slovic, Paul, Baruch Fischhoff, and Sarah Lichtenstein. 1980b. Perceived risk and quantitative safety goals for nuclear power. Transactions of the American Nuclear Society 35:400-401.

Slovic, Paul, Sarah Lichtenstein, and Baruch Fischhoff. 1979. Images of disaster: Perception and acceptance of risks from nuclear power. In Energy Risk Management, ed. G. Goodman and W. D. Rowe, 223-245. London: Academic Press.

Slovic, Paul, Sarah Lichtenstein, and Baruch Fischhoff. 1984. Modeling the societal impact of fatal accidents. Management Science 30 no. 4:464-474.

Sowby, F. D. 1965. Radiation and other risks. Health Physics 11:879-887.

Starr, Chauncey. 1969. Social benefit versus technological risk. Science 165:1232-1238.

Tversky, Amos, and Eric Johnson. 1983. Alternative representation of perceived risks. Unpublished manuscript.

Vlek, Charles, and Pieter Jan Stallen. 1981. Judging risks and benefits in the small and in the large. _Organizational Behavior and Human Performance_ 28:235-271.

von Winterfeldt, Detlof, Richard S. John, and Katrin Borcherding. 1981. Cognitive components of risk ratings. _Risk Analysis_ 1:277-287.

Wildavsky, Aaron. 1979. No risk is the highest risk of all. _American Scientist_ 67(January/February):32-37.

Wilson, Richard. 1979. Analyzing the daily risks of life. _Technology Review_ 81(February):41-46.

Part Two

Overview: Measuring Consequences

We have defined hazards as threats to humans and what they value. Part 1 explores the causal structure of hazards, the ways in which society copes, and the methods of classifying hazards according to physical, biological, and social characteristics. Part 1 also indicates that hazard perception falls into regular patterns that permit prediction of the degree of adjustment and regulation desired by various groups of well-educated people. Whereas Part 1 demonstrates the richness and complexity of hazard structure, its qualitative approach does not specify the magnitude of the burden that hazards place on society.

As a step toward quantification of this burden, Part 2 provides an analysis of aggregate mortality (chapter 6) and a summation of economic costs and losses attributable to hazards and their control (chapter 7). In terms of the causal structure of hazards (chapter 2), most of what is at issue here concerns quantitative measures of hazard consequences. The measures employed in the two chapters are related. The sum of costs and losses discussed in chapter 7 includes as one term the productivity losses incurred through hazard mortality estimated in chapter 6.

The approaches of chapters 6 and 7 are predicated on the idea that it is interesting and useful to know something about the size of the "universe," even when detailed scientific understanding of many specific hazards is on an unsound or incomplete footing. The chosen methods of analysis do not rely on summing individual, hazard-specific deaths and costs but employ broad categories and correlational analysis. Because of insufficient scientific understanding of cancer etiology, for example, chapter 6 relies on strong, positive correlations between cancer incidence and the level of employment in industry.

For the analysis of mortality, the results of this "global" approach are not unreasonable and fall midway between similar estimates made by others. Chapter 6 indicates that 20-30 percent of male and 10-20 percent of female mortality correlate with technological hazard. The bulk of this mortality involves chronic causes of death for which the fraction of mortality assigned to technology is not much more than an educated guess. For example, it is estimated that 0-40 percent of male and female cardiovascular mortality, and 25 percent of female and 40 percent of male cancer mortality, are associated with technology.

127

When this analysis is translated into lost productivity and added to other costs and losses in chapter 7, additional large uncertainties arise. In estimating the impact of air pollution, for example, indirect methods of accounting can determine the 1978 cost of property damage to within only a factor of 6. Perhaps the most significant shortcoming of the economic balance sheet of chapter 7 is that it does not, and cannot, incorporate nonhuman mortality and losses, variously described in chapter 6 as ecosystem impacts, loss in biomass, or species extinction.

Despite the uncertainty and incompleteness in accounting, the economic sum of chapter 7 leads to a plausible and meaningful result. Estimates of lost productivity arising from human mortality and injury, and other costs and losses experienced in coping with technological hazards, are 7.8-12.4 percent of the gross national product, or $179-283 billion in 1979. Half of this substantial burden does not appear in direct private and public expenditures, however, since the method involves assigning dollar values, in the form of lost productivity, to human death and injury.

The economic sum of chapter 7 thus divides about equally between societal control costs and the costs of damages and losses. Interestingly, the public share of control costs, 32-39 percent, is about as large as the public fraction of national expenditures. This conclusion should be contrasted to that obtained in studies of government regulation alone, which generally ascribe relatively large private control costs to relatively low-cost publicly mandated regulation.

The overall conclusion is bittersweet. On the one hand, in the last 100 years technology has produced substantial increases in life expectancy through eradication of infectious disease and the reduction of many hazards with acute consequences. This means that in an historical sense the problem is not getting worse. On the other hand, as technology has become more sophisticated, the processes and products of technology have become ever more varied and complex. In this situation hazards associated with chronic, difficult-to-measure consequences have become the dominant cause of concern, and may, for all we know, be growing.

6
Human and Nonhuman Mortality[1]

Robert C. Harriss, Christoph Hohenemser,
and Robert W. Kates

Hazards are threats to humans and what they value: life, well-being, material goods, and environment. Today hazards originating in both nature and technology are a major concern in developing and industrial nations alike. Coping with hazards involves a wide range of adjustments--from learning to live with hazard, to sharing the burden of hazard, to controlling and preventing death, injury, property loss, and damage to human and natural environments.

In chapters 2-5, descriptions of causal structure, management, classification, and perception together provide a qualitative framework for hazard analysis. In this chapter we begin to quantify the burden of hazard consequences by estimating the extent of human and nonhuman mortality associated with technological hazards. The approach taken is "global," and rather than relying on hazard-specific data, depends on broad scale accounting and correlational analysis. We begin with a basic distinction.

Natural and Technological Hazards

For the majority of the world's people, living in rural portions of the developing nations, the hazards which most concern them are ancient ones and are predominantly rooted in nature. These natural hazards arise most often in connection with agriculture, food supply, or settlement, and they constitute a major burden. For example, the losses from geophysical hazards (floods, droughts, earthquakes, and tropical cyclones) each year in the developing world involve an average of 250,000 deaths and $15 billion in damage and costs of prevention and mitigation (Burton, Kates, and White 1978). This is equivalent to 2 to 3 percent of the gross national product (GNP) of the affected countries. Losses from vermin, pests, and crop disease are widely regarded as a larger problem (Porter 1976) and involve destruction of as much as 50 percent of food crops (Pimentel 1978). And infectious disease, though declining, typically accounts for 10 to 25 percent of human mortality, concentrated among the very young (World Health Organization 1976).

In contrast, for industrialized nations, natural hazards are relatively a much smaller problem. Thus, for the United States, geophysical hazards produce less than one thousand fatalities per year, and property damage and costs of prevention and mitigation are on the order of 1 percent of GNP. Vermin, pests, and crop disease, while leading to serious losses, are kept in bounds by pesticides and other techniques, and infectious disease accounts for less than 5 percent of human mortality. While controlling natural hazards to this extent, however, the industrialized nations have not escaped unscathed.

Taking the place of the ancient hazards of flood, pestilence, and disease are new and often unexpected hazards, predominantly rooted in technology. As Table 1 shows, these hazards now have an impact as large or larger than the natural hazards they have replaced (Kates 1978).

For example, in 1975 the United States spent $40.6 billion, or 2.1 percent of GNP, on air, land, and water pollution (Council on Environmental Quality 1978); and in the same year the cost of auto accidents was estimated as $37 billion, or 1.9 percent of GNP (Faigin 1976). In our estimate (see below) the death toll associated with technological hazards involves 20-30 percent of male and

TABLE 1
Comparative hazard sources in U.S. and developing countries

| | PRINCIPAL CAUSAL AGENT[a] | | | |
| | NATURAL[b] | | TECHNOLOGICAL[c] | |
	Social cost[d] (% of GNP)	Mortality (% of total)	Social cost[d] (% of GNP)	Mortality (% of total)
United States	2-4	3-5	5-15	15-25
Developing countries	15-40[e]	10-25	n.a.[f]	n.a.[f]

[a]Nature and technology are both implicated in most hazards. The division that is made here is made by the principal causal agent, which, particularly for natural hazards, can usually be identified unambiguously.
[b]Consists of geophysical events (floods, drought, tropical cyclones, earthquakes and soil erosion): organisms that attack crops, forests, livestock; and bacteria and viruses which infect humans. In the U.S. the social cost of each of these sources is roughly equal.
[c]Based on a broad definition of technological causation, as discussed in the text.
[d]Social costs include property damage, losses of productivity from illness or death, and the costs of control adjustments for preventing damage, mitigating consequences, or sharing losses.
[e]Excludes estimates of productivity loss by illness, disablement, or death.
[f]No systematic study of technological hazards in developing countries is known to us, but we expect them to approach or exceed U.S. levels in heavily urbanized areas.

10-20 percent of all female mortality. Overall, expenditures on hazard management in the private and public sectors, when added to value of direct losses, are estimated by Tuller to be in the range $179-283 billion (7.8-12.4 percent of GNP) for 1979 (see chapter 7).

Technological hazards are thus a big business, comparable in scope to such major sectors of the national effort as social welfare programs, transportation, and national defense. And the impacts of technological hazards go well beyond mortality. Table 2 details the various groups, sectors, and environments affected, along with the dimensions of the consequences considered in our work on the assessment and management of technological hazards.

How can the full scope of technological hazards be evaluated? Only by determining the sum of all of the impacts and consequences outlined in Table 2. This is a formidable task, one which no group,

TABLE 2
Impacts of technological hazards

HAZARD EXPOSURE RECEPTORS	DIMENSIONS OF CONSEQUENCES
HUMAN POPULATIONS: individuals, groups, cohorts	well-being (diminution, loss) morbidity (acute, chronic, trans-generational) mortality (acute, chronic, trans-generational)
ECONOMY: activities, institutions, production	individual and collective loss cost of control adjustment cost of mitigation
SOCIETY: activities, institutions, values	activity disruption institutional breakdown value erosion
ENVIRONMENTS: natural	landscape transformation air and water quality loss recreational opportunities lost
built	community loss architectural deterioration
ECOSYSTEMS: population, species, communities	species extinction productivity reduction resistance/resilience diminution
natural	landscape transformation air and water quality loss recreational opportunities lost
built	community loss architectural deterioration

to our knowledge, has accomplished, or even attempted. In this chapter, we concentrate on human mortality and ecosystem impacts (particularly impacts on biological species and communities). Human mortality is based on well-defined data and, of all impacts, is most susceptible to quantification; the ecosystem impacts are at best difficult to judge and nearly impossible to quantify. These two impacts thus delimit the range of current scientific understanding within which other impacts and consequences fall.

Measuring Technological Hazard

Human death is the best defined of all hazard consequences. Even many impoverished societies keep reasonable mortality records and, for a large number of countries, including the United States, mortality statistics grouped according to "causes of death" are extensively tabulated by age, sex, and even race (World Health Organization 1976; National Center for Health Statistics 1978). It would seem, therefore, that there is a direct and obvious answer to the question, "How much death is due to technological hazard?"--that it is simply necessary to add the contribution of each "cause of death" and note the relative magnitude of the sum.

Unfortunately, once we have added the toll of transportation and occupational accidents and the impact of violence, this approach ends in a quagmire of uncertainty for at least three reasons: (1) death rarely has a single cause and, in most cases technology is at best a contributing factor; (2) when chronic disease, such as cancer and heart disease, is given as "cause of death," one can deduce little directly about the role of technology, since the root causes of chronic disease are known in only a small percentage of cases; (3) much death is not accurately classified according to "cause," even in some cases of accidents and violence. Mortality statistics are further clouded, because in many developing countries as much as 50 percent of all mortality is classified as of "unknown orgin," and even for developed countries practices in assigning causes vary widely (Preston 1976).

In this study of technological hazards, we have sought to circumvent these problems by what amounts to an "end run." Instead of obtaining percentage of mortality resulting from technology by direct calculation and summing, we make an indirect estimate through a two-step process. First, we estimate the percentage of mortality that is preventable in principle or, equivalently, involves external or nongenetic causes. In the literature, this is often called exogenously caused mortality.[2] Second, we estimate the percentage of technologically preventable mortality. In doing so, we recognize that externally caused or exogenous mortality sets an upper limit on technologically caused mortality, but that exogenous mortality can result from social, cultural ("life style"), and environmental factors as well as technological ones. This division of exogenous mortality (illustrated in Table 3) is not a clear-cut one. In an interrelated and mutually dependent society such as ours, most deaths have multiple causes.

TABLE 3
Classification of morbidity and mortality

CAUSE	EXAMPLES
ENDOGENOUS: causes reside predominantly within the individual	Aging; genetic defects arising from inherited genetic load.
EXOGENOUS: causes reside predominantly outside the individual	
Natural Environmental	Infection; background radiation-induced cancer; latitudinal skin cancer effects; natural catastrophes.
Social and Cultural	Diet-based disease such as cancer from betel nuts, cirrhosis of liver from alcohol, heart disease from overweight; smoking-related disease, some urban-related mortality, some violence, war death.
Technological	
diffuse effects	Pollution-related disease; some urban-related mortality.
specific technology	Transportation accidents; cancers from specific industrial chemicals such as benzene, asbestos, and vinyl chloride; gun accidents.

Exogenous Mortality

What fraction of mortality is exogenous, that is, preventable at least in principle? To answer this question, we first divide all of mortality into acute and chronic causes of death (Tables 4 and 5). Among acute causes of death, we include all those cases in which death is sudden and not preceded by a long period of illness. Among chronic causes of death, we include all those cases where death results from a long period of prior illness due to deterioration of one or more body functions. The division into acute and chronic causes is made because the analysis of the two cases is fundamentally different.

Acute Causes of Death. Except for congenital malformations leading to sudden death, a small percentage of infectious disease, and a percentage of accidents, suicides, and homicides associated with inherited deficiencies and psychotic illness, all acute causes of death are prima facie exogenous. Assignment of the exogenous percentage is therefore made at or near unity in most cases, as shown in Table 6.

TABLE 4
Acute mortality in the United States, 1972

CAUSE OF DEATH	MORTALITY DEATHS/100,000		MORTALITY (PERCENT OF TOTAL)	
	male	female	male	female
Infectious Disease	45.7	35.3	4.2	4.3
influenza	2.4	2.4		
pneumonia	31.9	23.7		
infection of the kidney	3.0	3.6		
enteritis	1.0	1.1		
infectious hepatitis	0.3	0.4		
other	4.4	3.4		
Deaths in Early Infancy	27.2	19.6	2.5	2.4
diseases of early childhood	19.5	13.2		
congenital abnormalities	7.7	6.4		
Transportation Accidents	43.1	15.7	4.0	1.9
automobile	39.6	15.1		
other	3.5	0.6		
Other Accidents	38.9	19.3	3.5	2.4
poisoning	3.7	1.6		
falls	8.4	7.7		
fire	4.0	2.5		
drowning	5.0	1.0		
firearms	2.1	0.3		
industrial machinery	5.1	0.5		
others	7.2	4.1		
Violence	32.9	10.5	3.0	1.2
suicide	17.5	6.8		
homicide	15.4	3.7		
Other Acute Causes	11.8	9.0	1.0	1.0
TOTAL ACUTE CAUSES	199.6	109.4	18.4	13.5
MALE-FEMALE AVERAGE			16.3	

Source: World Health Organization (1976).

TABLE 5
Mortality from chronic disease in the United States, 1972

CAUSE OF DEATH	MORTALITY DEATHS/100,000		MORTALITY (PERCENT OF TOTAL)	
	male	female	male	female
Cardiovascular Disease	554.5	459.3	51.3	56.5
hypertension	9.5	10.9		
ischemic heart disease	382.4	277.6		
cerebrovascular disease	94.0	110.5		
arteriosclerosis	29.2	26.7		
other cardiovascular	39.5	33.6		
Cancer	188.1	149.6	17.4	18.4
lung, trachea, bronchia	56.8	14.0		
colon	17.4	18.8		
breast	0.3	29.2		
lymphatic tissues	10.5	8.4		
prostate	18.0	----		
stomach	9.2	5.8		
leukemia	8.1	5.8		
uterus	----	6.0		
rectum	5.6	4.2		
mouth-pharynx	5.3	2.0		
other	56.9	55.4		
Chronic Liver Disease	37.7	31.8	3.5	3.9
diabetes	15.6	21.4		
cirrhosis	21.1	10.4		
Chronic Respiratory Disease	25.8	7.6	2.4	0.9
tuberculosis	2.5	0.9		
bronchitis, emphysema, asthma	23.3	6.7		
Other Chronic Disease	74.6	55.1	6.8	6.8
TOTAL CHRONIC DISEASE	880.7	703.4	81.6	86.5
MALE-FEMALE AVERAGE			83.7	

Source: World Health Organization (1976).

TABLE 6
Estimated exogenous and technologically involved deaths in the United States

CAUSE OF DEATH	ESTIMATED U.S. EXOGENOUS PERCENTAGE OF MORTALITY (percent)		ESTIMATED U.S. TECHNOLOGICAL COMPONENT OF MORTALITY (percent)		(annual deaths in thousands)	
	male	female	male	female	male	female
ACUTE MORTALITY						
Infectious disease[a]	90	90	0	0	0	0
Deaths in infancy[b]	50	50	5	5	1	1
Transportation accidents[c]	100	100	90	90	39	15
Other accidents[d]	100	100	70	50	28	11
Violence[e]	100	100	30	30	10	3
Other acute deaths[f]	100	100	70	50	8	5
CHRONIC MORTALITY						
Cardiovascular disease[g]	80	60	0–40	0–40	0–217	0–132
Cancer[h]	60	45	40	25	82	35
Chronic liver disease[i]	80		0	0	0	0
Chronic respiratory disease[j]	60	10	0–20	0–5	0–5	0
Other chronic disease[k]	70	70	25	25	19	15
ALL MORTALITY			17–30	11–21	182–318	85–167

[a] Exogenous percentage of 90 percent is based on the hypothesis that this amount of infectious disease is in principle preventable before genetic factors become dominant. Supportive of the hypothesis is the fact that the decline trend of infectious disease mortality is steep. The technological fraction of zero is based on the fact that infectious disease is usually prevented by technology, not enhanced.

[b] Currently the U.S. ranks 13th in the infant mortality and, even in the lowest nations, infant mortality is still declining. The estimate for the exogenous percentage is meant to reflect these facts qualitatively. The technological percentage is low because infant deaths are caused largely by disease.

[c] Transportation accidents are prima facie 100 percent externally caused. The technological percentage given includes all deaths except those that are estimated to be predominantly homicidal and suicidal.

[d] Other accidents include numerous categories, as shown in Table 4. All are by definition externally caused. Some, like drowning and falls, are primarily rooted in culture and society, not technology, and hence these are excluded in estimating the technological percentage.

[e] Although nearly all violence is committed with the help of technological devices, and this suggests 100 percent exogenous causation, there is little evidence that violence is prevented by modification of technology. Rather, violence is rooted in culture and society. The assignment of a modest technological percentage reflects this fact.

[f] Other acute deaths involve many causes but relatively small numbers. The values given represent the average behavior of other acute deaths.

[g] The exogenous percentage is based on 36-nation comparisons as illustrated in Figure 1. The technological percentage is uncertain, yielding 40 percent based on cross-national plots similar to Figure 2, the difference between the U.S. rate and some theoretical rate without technology (0 percent), yet yielding near zero based on state-by-state comparison within the U.S. similar to Figure 3. Since much of cardiovascular epidemiology points toward diet and stress, we are inclined to believe the lower technological percentage.

[h] The exogenous fraction is based on Figure 1, the technological fraction on Figure 2 and the support given this by Figure 3 as well as the available literature on cancer epidemiology.

[i] The exogenous percentage is based on data similar to Figure 1. The low technological percentage is based on the predominant role of diet and alcohol in liver disease epidemiology.

[j] The exogenous percentage is based on analysis similar to Figure 1. The technological fraction is based on the literature describing the urban-rural difference in epidemiological studies of smoking-related disease.

[k] Other chronic diseases involve a large variety of causes, but rather small total mortality. Percentages assigned here are guesses based on the average behavior of the chronic diseases which we have analyzed.

Chronic Causes of Death. For chronic causes of death we obtain the exogenous percentage by a comparison of the mortality statistics reported by thirty-six nations to the World Health Organization (1976). The nations selected are believed to have sufficiently reliable statistics for our purposes; all have mortality rates for "unknown causes" amounting to less than 10 percent of total mortality. From the thirty-six nation data, the lowest age-specific mortality rate was chosen and used as the "base case." Exogenous mortality for each nation was then defined operationally as the excess mortality observed in each relative to base-case mortality.

Several problems with this definition make it necessary to regard it as only an approximate estimate of true exogenous mortality. Thus, use of the definition implicitly assumes that the genetic disposition toward mortality of various populations is identical. This is not always the case. Some cancers, for example, appear to have a genetic basis. On the other hand, when populations migrate, they usually take on the mortality patterns of their new home, thus indicating the predominance of external factors.

In addition, our method for obtaining the exogenous mortality rate is critically dependent on the validity of the base case. Our definition will tend to underestimate true exogenous mortality if some base-case mortality is preventable in principle; and it will tend to overestimate true exogenous mortality if the base case involves serious under-recording of certain chronic causes of death. Fortunately these latter effects, both of which are surely present, will at least partially cancel each other out.

Figure 1 illustrates the kind of data used: age-specific cardiovascular and cancer mortality for males and females in selected countries, including the lowest and highest mortality cases. Male and female exogenous percentages deduced from this data were 80 and 60 percent for cardiovascular disease, and 60 and 45 percent for cancer, respectively. Exogenous percentages for all causes of death are summarized in Table 6.

Technological Mortality

What percentage of exogenous mortality is associated primarily with technology, rather than with environment and culture? This is a much more difficult question, with a considerably more uncertain answer than in the case of exogenous mortality per se. There is no simple argument that allows approximate separation of the technological percentage. Our present best estimate must therefore be something of a guess, though, we hope, a good one. In order to make this guess, we again treat acute and chronic causes of death separately.

Acute Mortality. Infectious disease, though influenced by the level of technology, is largely environmental and cultural in origin. To the extent that technology is involved, it usually leads to a reduction of disease rather than increased hazard. In contrast, accidents, homicide, and suicide are highly associated with technology and culture and only marginally with the natural environment. Our estimate of the technologically involved percentage of acute mortality thus ranges from zero percent in the case of infectious disease to 90 percent in the case of transportation accidents (see Table 6). For cancer, similar results have been obtained by other

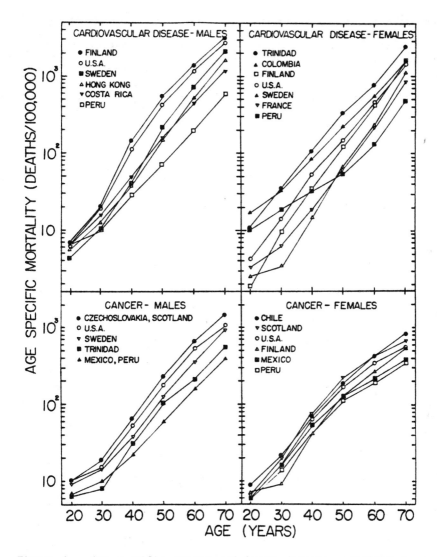

Figure 1. Age-specific cancer and heart disease mortality in selected countries for males and females for 1972-1973. The countries
were selected because they are believed to have reliable statistics
and because they represent the full range of recorded mortality,
from lowest to highest. The difference between the lowest and U.S.
mortality was used to estimate the exogenous fraction of mortality
for the U.S., as indicated in Table 6. Note that the plots shown
here utilize a logarithmic scale for mortality. This is a convenient device for numbers that vary over a very large range and has
the effect of giving equal intervals to each factor of ten. The
source for the data is the World Health Organization (1976).

researchers (Higginson 1976;Epstein 1976;Council on Environmental Quality 1977).

Chronic Mortality. We have already noted in our discussion of exogenous mortality that direct assignment of cause in the case of chronic disease is usually not possible. For estimating the technological component of exogenous mortality we again use an indirect method, based on national and international comparisons. Our approach is to look for correlations of chronic disease mortality with certain indicators of technology, such as per capita GNP, per capita energy consumption, and percent of labor in manufacturing. If chronic disease increases with level of technology, this analysis yields the equivalent of a "dose-effect" relation: it permits the determination of the change in mortality occasioned by a given change in level of technology. Unlike high quality dose-effect relations in the field of toxic substance epidemiology, the exposed populations in this case are poorly controlled for factors other than level of technology. Hence, one must expect a certain amount of scatter in mortality at a given level of technology.

The Case of Cancer. The incidence of cancer can be used as an illustration of this type of analysis. International "dose-effect" relations for men and women are shown in Figure 2; equivalent relations for the United States, for both blacks and whites, are shown in Figure 3. In each case, the mortality in 1972-1973 is plotted against percent of the labor force in manufacturing in 1940, thus allowing for the latency of cancer. Our interpretations of the observed relations are as follows:

- Internationally, cancer in males varies widely and shows an average increase of a factor of 2.7 and 1.7 for males and females as the level of technology varies from lowest to highest. Particularly for males, the scatter is very large, indicating that there are many other causes at work.
- For whites within the United States, the international pattern is repeated, though with smaller increases and less scatter. Thus mortality increases by an average factor of 1.5 and 1.4 for males and females as the level of technology rises from lowest to highest.
- For blacks within the United States, the pattern is significantly different. For black males, the increase in mortality is an average factor of 2.0 as technology varies from lowest to highest. This is a significantly greater increase than that for whites. For black females, on the other hand, no significant effect is seen, though scatter is large and average values are higher than for white females.

Thus, although there are some puzzles, a reasonably consistent picture emerges. Cancer mortality, as one would expect from the epidemiological literature (Fraumeni 1975), has an appreciable technological component. Using the international data shown in Figure 2, we estimate the difference between the U.S. rate and some theoretical rate without technology (zero percent in manufacturing). This gives a conservative estimate of the technological component of at least 40 percent for men and 25 percent for women in the United

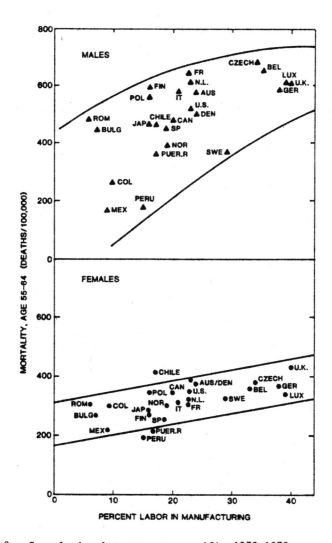

Figure 2. Correlation between age-specific 1972–1973 cancer mortality and percent labor in manufacturing in 1940 for nations believed
to have reliable mortality statistics. Though the data exhibit wide
scatter, both males (top) and females (bottom) show increasing
cancer mortality with increasing industrialization. The scatter
indicates causes for cancer other than industrialization. The consistent increase of cancer mortality probably implicates industrialization as one of the causes of cancer. The choice of 1940 allows
for the known, approximately 30-year lag between exposure to carcinogens and the occurrence of cancer. Note that, consistent with
their greater participation in industry, males show a bigger increase than females. These data were used to estimate the fraction
of technologically involved mortality given in Table 6. Source of
the mortality data: World Health Organization (1976); source of the
percent labor in manufacturing: Woytinsky and Woytinsky (1953).

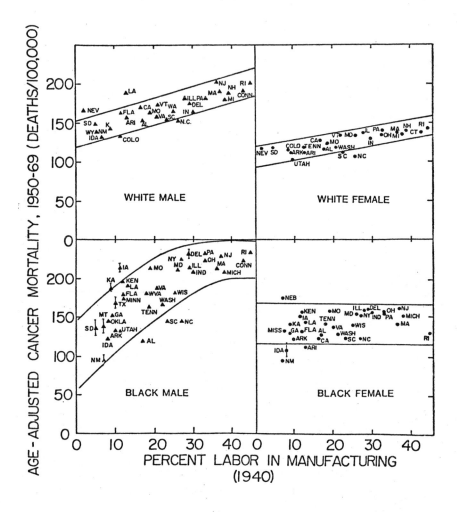

Figure 3. Correlation between average 1950–1969 age–adjusted cancer mortality and percent labor in manufacturing in 1940 for states within the United States. As might be expected from the greater homogeneity of the United States, the scatter is considerably reduced relative to international data shown in Figure 2. At the same time, the pattern of increasing mortality with increasing industrial exposure is repeated. Again men show a more pronounced increase than women and, in addition, black men show a bigger increase than white men. The only surprising aspect of the data is that black females show no apparent increase. Source: U.S. Cancer Mortality by County, 1950–69 (U.S. National Cancer Institute 1974).

States. Thus, very roughly speaking, about half of exogenous cancer mortality is the result of technology, the rest of social and cultural causes. Similar results are obtained if the per capita energy consumption is used as an indicator of technological development.

We do not wish to claim that energy consumption or industrial employment causes cancer per se. Correlations such as those shown in Figures 2 and 3 are too weak a tool for this purpose. When correlations with mortality exist for several indicators across a wide range of populations and cultures, however, it is likely that the results are not accidental but are evidence of a number of factors that form links in the causal chains leading directly or indirectly to observed chronic disease mortality. Sometimes these links are fairly simple and well-established; for example, coal mining leads to deposit of fine coal dust in deep lung cavities and, through obstruction of these, reduces lung function (black lung disease). In other cases the links are highly complex, such as the incompletely understood connections between diet and heart disease. It is the task of medical science, particularly epidemiology, to identify and describe these specific links, and it is the task of hazard management to control them. Our purpose here is to explore the magnitude of the problem and for this our correlations of disease with general indicators of technology are adequate and appropriate.

Using a method similar to that employed in the cancer illustration above, we have estimated for other chronic disease the percentage of mortality that is technologically related (Table 6). It is necessary to stress that by technologically related percentage we mean mortality, which is in principle preventable by adjustments in technology. This does not exclude the involvement of other causative factors such as genetics, cultural milieu, life-style, and natural environmental conditions as contributing causes. To compensate, and to be conservative, we exclude smoking and diet as technological causes, even though technologies have figured highly in the consumption of cigarettes (as opposed to the less hazardous tobacco forms) or in the availability of low-cost meat and dairy products (implicated in cardiovascular disease).

The Cost of Technological Mortality

Estimates of mortality and morbidity costs for various causes of death are available in the literature (Rice, Feldman, and White 1976). These estimates indicate the dollar value of medical costs and of the cost of lost productivity. Such estimates do not place any dollar value on life and suffering as such, since this necessarily depends on diverse personal and societal ethical judgments that are widely held to be beyond economic valuation. However, such estimates are important because they define the magnitude of the economic problem of lost life and illness and in this way serve to indicate the savings that can be realized if mortality and illness are prevented.

Using the percentages of technologically involved deaths given in Table 6 and the estimated values of life shortening applicable to each cause of death, we find that the total annual loss due to technological hazard mortality is approximately seven to ten million person-years, about two-thirds of which occurs in males. Using the methodology developed by Rice, Feldman, and White (1976) for translating this into medical and lost productivity costs, we find that an annual loss of $50 to $75 billion (1975 dollars) is due to technologically involved mortality and related morbidity. Interestingly, accidents and violence, though they constitute only 10 and 6 percent of male and female mortality, respectively, account for 40 to 60 percent of the costs. This is because of the relatively higher technological percentages and larger life-shortening effects in the case of accidents and violence.

Technological Impacts on Ecosystems

In contrast to human mortality, the ecosystem impacts on biological communities, while perhaps the most important of the dimensions of technological hazard in the long run, are also the most difficult to quantify. Here there are no world-wide, nearly all-inclusive accounting systems such as death certificates. And instead of dealing with one dominant species, we are dealing with literally millions of species related by a complex and often fragile system of interdependence. How can the impacts of technology on this system be defined? Two possible measures of ecosystem impacts by technological hazards (recall Table 2) are species extinction and ecosystem productivity. Both of these measures are in principle quantifiable. Yet each has less specific meaning to humans and what they value than does human mortality. Each is separately considered below.

Species Extinction

Species extinction is the most drastic and inclusive form of wildlife mortality. Like human mortality this can occur naturally, independently of any technological effects. As in the case of human mortality, we are interested here in the percentage of species extinction that is of technological origin. As before, we divide the problem by asking two questions:

1. What is the rate of exogenous species extinction, that is, the percentage of cases for which the underlying causes are not of predominantly natural origin?
2. What is the rate of technologically involved extinction, that is, the percentage of exogenous extinction that is predominantly related to technological causes?

One approach to the first question is through the historical record. As shown in Figure 4, the world-wide rate of vertebrate extinction has speeded up considerably during the last hundred years, culminating in a current rate that is at least ten times the "baseline" or evolutionary rate observed 300 years ago (Ehrlich, Ehrlich, and Holdren 1977). As shown in Table 7, one in ten species of native, higher plants in the United States is currently endangered, threatened with becoming endangered, or recently extinct; in Hawaii, nearly half of the total diversity of native vegetation is similarly involved (Council on Environmental Quality 1974;Brunnel and Brunnel 1967;Uetz and Johnson 1974;Fisher, Simon, and Vincent 1969).

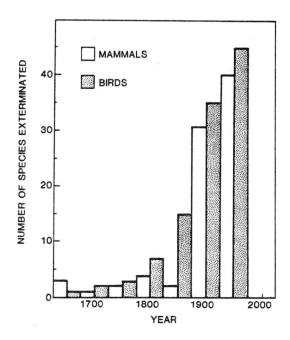

Figure 4. The number of exterminated mammal forms (white bars) and bird forms (shaded bars) eliminated over the last 300 years. Each bar represents a 50-year period. Source of the data is the National Center for Health Statistics (1978).

Another approach to estimating the rate of exogenous extinction is through direct classification of species extinction according to cause. Using available data (Fisher, Simon, and Vincent 1969) on extinction and rarity for birds and mammals since 1800, we have obtained the division into exogenous and natural causes, as shown in Table 8. Thus, for the period studied, more than two-thirds of extinction and rarity have specifically non-natural causes.

How much of exogenous extinction is of distinctly technological origin? This question is unfortunately unanswerable in terms of any well-defined analytical approach. Technology certainly plays an important role in hunting and in much of physical habitat modification, but we do not have the data for a case-by-case review of recorded extinction. In the absence of such detailed data, we conservatively estimate the technological percentage of exogenous species extinction at approximately one-half, with the remainder being largely of cultural character.

Whatever the division between technological and cultural causes may be, it is clear that the rates of exogenous extinction currently being observed are much faster than the normal evolutionary process of replacement. Nor is it possible to insure adequately against such loss in zoos, botanical gardens, and other protected environments (Ehrlich, Ehrlich, and Holdren 1977). Ecological theory, furthermore, suggests that wildlife mortality of the magnitude currently being observed can lead to significant diminution and loss

146

TABLE 7
Endangered, threatened and extinct species of native higher plants
in the U.S.

STATUS	CONTINENTAL UNITED STATES		HAWAII	
	species, sub-species and varieties	percent	species, sub-species and varieties	percent
Total native higher plants	20,000	100.0	2,200	100.0
Endangered[a]	761	3.8	639	29.0
Threatened[b]	1,238	6.1	194	8.8
Extinct[c]	100	0.5	255	11.6
TOTAL	2,099	10.4	1,088	48.9

Source: Smithsonian Institution (1975).

[a]Endangered is defined as in danger of becoming extinct throughout all or a significant portion of their natural range.
[b]Threatened is defined as likely to become endangered in the foreseeable future.
[c]Extinct is defined as limited to recently (or possibly) extinct species only: they cannot be found after repeated searches in the localities where they were formerly observed or other likely places. Some of the latter appear to be extinct in the wild but are still preserved in cultivation.

of ecosystem productivity and resilience, with occasionally cata-strophic consequences.

We wish to emphasize that counting species by itself is inade-quate for defining the impact of technology on ecosystems. It is not enough to have a catalog of characters to predict the outcome of an evolutionary play. What is needed is some measure of the effec-tiveness with which ecosystems use energy and how well an ecosystem is able to recover from a stressed condition (resilience). Impor-tant new concepts related to ecosystem energy analysis (Odum and Odum 1976) and ecosystem resilience (Fiering and Holling 1974) are currently undergoing intensive study in the scientific community. Until these provide well-defined indicators, however, it seems prudent to use crude indicators, such as species extinction, as warning signals of potential hazard.

Ecosystem Productivity

As a second measure of ecosystem impacts we consider productiv-ity, or the ability of ecosystems to produce organic material from inorganic substrate and sunlight. In so doing, we limit ourselves

TABLE 8

Classification of causes of extinction and rarity for birds and
mammals since 1800 on a worldwide scale

CAUSE OF EXTINCTION	BIRDS (%)	MAMMALS (%)
NATURAL CAUSES	24	25
EXOGENOUS CAUSES		
Acute (hunting)	42	33
Chronic		
habitat disruption (physical)	15	19
habitat modification	19	23
(biological and chemical)		
TOTAL	100	100

CAUSE OF RARITY	BIRDS (%)	MAMMALS (%)
NATURAL CAUSES	32	14
EXOGENOUS CAUSES		
Acute (hunting)	24	43
Chronic		
habitat disruption (physical)	30	29
habitat modification	14	14
(biological and chemical)		
TOTAL	100	100

Source: Recalculated from Fisher, Simon, and Vincent (1969).

to the changing magnitude of the land biomass--that is, the organic
material of biological origin found on land. Land biomass is subject to natural variability arising from such factors as weather and
disease; it also responds to the expansion of timbering, agriculture, urbanization, and similar pressures from humans. The question
of biomass impacts can therefore, as before, be divided into natural
and exogenous effects.

Global changes in land biomass have recently been explored in
connection with studies of the world carbon cycle (Bolin 1977;
Woodwell et al. 1978). These studies show a net annual decline in
global land biomass (albeit with great uncertainty) amounting to 0.2
to 2 percent. The causes of change are largely exogenous and, as
seen in Table 9, involve decline and destruction of major land plant
communities in areas of maximum population pressure. Among the
communities destroyed, tropical forests are of particular concern
because it is not clear that reforestation can take place in some
lateritic soils. A detailed study of tropical forests estimates
that 0.3 to 0.6 percent of the total is being destroyed each year
(Sommer 1976).

TABLE 9
Estimates of current net loss of major land plant biomass, as
reflected by the release of carbon into the atmosphere

PLANT COMMUNITY	CARBON RELEASED (BILLION TONS/YR)	
	Average	Range
Tropical forests	3.5	1-7
Temperate forests	1.4	0.5-3
Boreal forests	0.8	0-2
Other vegetation	0.2	0-1
Detritus and humus	2.0	0.5-5

Source: Modified from data in Woodwell et al. (1978).

In addition to direct losses in ecosystem productivity from deforestation, indirect impacts on drainage basins, resulting from major changes in hydrologic and chemical cycles, can also diminish long-term productivity of the total ecosystem. For example, replacing biomass and nutrients lost in harvesting northern hardwoods may take sixty to eighty years (Likens et al. 1978).

As with the case of species extinction, exogenous decline of land biomass is of specifically technological as well as cultural origin. Because the bulk of the large changes now being seen, particularly in tropical forests, involves the application of high technology, we believe the technological component of biomass decline to be as high as 75 percent of the total.

Technological Hazards in Historical Perspective

Our discussion so far has focused on present technological hazard impacts. Except in the case of species extinction, we have made no effort to look at the historical record. Industrial development in the West is now 300 to 400 years old, and much of what has occurred in the past fifty years has been termed "post-industrial." Historical experience with technology is therefore extensive, and it is thus interesting to ask whether the problem of technological hazards is getting worse.

Human Mortality

In regard to human mortality, the benefits of technology appear to have been large and dramatic. As already noted, they include the near elimination of the worst of natural hazards—infectious disease. This development is largely responsible for the fact that, since 1850, when the United States had a highly dispersed agricultural population, life expectancy has shown a near doubling at birth, a 30 to 50 percent increase at midlife, and a modest increase at age sixty. Technology has also led to a food supply system that is so productive that few in the industrialized world need fear even slight deprivation in relation to this basic need.

In addition, hazards of technology were undoubtedly higher in earlier, less fully managed stages of industrial development. Thus occupational mortality, at least of the acute variety, has shown a continuing and steady decline, as shown in Figure 5; and large technological disasters apparently peaked during 1900-1925 (National Safety Council 1977). If evidence from literature is desired, one needs only recall the novels of Charles Dickens and D.H. Lawrence, which contain accounts of environmental pollution and human exploitation in an industrial setting that find few parallels in the modern age.

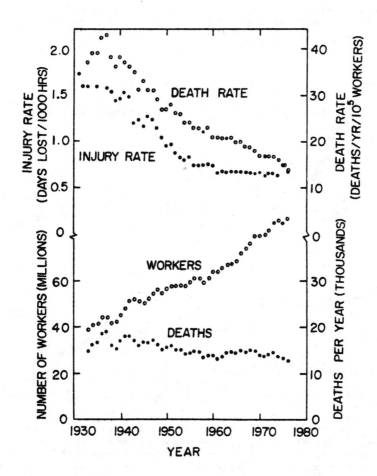

Figure 5. Historical variation of occupational death and injury rates in the United States for 1930-1976. Due to varying recording practices, injury data are considered to be only roughly correct. Source of the data is the National Safety Council (1977).

Thus, at worst the present problem may be that the positive effects of technology have for some time now reached their maximum effect on human mortality, whereas the hazards of technology continue partially unchecked, affecting particularly the chronic causes of death that currently account for 85 percent of mortality in the United States. Supportive of this view is the fact that male life expectancy has not increased since 1950 and has even shown a slight decline.

But this view may be too pessimistic. Even the apparent increase in chronic disease, which forms the principal evidence for unchecked technological hazard mortality, may be erroneously interpreted. Thus, as shown in Figure 6, along with most other causes of death, the age-adjusted mortality from heart disease is declining; and increasing cancer mortality can in large part be explained by the delayed effect of earlier increases in smoking. In addition, there is indirect evidence (Preston 1976) that certain chronic diseases were seriously under-reported in earlier parts of this century. Therefore, the actual cancer and heart disease mortality

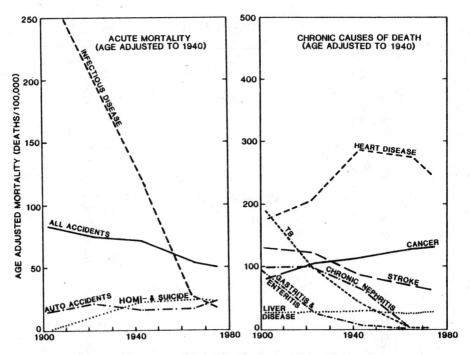

Figure 6. Historical variation of age-adjusted causes of death in the United States from 1900 to present. Among acute causes of death, note the sharp decline of infectious disease and the rise in auto-accident mortality; among chronic causes of death, note the decline of most causes except for cancer and cardiovascular disease. Even the latter shows a peaking in 1940, followed by a subsequent decline. Sources of the data is Spiegelman and Erhardt (1974).

rates shown from 1900-1940 were probably higher and the overall increase since 1900 lower than shown in Figure 6.

In summary, we believe the burden of technological hazard mortality is not currently rising. Rather, it is clear that in the United States the last century has brought three things: (1) a longer life through elimination of old ways of death, which were largely acute and rooted in natural hazards, (2) an increase in chronic causes of death, which are rooted significantly in technology, providing therefore (3) a continuing burden of death, close to half of which results from accidents and violence and the remainder from various chronic diseases.

Ecosystem Impacts

Beyond species extinction and productivity decline, what are the long-term trends in technological hazard impacts on ecosystems?

On the positive side of the ledger, it is clear that massive, local releases of pollutants to the environment, as exemplified by the London killer smog, Minamata disease, and fish-killing concentrations of pesticides in rivers, are now less frequent. Trends in air and water quality indicate that, after massive investments, environmental quality in the heavily populated and industrialized areas of the United States is generally improving (Environmental Protection Agency 1977a;1977b). Thus strong control programs for particulates and sulfur dioxide have reduced emissions to the point that very few urban regions are now experiencing violations of standards for these pollutants. Fish have returned to western Connecticut's Naugatuck River, even in areas where no aquatic life could survive in the 1950s (Council on Environmental Quality 1978). Interestingly, almost all of the major ecological hazards which have been identified and brought under control share two attributes that determine the nature of the hazard management process—they originate from an easily identifiable point source, and they are amenable to control by technological fixes of the source.

On the negative side of the ledger, it is equally clear that widespread releases of pollutants in relatively low concentrations are degrading aquatic and terrestrial ecosystems at an unmitigated or even increasing rate. Calculated ratios of manmade to natural fluxes of heavy metals, for example, indicate that natural cycles of mercury, lead, antimony, and selenium are being significantly altered by human activities (Stumm 1977). The input of mercury to the global atmosphere from industrial and fossil fuel emissions exceeds the natural flux eighty-fold, and the ratio of man-made to natural flux is large for a number of other cases (Table 10). This explains, in part, why toxic metal pollution was cited by thirty-five of forty-one states that reported water quality problems to the Environment Protection Agency in 1976 (Council on Environmental Quality 1978).

Similarly, persistent pesticides consisting of chlorinated hydrocarbons, though banned for some time because of potentially harmful ecosystem impacts, are found with a 68 percent detection rate in water and sediment samples in Houston, Texas (Council on Environmental Quality 1978). And DDT, although controlled in the U.S., is increasingly being produced for global sale in developing countries (Goldberg 1976).

TABLE 10
Global average ratios between manmade and natural flux of selected
heavy metals in the environment

ELEMENT	RATIO OF MANMADE TO NATURAL FLUX
Nickel	0.9
Vandium	1.3
Copper	2.3
Arsenic	3.3
Tin	3.5
Zinc	4.6
Cadmium	5.2
Selenium	14
Antimony	28
Molybdenum	29
Lead	70
Mercury	80

Source: Modified from Garrels, Mackenzie, and Hunt (1975).

Finally, acid rain, resulting from regional deterioration of
air quality in areas downwind from urban centers, is having a number
of effects. One of the most remarkable and potentially hazardous of
these is the fact that it apparently results in a complete shift in
forest floor mineral cycling processes which may eventually lead to
problems with nutrient availability and metal toxicity as well as
direct damage to leaf tissue (Cronan et al. 1978;Seliga and
Dochinger 1978).

Thus, for ecosystems, as for human mortality, we observe a
change from acute to chronic effects, from easily understood to
complex causal structure. Much of what is happening in ecosystems
is so incompletely understood that no clearcut directives can flow
from scientific work to hazard management. All that science can
currently hope to provide are warnings about what may possibly
happen.

The Challenge

Hazards arising explicitly or implicitly out of technological
practices have, in the industrialized world, significantly surpassed
natural hazards in impact, cost, and general importance. At present
in the United States, technological hazards account by our estimate
for 15 to 25 percent of human mortality, with associated economic
costs and losses of $50 to $75 billion annually. About half of
these costs and losses are associated with accidents and violence,
the remainder with various forms of chronic disease. Ecosystem
impacts, though difficult to define fully, are indicated by a number
of danger signals, such as significant exogenous species extinction,
productivity losses, and high concentrations of man-made toxic
chemicals in the environment.

Overall, the burden of risk assessment, hazard management,
coping, and adjustment may be as high as $280 billion per year (1979

dollars), or 12.4 percent of GNP (chapter 7). So far, the principal result of this effort has been the elimination of numerous acute effects, such as infectious disease and point-source pollution, with little progress in stemming the tide of chronic disease and ecosystem impacts.

We conclude, therefore, that, although the problem of technological hazards is on balance not getting worse, the main success of hazard management has been with the relatively more accessible part of the problem. And, whereas this part of the problem is by no means under control, as indicated by the continuing burden of violence and accidents, the principal challenge for the future involves hazards that have indistinct causes and a broad distribution of impacts. Coping with technological hazards is and will continue to be one of the major social issues of our time.

ACKNOWLEDGMENTS

We thank Robert Goble, Thomas Hollocher, Jeanne Kasperson, Roger Kasperson, and Jim Tuller, all colleagues in the Clark Hazard Assessment Group, for providing us with interesting insights and specific information. The research of the group was supported in part by the National Science Foundation under grant number ENV 77 15334. Any opinions, findings, conclusions or recommendations expressed herein are those of the authors and do not necessarily reflect the view of the National Science Foundation. We are also most grateful to Dennis Chinoy, Thomas Hollocher, Henry Kissman, Leslie Lipworth, David Pimentel, and Chris Whipple who served as reviewers and provided helpful commentary.

NOTES

1. This chapter originally appeared as "Our Hazardous Environment." Except for minor revisions appropriate to this volume, the article is reproduced with permission from Environment 20 no. 7 (September 1978):6-15,38-41, a publication of the Helen Dwight Reid Educational Foundation (HELDREF).

2. The Latin meaning of "exogenous" is "of external origin." As used in our discussion "exogenous mortality" does not necessarily exclude any genetic involvement. Rather, it refers to that fraction of mortality which, in a purely statistical sense, can be altered by changing external conditions. Genetic factors, including inherited susceptibility to a particular disease, can easily be active in this context. One needs only think of the initiation of disease in an individual case as a combination of genetic predisposition and external factors.

154

REFERENCES

Bolin, Bert. 1977. Changes in land biota and their importance for the carbon cycle. Science 196:613–615.

Brunnel, F., and P. Brunnel. 1967. Extinct and vanishing animals. New York: Springer Verlag.

Burton, Ian, Robert W. Kates, and Gilbert F. White. 1978. The environment as hazard. New York: Oxford University Press.

Council on Environmental Quality. 1974. Environmental quality 1974. Washington: Government Printing Office.

Council on Environmental Quality. 1976. Environmental quality 1976. Washington: Government Printing Office.

Council on Environmental Quality. 1978. Environmental quality 1978. Washington: Government Printing Office.

Cronan, Christopher S., William A. Reiners, Robert C. Reynolds, and Gerald E. Lang. 1978. Forest floor leaching: Contributions from mineral, organic, and carbonic acids in New Hampshire subalpine forests. Science 200:309–311.

Ehrlich, Paul A., Anne H. Ehrlich, and John P. Holdren. 1977. Ecoscience: Population, resources, and environment. San Francisco: W.H. Freeman.

Environmental Protection Agency. 1977a. National water quality inventory: 1976 report to Congress. EPA–440/9–76–024. Washington: Environmental Protection Agency.

Environmental Protection Agency. 1977b. National air quality and emission trends report: 1976. EPA–450/77–022. Washington: Environmental Protection Agency.

Epstein, Samuel S. 1976. The political and economic basis of cancer. Technology Review 78 no. 8:34–39, 42–43.

Faigin, Barbara M. 1976. 1975 societal costs of motor vehicle accidents. DOT-HS 802–119. Washington: Department of Transportation.

Fiering, M.B., and C.S. Holling. 1974. Management standards for perturbed ecosystems. Agro-Ecosystems 1:301–321.

Fisher, J., N. Simon, and J. Vincent. 1969. Wildlife in danger. New York: Viking Penguin.

Fraumeni, Joseph F. Jr., ed. 1975. Persons at high risk of cancer: An approach to cancer etiology and control. New York: Academic Press.

Garrels, R.M., F.T. Mackenzie, and C. Hunt. 1975. Chemical cycles and the global environment. Los Altos, Calif.: Kaufmann.

Goldberg, E.D. 1976. The health of the oceans. Paris: UNESCO.

Higginson, John. 1976. A hazardous society? Individual versus community responsibility in cancer prevention. American Journal of Public Health 66:359–366.

Kates, Robert W. 1978. Comparative hazard sources in the U.S. and developing countries. Unpublished working paper, Hazard Assessment Group, Clark University. Worcester, Mass.: Center for Technology, Environment and Development (CENTED), Clark University.

Likens, G.E., F.H. Bormann, R.S. Pierce, and W.A. Reiners. 1978. Recovery of a deforested ecosystem. Science 199:492–496.

National Center for Health Statistics. 1978. Vital Statistics of the United States. Washington: Government Printing Office.

National Safety Council. 1977. Accident Facts 1977. Chicago: National Safety Council.

Odum, H.T., and E.C. Odum. 1976. Energy basis for man and nature. New York: McGraw Hill.

Pimentel, David. 1978. Private conversation.

Porter, P.W. 1976. Agricultural development and agricultural vermin in Tanzania. Paper presented at the annual meeting of the American Association for the Advancement of Science, Boston, February.

Preston, Samuel H. 1976. Mortality patterns in national populations. New York: Academic Press.

Rice, Dorothy J., Jacob J. Feldman, and Kerr L. White. The current burden of illness in the United States. Occasional paper. Washington: Institute of Medicine, National Academy of Sciences.

Seliga, T.A., and L.S. Dochinger, eds. 1976. Proceedings of the first international symposium on acid precipitation and the forest ecosystem. Washington: Dept. of Agriculture.

Smithsonian Institution. 1975. Report on endangered and threatened plant species in the United States. Washington: Government Printing Office.

Sommer, A. 1976. Attempt at an assessment of the world's tropical moist forests. Unasylva 28 nos. 112 and 113:5-24.

Spiegelman, Mortimer, and Carl L. Erhardt. 1974. Mortality in the United States by cause. In Mortality and morbidity in the United States, ed. Carl L. Erhardt and Joyce E. Berlin, 21-38. Cambridge, Mass.: Harvard University Press.

Stumm, Werner, ed. 1977. Global chemical cycles and their alterations by man. Berlin: Abakon Verlagsgesellschaft.

Uetz, George, and Donald L. Johnson. 1974. Breaking the web. Environment 16 no. 10:31-39.

U.S. National Cancer Institute. 1974. U.S. cancer mortality by county, 1950-69. By Thomas W. Mason and Frank W. McKay. DHEW-NIH 74-615. Washington: Government Printing Office.

Woodwell, G.M., R.H. Whittaker, W.A. Reiners, G.E. Likens, C.C. Delwiche, and D.B. Botkin. 1978. The biota and world carbon budget. Science 199:141-149.

World Health Organization. 1976. World health statistics annual. Vol. 1. Geneva: World Health Organization.

Woytinski, Wladimir S., and Emma S. Woytinsky. 1953. World population and production: Trends and outlook. New York: Twentieth Century Fund.

7
Economic Costs and Losses

James Tuller

At a time when public concern over technological hazard appears to be growing, a rising chorus complains that the costs of controlling technology have become excessive. It is not a coincidence that government regulation, insurance and liability costs, and health care expenditures all have been criticized, since each of these efforts serves in part to prevent or mitigate hazard consequences. Unfortunately, our knowledge of the extent and distribution of economic losses and the costs of reducing these risks is incomplete. In what follows, I review what is known about this burden as it affects the United States.

It is easy to draw misleading conclusions about the real distribution of economic burdens, particularly if one views risk-reduction strategies in isolation. For example, one recent study concludes that for every dollar spent by federal agencies to write and enforce regulations, private business must spend 20 dollars to comply (Weidenbaum 1978). Whereas it may seem that businesses and, ultimately, consumers must pay an extremely large and perhaps unfair share of these costs, it would be incorrect to conclude that all hazard management costs have a similar distribution, or even that all compliance costs are in fact attributable to hazard control. The large share of private sector costs revealed by studies of regulation is more an artifact of study designs than a feature of social reality. Any study of **only** regulation is likely to examine only those government programs that generate substantial private compliance costs and to ignore programs that may have the same objectives as regulation.

In addition to avoiding such errors, a comprehensive assessment of economic costs should include as many of their economic impacts as can be estimated. It should also at least identify impacts, such as psychological damage, for which no estimates are available. Among adverse impacts, the costs associated with mortality, morbidity, property damage, and with the effort to prevent these consequences are the best documented by budget analysis, census data, and other surveys. This study relies largely on such sources.

Few hazards can be characterized as exclusively natural, social, or technological in origin, since all these elements interact in most real situations. Because this volume focusses on technological hazards, however, the analysis that follows excludes hazards derived from extreme natural events and social problems.

158

Thus, it excludes most aspects of such hazards as crime, drug abuse, floods, infectious disease, residential sewage disposal, and war and preparation for war. Especially the last eliminates a burden as large as that of all other hazards combined!

Given the uneven quality of information used in estimating costs, I estimate ranges of costs for different categories. These ranges do not represent statistical confidence limits, but they result from problems of definition, reliability, aggregation, and especially interpretation. My goal has been to provide a range of reasonable estimates within the varied scope of interpretations permitted by the study's definition of technological hazard and the joint character of most costs and losses. This joint character is a consequence of the fact that many budgeted items have one or more social purposes in addition to hazard management.

To organize the allocation of costs and losses, I employ the causal structure of hazard defined in chapter 2. According to this, hazards are a sequence of events that originate in human needs and wants, evolve to the choice of technology, initiating events, release of energy and materials, exposure, and consequences. The data used in allocating costs and losses cannot be disaggregated into the entire causal structure. For the present purpose, I combine stages of the causal structure into four useful categories: (1) choice of technology; (2) prevention of events and consequences (which includes initiating events, releases, exposure, and losses); (3) consequence mitigation; and (4) the bearing of losses and damages not otherwise prevented or mitigated. Table 6, which summarizes my findings, employs these four categories.

Issues in Cost and Loss Estimation

To introduce my estimates of economic costs and losses, I review issues that affect these estimates using the four categories of hazard control. My analysis begins with the fourth category.

Losses and Damages

Uncontrolled hazard consequences create economic losses associated with lost productivity due to illness and death and with damage to property and environment. Since losses caused by sickness and death are heavy, any estimate of the economic burden of technological hazards will be quite sensitive to the proportions of illness and death attributable to technological hazards. The proportions used here derive from estimates of technologically associated mortality as developed in chapter 6. Although morbidity/mortality ratios may differ significantly from hazard to hazard, I assume for simplicity that morbidity proportions are the same as the mortality proportions discussed in chapter 6.

To compute actual losses, the morbidity and mortality proportions were submitted to the methodology of Rice, Feldman, and White (1976), who calculate the cost of lost productivity attributable to each of several standard diagnoses of illness and death in the United States.

The effects of chronic hazards such as air and water pollution are more difficult to establish. Although health effects of these chronic hazards are included in the proportions of technologically associated mortality and morbidity, property damage is not. Numerous studies have estimated property damages caused by pollution or the benefits of controlling pollution to reduce such damage. Scientific uncertainty and differences in methods, goals, and definitions account for the wide variation in estimates. A review (Freeman 1979) prepared for the Council on Environmental Quality attempts to reconcile some of these differences. I have used this review and others (Abel 1975;U.S. Congress 1975;Heintz, Hershaft, and Horak 1976;Waddell 1974) to prepare the cost estimates of property damages.

Fires and accidents are acute events that ordinarily result in specific claims subject to insurance company investigation. Such claims are a good measure of economic losses, but the cost of administering the insurance system is more appropriately charged to consequence mitigation (see below). Health care costs might also be viewed as losses or damages, but in the context of the categories selected for study, they are also more appropriately assigned to consequence mitigation.

Consequence Mitigation

After a hazard has led to experienced injury or damage, there are still many possibilities for intervention. These interventions, both of which qualify as consequence mitigation, may compensate for experienced losses, or they may seek to prevent further, more serious consequences. Administrative expenses of insurance used to distribute damages are an appropriate example of the former; health care costs are an example of the latter.

One may well ask whether expenditures such as the Federal Supplementary Security Income Program are more a form of consequence mitigation than a redistribution of damages and losses. This question carries particular significance since the location of these efforts within the hazard structure categories affects the estimation procedure itself. Public mitigation efforts are estimated directly from government budgets, whereas losses and damages are inferred from earnings rates and vital statistics. If these expenditures are redistributed losses that have been misclassified as mitigation, they would be counted twice.

If government disability programs are strictly a form of social insurance that prevents the economic devastation of the disabled, then only the administrative expense of these programs (about 7 percent of the budgetary cost) should count as mitigation. Evidence shows that these programs go further. Disability programs, funded and administered separately from other public assistance programs, provide a higher level of benefits and a wide variety of services. Significantly, many program participants experience difficulty in returning to employment because their earnings would be more than offset by the reduction in their benefits. In short, the programs provide services that exceed lost earnings.

A strict concern with effects on national income would take care to avoid any double counting of transfer payments. Yet it is informative for purposes of general policy guidance to review the trend of actual expenditure patterns and budgetary priorities. Therefore, I include disability programs in mitigation estimates, even though this may create some duplication of costs.

Prevention of Events and Consequences

In a typical case, a government agency might order an industrial plant to reduce or eliminate emissions of pollutants. The agency incurs administrative costs in promulgating and enforcing the order, whereas the industry incurs compliance costs. Existing management costs are plausibly taken as baseline costs. Although this approach to defining the cost of consequence prevention is frequently used, it can be misleading. In order to protect itself from other kinds of social controls (e.g., lawsuits), an industry might "voluntarily" produce safer products than those required by government standards. In other cases, business or professional organizations may undertake self-regulation in order to forestall what they fear would be excessive government regulation.

Baseline and incremental control costs also are difficult to identify because of the dynamic nature of hazard management. In general, mandated controls become absorbed into everyday practice. It can be difficult to determine which period represents the baseline and which set of regulations constitutes the increment. Special safety requirements of all types become assimilated and provide economic benefits that are only indirectly related to their roles in controlling hazards. The first wagons probably had no brakes. Now vehicles without brakes are unthinkable. A better example is the improvement of railroad braking and signalling devices, which not only reduced the frequency of collisions but also reduced the headway distance between trains; this, in turn, increased the capacity and economic value of the track (Broadbent 1962). What portion of the cost of railroad brakes should be allocated to hazard control? The answer is probably "little or none."

To avoid such questions many studies of safety costs focus strictly on incremental costs and seek to estimate only the additional cost of new, improved brakes. Yet even these incremental costs can generate new economic benefits unrelated to hazard control. Hence, the degree to which they are simply ordinary investments, as opposed to risk-reduction efforts, is unclear. Ashford et al. (1980) have published a review of these kinds of problems.

Most recent surveys of compliance costs have selected a period of incremental comparison that takes advantage of a real discontinuity in United States hazard management policy. Between 1968 and 1972 Congress enacted an important series of laws that resulted in substantially more stringent regulation of motor vehicle safety, pollution emissions, and occupational hazards. For quantifying compliance costs of hazard regulation, I have used two studies that utilize this incremental-cost approach (Denison 1978;McGraw Hill 1980).

An alternative method of estimating incremental costs uses current practice as a baseline and, indifferent to safety and health considerations, estimates the hypothetical minimal costs of a technology (Cremeans and Segal 1975,9). Since no designer is truly indifferent to all safety and health considerations, however, the selection of a baseline is not entirely avoidable.

Choice of Technology

Hazard management sometimes surfaces as technology sponsorship, which may entail both the choice of less hazardous technologies and efforts to develop and disseminate them. Some hazard managers will adopt an intrinsically safer, new technology rather than correct the hazards of an existing system. Such approaches carry the potential for dramatic improvements in public health and safety, yet these efforts are easy to overlook because they do not follow the typical pattern of regulation and compliance. Choices of new technology may be encouraged by major government initiatives motivated by broad social and economic considerations, including among other things safety benefits. In the course of developing the interstate system to reduce congestion and decrease travel times, for example, the federal government introduced measures (e.g., divided highways) that strongly improved safety. Therefore, some, but certainly not all costs of the system may reasonably be allocated to hazard control.

The problems of allocating the costs of new technologies to hazard control are clearly quite similar to those involved in estimating regulatory compliance costs. Well-controlled technologies tend to be both efficient (i.e., economical) and safe, whereas poorly controlled technologies can be dangerous and expensive. To disaggregate the dual functions of economic efficiency and safety, one may well ask: "How might things be done differently to reduce costs while continuing to accept higher risks?" This question lends itself naturally to an alternative cost analysis, similar to that discussed in relation to regulation.

In this chapter, wherever possible, I have compared costs of new technologies with costs of alternative systems designed to yield the same economic benefits, except for those benefits associated with hazard reduction. The difficulty of this approach to estimating hazard control costs via technology choice is the weakest part of my analysis and leads to significant underestimates of managerial effort in the private sector.

Estimated Losses and Damages

The costs of losses and damages associated with technological hazards, as summarized in Table 1, total $80-150 billion. The large estimated range stems primarily from the uncertainties in current efforts to measure pollution property damages. In the following I discuss separately each of the items in Table 1.

Mortality and Morbidity

To estimate morbidity and mortality, I begin with the present value of lost productivity, discounted at a rate of 4 percent, for all illnesses and deaths as distributed by diagnoses, for the year 1975 (Rice, Feldman, and White 1976). In combining this distribution with cause-specific estimates of technology-associated mortality (as reported in chapter 6), one finds that technological hazards are implicated in 28-41 percent of mortality costs and in 21-29 percent of morbidity costs. To obtain the results stated in Table 1, it was necessary to project the results of Rice, Feldman, and White backward to 1974 and forward to 1979.1

TABLE 1
Losses and damages from technological hazards by consequence type,
1974 and 1979

CONSEQUENCE TYPE	COST (IN BILLIONS OF CURRENT DOLLARS)	
	1974	1979
Mortality	23.8–34.9	37.6–49.1
Morbidity	11.7–16.1	18.2–25.2
Motor Vehicle Accidents (Property Damage)	6.0	11.5*
Fires (Property Damage)	2.2–2.4	3.2–3.4
Pollution Property Damage		
Air	4.6–26.8	6.4–37.0
Water	5.0–17.4	6.9–24.0

* 1978 data adjusted for inflation.
Totals may not add due to rounding.

Motor Vehicles and Fires

The costs of property damages resulting from fires and motor
vehicle accidents have been derived from insurance company reports.
Fire losses of $3.1–3.4 billion and motor vehicle damages of $11.5
billion exclude the costs of fires and accidents thought to be intentional (Insurance Information Institute 1979). Many industrial
accidents involve fires and are included under fire losses. The
National Safety Council (1978) reports a significant loss each year
from indirect costs of occupational accidents. These amounted to
$10.8 billion in 1978. Since the basis for calculating such figures
could not be ascertained but probably overlapped to a significant
degree with consequence mitigation, this account of losses and
damages excludes these estimates.

Air and Water Pollution Property Damage

The many attempts to calculate the economic costs of air and
water pollution and the costs or benefits of controlling these pollutants often produce seemingly contradictory results because of
differences in their objectives, assumptions, and methods. Freeman
(1979) has attempted to reconcile these differences to correct for
some of the shortcomings that have been identified in individual
studies. He intended to restrict the range of estimated benefits or
damages to values that reflect uncertainties concerning only the
dose-response effects of pollution upon health and property and to
eliminate estimates of the economic value of preventing death or the

efficacy of actual or planned pollution-control efforts.[2] In estimating air and water pollution damage costs, I have used Freeman's analysis as a starting point and made adjustments as needed for my target years of 1974 and 1979.[3]

In the case of air pollution Freeman reviewed several measures of air quality for the period since 1970. He concluded that a 20 percent reduction in sulfur dioxide and total suspended particulates was the best estimate of post-1970 air-quality improvements. These reductions generally occurred most rapidly in the years immediately following 1970, with most or all of the 20 percent reductions having been achieved by 1974 (Freeman 1979,62). Assuming that 1970 pollution levels prevailed in 1978, Freeman estimates 1978 damages (or potential benefits) in four categories.[4] By deducting the actual from the potential benefits, one may obtain the 1978 damages. Adjusting for inflation this leads to my estimate of $6.4-37 billion in air pollution damages for 1979.[5]

Various indicators of water quality have fluctuated without showing significant improvement or deterioration since 1970. The implementation of recent federal water-pollution requirements is scheduled for completion by 1983. For these and other reasons Freeman chose to estimate those benefits that will be achieved in 1985 as a result of compliance with the "best available treatment technology" provisions of the 1972 amendments to the Federal Water Pollution Control Act. Most of these benefits would result from increased recreational opportunities. The understanding of these opportunities and the ability to predict their economic consequences or estimate their value is not as advanced as the ability to estimate the economic effects of air pollution. Therefore, the empirical basis for the water pollution damage estimates is less sound than that for the air pollution results.

Freeman's results rely in part upon estimates of the total damages of water pollution:

> There are at least two reasons why estimates of damages might be taken as upper-bound estimates of the benefits of full compliance with the 1983 standards. First, these standards call for strict control of discharges and in some cases will result in the elimination of certain types of discharges. And second, if...the first clean-up efforts produce the largest benefits, the bulk of benefits will have been achieved (and damages avoided) as the 1983 standards are approached and attained. (Freeman 1979,134)

Since not all pollutants will be controlled, benefit estimates will tend to be less than total damages. Since these estimates are based upon the increase in population and income levels expected by 1985, however, they would tend to overestimate the benefits that would achieved in 1974 or 1979. Despite these uncertainties, Freeman's results remain the best information available on water pollution damages and are consistent with another recent comprehensive survey (Heintz, Hershaft, and Horak 1976). Using them as a basis, my estimates of nonhealth water pollution damages are $5.0-17.4 and $6.9-24.0 billion for 1974 and 1979, respectively.

Private Control Costs

To discuss the cost of technological hazard control, I consider private and public controls separately. Private controls, summarized in Table 2, account for a total expenditure of $67–80 billion for 1979. This total comprises separate subtotals for the four categories of hazard control.

TABLE 2
Private costs to control technological hazards, by control action (1974 and 1979)

CONTROL ACTION	COST (IN BILLIONS OF CURRENT DOLLARS)	
	1974	1979
Consequence Mitigation		
Medical Costs (includes insurance)	13.8–19.1	27.1–37.5
Insurance, Administrative Expense (nonmedical)	2.8	5.6
Total Consequence Mitigation	16.6–21.9	32.7–43.1
Prevention of Events and Consequences		
Motor Vehicle Safety*	0.7–2.6	0.1–3.6
Occupational Safety	0.3–2.9	2.9
Pollution Abatement	12.5–18.1	31.7–31.9
Total Prevention Costs	13.6–23.6	34.6–37.4
Choice of Technology	n o t e s t i m a t e d	
Total Private Control Costs	30.2–45.5	67.4–80.4

* 1978 data adjusted for inflation.

Consequence Mitigation

These costs include health care and administrative expense of insurance used to redistribute the damages caused by hazards. As Table 2 shows, they totalled $32.17–43.1 billion in 1979.

Medical costs were derived in a manner similar to that used in productivity loss estimates described above; that is, by combining fractions of technological-hazard-associated mortality and morbidity with estimates of private health care expenditures. Because Rice, Feldman, and White (1976) count only health-care expenditure related to specific diagnoses, I employed instead estimates by Gibson (1979) for the **total** cost of private health care expenditure.6   As a result I project health-care-related expenditures of $27.1-37.5 billion for 1979.

Administrative insurance expenses are the difference between premiums paid and claims, as adjusted for coverage in force. It is the cost of redistributing losses rather than the losses themselves. Nonhealth-care insurance expenses were estimated to be $5.6 billion in 1979.

Prevention of Events and Consequences

Many of the costs in this category are difficult to identify because of the problems of distinguishing safety-related and ordinary expenditures. In the last decade, however, new federal initiatives in the areas of motor vehicle safety, occupational health, and pollution abatement have encouraged efforts to measure hazard management costs in these important areas. As shown in Table 2, total private costs for these control measures are estimated to be $34.6-37.4 billion in 1979. These costs are obviously underestimates of private prevention efforts since they cover only three hazard areas.

For automobile hazards, a good source of data is the Bureau of Labor Statistics (BLS), which has surveyed for a number of years, the average cost of safety and other improvements required in passenger cars. Only the incremental costs for the current model year, which are identified as compliance costs, can be accepted with a satisfactory degree of confidence. There is no way to tell from the BLS survey whether devices introduced to meet safety standards in previous years are still present on vehicles produced in later years, and at what cost. Some studies of compliance costs have accumulated the BLS data in yearly increments (Denison 1978). Such a procedure results in a projection of $3.6 billion in 1979, but that almost certainly constitutes a vast overestimate. The incremental single-year compliance costs were only $61 million in 1979.

For occupational health costs, a useful source is the McGraw Hill annual survey of business expenditures on plants and equipment (McGraw Hill 1980). There is some question, however, whether the results of this survey, published since 1972, represent costs of complying with the Occupational Safety and Health Administration (OSHA) regulations. According to Denison (1978,36), "...the trend of capital outlays from 1972 to 1975 suggests a different interpretation: that nearly all of the reported expenditures would have been made in the absence of the new legislation." Denison has employed a method for estimating the proportion of the McGraw Hill data that is attributable to the OSHA requirements. Reported safety spending declined during 1979, however, and application of the Denison procedure to that year produces negative compliance costs. In Table 2, I report outlays by Denison as OSHA compliance outlays and use them as lower bounds for the 1974 data. The unmodified McGraw Hill results serve as the upper bound. Only the McGraw Hill estimate has been reported for 1979 (McGraw Hill 1980).

Annual estimates for pollution abatement are prepared by the Bureau of Economic Analysis and the Council on Environmental Quality (1975 and 1979;Rutledge and Trevathan 1980). At this writing, the latest available results were for 1978 and were projected to 1979 by adjusting for real growth and inflation. Based on this, I estimate pollution-abatement costs at $31.7-31.9 billion for 1979.

Choice of Technology

The private sector lacks obvious counterparts to the examples of government-subsidized new technologies whose introduction is encouraged because of superior safety characteristics. Undoubtedly, market demands for improved safety have accelerated the introduction of many new technologies. But new technologies may also improve safety, incidental to other advantages, a condition only recognizable with hindsight. Automobiles helped reduce infectious disease hazards associated with animal transport; oil use greatly reduced the air pollution hazards of coal burning; and the jet proved to be safer and more reliable than piston-engine aircraft. Although some portion of the private resources used to develop and produce new technologies can be considered as expenditures to reduce hazards, I have attempted no estimate of this amount.

Public Control Costs

Compared to private control costs, public sector control costs are well documented, largely through budgets. In the following, I consider federal and state expenditures separately.

Federal Hazard Control Costs

Federal expenditures to control technological hazards were estimated from detailed reviews of actual obligations reported in the Appendix to the Budget of the U.S. Government. Because of the consistency, accuracy, and detail of the budget documents, the estimates of federal expenditures are reliable, especially for year-to-year comparison, within the context of the basic uncertainties and questions of interpretation in the allocation of costs which are always present in a study of this type. As Table 3 indicates, federal control costs are estimated at $21.7-34.8 billion for fiscal year 1979. This breaks down into three categories.

Consequence Mitigation. Since the introduction of major federal health care and social welfare programs in the 1960s, expenditures in this category have become an increasingly large component of federal hazard-management efforts. The 1979 costs of $14.2-23.3 billion include both health care programs and $4.9-8.1 billion in disability aid.

Prevention of Events and Consequences. This category includes those costs that correspond most closely to what have been called the federal costs of regulation. The results for fiscal year 1979 of $5.3-6.9 billion and the trend from the fiscal year 1974 results are consistent with other studies of regulatory costs (DeFina 1977;U.S. Congress 1976).

Some expenditures that ought to be included here are covered elsewhere. In those cases where the government is a final consumer or purchaser of products, such as automobile safety and emission devices, the costs have been included under private expenditures.

Choice of Technology. Expenditures during 1979 of $2.2-4.6 billion are concentrated primarily in investments related to energy

TABLE 3
Federal costs to control technological hazards, by control action

CONTROL ACTION	COST (IN BILLIONS OF CURRENT DOLLARS)		
	FY 1964	FY 1974	FY 1979
Consequence Mitigation	0.7-1.2	6.5-10.6	14.2-23.3
Prevention of Events and Consequences	1.0-1.7	2.5-4.0	5.3-6.9
Choice of Technology	0.5-1.8	0.6-2.3	2.2-4.6
TOTAL	2.2-4.7	9.7-16.9	21.7-34.8

CONTROL ACTION	COST (IN BILLIONS OF 1972 DOLLARS)		
	FY 1964	FY 1974	FY 1979
Consequence Mitigation	1.0-1.7	5.8-9.5	9.0-14.7
Prevention of Events and Consequences	1.3-2.3	2.3-3.6	3.3-4.4
Choice of Technology	0.7-2.4	0.5-2.0	1.4-2.9
TOTAL	3.0-6.5	8.6-15.1	13.7-22.0

and transportation. The results are based upon alternative cost estimates of additional expenditures incurred to achieve a lower level of risk than might have been possible with less costly alternatives.

Trends in Federal Expenditure

For three years (1964, 1974, and 1979), Table 4 provides a more detailed account of federal hazard related expenditures in traditional programmatic categories. Year-to-year variations in these categories should be interpreted with caution, since they might be an artifact of changing explanatory details in the budget documents. Nonetheless, they, along with the summary figures in Table 3, make it possible to track the federal government's effort over a longer period. Beginning in 1964, at a lower level of environmental and hazard consciousness, federal expenditures have increased between 6 percent and 9 percent per year and have undergone a definite change in character. Major technology-sponsorship initiatives, such as the Interstate Highway Program, have increased only modestly in 15

TABLE 4
Federal efforts to control technological hazards, by program
category

PROGRAM CATEGORY	COST (IN MILLIONS OF 1972 DOLLARS)		
	Fiscal Year 1964	Fiscal Year 1974	Fiscal Year 1979
Air Transport	14-217	22-343	19-403
Water Transport	719-728	553-589	514
Highway Transport	579-1,734	641-1,609	1,393-2,326
Railroad Transport	4	14-27	44-84
Food Quality and Pesticides	17-190	93-400	394-504
Soil Conservation	202-237	201-221	212-284
Marine Environment, Fish & Wildlife	0.3-209	76-286	234-433
Air Pollution	15-18	130-169	200
Water Pollution	6-48	50-433	298-304
Intermodal Transport	0.4-5	11-40	33-158
Nuclear/Radiation	182-653	283-454	268-593
Occupational Health	24	156	150
Disability Programs	381-661	2,105-3,434	3,068-5,031
Health Research	145-366	351-378	394-500
Health Care & Resources	416-719	3,373-5,684	5,442-9,055
Basic Science	59-74	71-88	76-108
Banks & Exchanges	86-245	275-314	51-113
Miscellaneous	69-362	214-522	855-976

years, whereas health-care programs show an extremely strong tenfold
increase. Growth in spending to prevent events and consequences
fell somewhere between the two, increasing about as fast, or only
slightly faster than, the total federal budget. Even allowing for
the rapid growth in real health costs, perhaps independent of hazard
management, a larger proportion of federal effort still favors
treatment over prevention. Despite the widespread perception of

rising expenditure for coping with technological hazards, the direct managerial component for preventing events or exposure has been growing only at the same rate as other government expenditures.

State and Local Hazard Control Costs

Information on the expenditures of the 70,000 to 80,000 state and local governments is much less detailed than that available for the federal government. Estimated expenditures for technological hazards of $11.0-17.3 billion in fiscal year 1978 are based upon the annual Bureau of the Census survey of government finances, Social Security Administration statistics on social programs, and the Bureau of Economic Analysis estimates of pollution-control expenditures. Results are summarized in Table 5. A breakdown into three categories follows.

Consequence Mitigation. These expenditures of $5.4-7.5 billion consist primarily of health-care costs; disability programs account for only $0.2 billion of the total. Health-care costs include some preventive public health-care programs that are not a form of mitigation.

Prevention of Hazard Consequences. Prevention costs in 1979 are estimated at $5.7-9.8 billion. Only three major program

TABLE 5
State and local costs to control technological hazards, by control action

CONTROL ACTION	COST (IN BILLIONS OF CURRENT DOLLARS)		
	FY 1964	FY 1974	FY 1978
Consequence Mitigation			
Health Care	1.0-1.4	3.3-4.6	5.2-7.2
Disability Aid	0.2-0.3	0.1	0.2
Total Consequence Mitigation	1.2-1.7	3.4-4.7	5.4-7.5
Prevention of Events and Consequences			
Fire Protection	0.9	2.1-2.3	3.4-3.6
Transportation	0.3-1.1	1.2-4.1	1.5-5.0
Pollution Control	0.5-0.9	0.8-1.2	0.8-1.2
Total Prevention Costs	1.7-2.9	4.1-7.6	5.7-9.8
TOTAL	2.9-4.6	7.6-12.3	11.0-17.3

TABLE 6
1979 Societal costs and losses from technological hazards, by sector

| SECTOR | Choice of Technology | COST BY CATEGORY (IN BILLIONS OF 1972 DOLLARS) | | | Damages and Losses |
		Prevention of Events & Consequences	Consequence Mitigation	Total Control Costs	
Private	NA	34.6–37.4	32.7–43.1	67.4–80.4	
Federal	2.2–4.6	5.3–6.9	14.2–23.3	21.7–34.8	
State & Local*	NA	5.7–9.8	5.4–7.5	11.0–17.3	
All Public	2.2–4.6	11.1–16.7	19.6–30.8	31.7–52.1	
TOTAL	2.2–4.6	45.7–54.1	52.3–73.9	99.1–132.5	79.8–150.2

* 1978 data
(Totals may not add due to rounding)

categories are presented: fire protection, transportation, and pollution control (including sanitation). As in the case of federal costs, some expenditures occurring in the form of consumer product purchases have been listed under private costs. Transportation expenditures are estimated from functional expenditures classifications in the Federal Highway Administration's Highway Statistics (1979) and from the government finances survey of the U.S. Bureau of the Census (1979a, 1979b, 1980).

Choice of Technology. These costs have not been estimated, although some highway construction funds counted as prevention costs might qualify as sponsorship expenditures.

Summary and Conclusion

The societal costs and losses from technological hazards are summarized in Table 6 for the year 1979. The table combines the results of the previous sections. Despite the inherent uncertainties already discussed, significant and relatively unambiguous conclusions emerge from the analysis.[7]

The absolute economic burden associated with technological hazards is large but not overwhelming. Its magnitude of $179-283 billion for 1979 represents 7.8-12.4 percent of the Gross National Product, a fraction that has not changed significantly in the five-year period for which I have made comparative estimates (for 1973-74, the fraction is 7.0-12.6 percent).

The costs of controlling technological hazards (column four, Table 6) are about as high as damages and losses (column five, Table 6). Damage costs break almost equally between health and property effects.

The public share of control costs for technological hazards was 32-39 percent of the total in 1979. Despite probable underestimation of some private costs, these results suggest that the public share of technological hazard control is about as large as the public share of total national expenditures. This conclusion is quite different from implications that might be drawn from studies of regulatory costs alone (Weidenbaum 1978).

Is 7-12 percent of the Gross National Product a high, low, or reasonable cost for society's hazard management effort? Given the structure of costs that I have explained, large fluctuations in this statistic are unlikely. A period of heavy losses would bring real declines in the amount of property and productivity at risk. Similarly, increases in spending to reduce losses also increases control costs, causing the total burden to rise again. Yet, if it is reasonable to expect that the economic burden of technological hazards might soon be reduced by as much as an order of magnitude, it is unwarranted to conclude that smaller fluctuations in this burden would be of minor importance.

The answer to the question "are costs reasonable?" must therefore be: the costs are large enough that even marginal improvements in hazard management efficiency can have substantial economic impacts. This is a welcome economic incentive to reduce the hazard burden, where even small gains can have a great value in human terms.

ACKNOWLEDGMENTS

I thank Robert W. Kates for his generous assistance. Valuable help and suggestions were offered by Christoph Hohenemser, Branden Johnson, Jeanne Kasperson, and Roger Kasperson. Portions of this research were supported by the National Science Foundation under grant number ENV 77-15334, and by grants from the Jesse H. Noyes Foundation.

NOTES

1. The results generated by the method of Rice, Feldman, and White (1976) are roughly proportional to the number of deaths, days of disability, labor force participation rates, and mean earnings rates associated with the various sex, race, and age groups in the population. The average changes in these values for the population as a whole were used to project the 1975 results to 1974 and 1979. Although this adjustment may introduce errors if the distribution of values among the population groups changes, the short time period of the projection tends to minimize this problem.

2. Freeman (1979) calculated previously reported estimates of costs or benefits in dollar values at the 1978 price level. In addition, estimates of the benefits of post-1970 pollution controls were based upon standardized assumptions about the degree of improvement achieved in air and water quality. Finally, in the case of health effects, a single estimate of the value of a "statistical life" was employed.

3. To adjust for inflation, Freeman has used the Consumer Price Index rather than the GNP Implicit Price Deflator (U.S. Bureau of the Census 1979b,476) that I have used elsewhere in this study. Although these indicators can be significantly different in some periods, in this instance the use of either index yields results that differ by no more than 1 percent.

4. To the extent that property damage categories may overlap, some double counting is possible where independent estimation methods assess the same damage costs. Freeman (1979) suspects that this is a serious problem only in the case of residential property values and indicates that only 30 percent of the damages in this category should be included in an estimate of total costs.

5. In his summary table Freeman (1979,117-118) declines to specify the estimated range for potential damages at 1970 pollution levels to property values from stationary-source pollutants, "because of uncertainties about the form of the dose-response function at lower pollution levels." An estimated range of $4.7-34.5 billion is the basis for his calculation of realized benefits, which **are** reported in the table for this category. Since this fact is explicit in the text (Freeman 1979,107-109),

these values have been used to estimate total air pollution property damages.

6. Rice, Feldman, and White (1976) count only personal health-care expenditures allocated to specific diagnoses. It is true that only these costs can be definitely associated with specific illnesses; to exclude other medical costs from the economic burden of illness, however, implies that these other expenditures are unrelated to health and disease. Since the goal of this study is an overall estimate of costs attributable to technological hazards, rather than a detailed allocation of costs to specific illnesses, I have used the estimation of total health-care expenditures given by Gibson (1979).

7. Given the wide range exhibited in estimates of certain categories of expenditures, it is reasonable to question the number of significant figures or to seek "confidence limits" for individual estimates. But these estimates are unlike those made of a physical parameter under conditions of believed certainty. Rather, these estimates depend upon issues of interpretation for which it is inappropriate to specify anything like a statistical confidence limit. It may well be possible to determine accurately federal expenditures on antismoking education, but is cigarette smoking—excluding analysis—a technological hazard? The technologies of low-cost mass production and of mass merchandising surely affect the widespread use of cigarettes. This example points up the role of uncertainty that derives from interpretation both of the hazard itself and of the intent and effects of programs that address the hazard—and not from inaccuracies in financial accounting.

REFERENCES

Abel, Fred H., Dennis P. Tihansky, and Richard G. Walsh. 1975. National benefits of water pollution control. Washington: U.S. Environmental Protection Agency.
Ashford, Nicholas A., et al. 1980. Benefits of environmental, health, and safety regulation. Prepared for the Committee on Governmental Affairs, U.S. Senate, by the Center for Policy Alternatives, Massachusetts Institute of Technology. Washington: Government Printing Office.
Broadbent, H.K. 1962. Fundamentals of railway braking. In Proceedings of the Convention on Railway Braking. London: Institute of Mechanical Engineers.
Council on Environmental Quality. 1975. Environmental quality 1975. Washington: Government Printing Office.
Council on Environmental Quality. 1979. Environmental quality 1979. Washington: Government Printing Office.
Cremeans, John E., and Frank W. Segal. 1975. National expenditures for pollution abatement and control, 1972. Survey of Current Business 55 no. 2:8-11,35.

DeFina, Robert. 1977. Public and private expenditures for federal regulation of business. St. Louis: Center for the Study of American Business, Washington University.

Denison, Edward F. 1978. Effects of selected changes in the institutional and human environment upon output per unit of input. Survey of Current Business 58 no. 1:21-44.

Federal Highway Administration. 1979. Highway statistics. Washington: Government Printing Office.

Freeman, A. Myrick. 1979. The benefit of air and water pollution control: A review and synthesis of recent estimates. Washington: U.S. Council on Environmental Quality.

Gibson, Robert M. 1979. National health expenditures, 1978. Health Care Financing Review 1 (Summer):1-36.

Heintz, H.T., A. Hershaft, and G.C. Horak. 1976. National damages of air and water pollution. Washington: Environmental Protection Agency.

Insurance Information Institute. 1979. Insurance Facts (annual). New York: Insurance Information Institute.

McGraw-Hill. 1980. Annual McGraw-Hill survey: Business plans for research and development expenditures.

National Safety Council. 1978. Accident facts 1978. Chicago: National Safety Council.

Rice, Dorothy, Jacob J. Feldman, and Kerr L. White. 1976. The current burden of illness in the United States. Occasional Paper, Institute of Medicine. Washington: National Academy of Sciences, 27 October.

Rutledge, Gary L., and Susan S. Trevathan. 1980. Pollution abatement and control expenditures, 1972-1978. Survey of Current Business 60 no. 2:27-33.

U.S. Bureau of the Census. 1979a. Census of governments, 1977. Washington: Government Printing Office.

U.S. Bureau of the Census. 1979b Statistical abstract of the United States, 1979. Washington: Government Printing Office.

U.S. Bureau of the Census. 1980. Government finances, 1977-1978. Washington: Government Printing Office.

U.S. Congress. 1975. Committee on Public Works. Air quality and stationary source emission control. Committee Print. 94th Cong., 1st sess. Washington: Government Printing Office.

U.S. Congress. 1976. Committee on Interstate and Foreign Commerce, Subcommittee on Oversight and Investigations. The number of federal employees engaged in regulatory activities. Committee Print. Washington: Government Printing Office.

Waddell, T.E. 1974. The economic damages of air pollution (report prepared for the U.S. Environmental Protection Agency). Washington: Government Printing Office.

Weidenbaum, Murray L. 1978. The impacts of government regulation. St. Louis: Center for the Study of American Business, Washington University.

Part Three

Overview: Assessing Hazards

The five chapters of Part 3 concern the early stages in the cycle of hazard management--hazard assessment and control analysis (recall Fig. 1, chapter 3). To provide a foundation in experience, chapters 8-10 describe in detail three specific hazards: automobile accidents, airborne mercury, and nuclear power. Following these excursions into specific hazard experience, chapter 11 addresses the general problems of hazard assessment and its "ideologies," and chapter 12 reviews the methods used in judging risk tolerability. Each of these two chapters significantly expands corresponding treatments in chapter 3, and each gains meaning from the experience given for specific hazards in chapters 8-10.

The sample is not wholly representative of the full range of technological hazards. Nevertheless, the three hazards studied provide a varied and interesting cross-section of the problems encountered in hazard assessment and control analysis--particularly in judging tolerable risk. In terms of the taxonomy of chapter 4, automobile accidents (chapter 8) represent the class of "common killers," defined as uniquely high in the factor **mortality**. Airborne mercury (chapter 9) represents the class of "persistent teratogens," defined as extreme in the factor **delay**. Nuclear power (chapter 10) represents the special class of hazards for which two or more factors take on extreme values: nuclear power ranks high both in the factor **delay** and in the factor **catastrophic**.

Auto accidents have few rivals among common killers in the degree to which risk structure and methods of risk reduction are understood and directly relatable to the laws of statistics and physics. Chapter 8 uses the statistics of auto accident risk to derive a comprehensive individual prescription for risk reduction; it also employs the laws of Newtonian mechanics to derive several broad guidelines for effective hazard reduction. In view of this, it is perhaps surprising that society continues to suffer an annual toll of about 50,000 deaths and more than 2 million injuries. The resolution of this puzzle, it turns out, lies not in the shortcomings of hazard assessment, but to a significant extent in the level of risk that the public willingly tolerates. Other roots of the problem lie in the politics of regulation, details of which are discussed in chapter 14 of this volume.

Airborne mercury (chapter 9), compared to auto accidents, involves a set of consequences for which the risk is less well

175

understood. Nevertheless, among the persistent teratogens (a group that includes most toxic chemicals), airborne mercury is rich with experience, involving early outbreaks of acute disease, a broad, quantitatively defined basis for acute human effects, and fairly well-defined pathways for dispersal and exposure. Despite multiple warnings and a substantial effort to set standards by a variety of groups, mercury production has not declined, and the environment is littered with hot spots in which mercury levels exceed standards. Of particular concern are consistently high mercury levels in the workplaces of certain industries. These levels equal or exceed the level where explicit symptoms of mercury poisoning are observable by the best diagnostic methods. If the assessment of the airborne mercury hazard has a soft underbelly, it is that levels currently experienced are sufficiently low so that their effects are unlikely to be detected directly in affected populations. This parallels the pattern for most "low-level" toxic pollutants and represents what is surely a generic problem with this class of materials.

Nuclear power in recent years has received more notoriety than any other technology, despite a record of safety that is far better than many technologies, including both automobiles and mercury. As chapter 10 indicates, the distrust of nuclear power has multiple causes, which include: (1) a troubled history that often associated nuclear power with nuclear weapons; (2) intractable problems of hazard assessment that combine low-probability, high-consequence risks with delayed impacts that stretch well beyond the present generation; and (3) a continuing, rancorous conflict, about the regulation and safety of the technology, which typically pits one part of the scientific community against another. Nevertheless, as chapter 10 shows, the uncertainties in assessing the risks of nuclear power are not so great as to preclude a quantitative analysis of expected mortality and a comparison to other technological risks. In these terms, nuclear power looks more favorable than other energy technologies. Hence the distrust of nuclear power must lie outside the realm of traditional probabilistic risk assessment. Chapter 10 suggests that the association with weapons and the ineptitude of past regulation and management go a long way toward explaining the case. An additional explanation, not discussed in chapter 10, can be found in analyses of perceived risk (chapters 4 and 5). Such analyses show that nuclear power combines a number of risk qualities that contribute strongly to a high level of perceived risk but are not included in the idea of risk defined as probability of dying.

With the experience of auto accidents, airborne mercury, and nuclear power as bases, chapter 11 dissects the hazard assessment process and describes the competing "ideologies" that underlie this process. The treatment of hazard assessment expands the material already given in chapter 3, particularly as regards hazard identification and risk estimation. The treatment of ideology makes the point that, as in other cases of policy analysis, hazard assessors fall into groups according to the assumptions they make. Thus author Robert Kates proposes three "schools" of ideologists, whom he labels "count-the-bodies," "tip-of-the-iceberg," and "worry-beads" assessors. Each group brings to the task of hazard assessment its own heuristics, which may result in serious over- or under-estimates of the risks attendant on a given hazard.

To conclude Part 3, chapter 12 reviews the methods of arriving at judgments of risk tolerability and expands considerably the treatment of these issues given in chapter 3. Even a cursory reading of chapter 12 leads easily to the conclusion that between hazard assessment and judging tolerability (usually the first step in control analysis), the latter is by far the more confounded. Hazard assessment begins with a shared scientific paradigm, and, ideologies notwithstanding, only in its final stages runs up against problems of social valuation. In contrast, judging risk tolerability, by whatever method, immediately immerses the assessor or manager in a quagmire of conflicting and inconsistent values. Because these values are most often implicit, the first task for analysis is to untangle the web of assumptions and to make them explicit. Chapter 12 approaches this goal by analyzing four nominal methods for judging risk tolerability: (1) risk/benefit analysis; (2) revealed preferences; (3) expressed preferences; and (4) natural standards. Though each method provides useful insights, the authors conclude that none is adequate by itself. Rather than offer a design for an alternative approach that avoids individual shortcomings of the four methods, they suggest using the four methods in various combinations, or, put another way, "muddling through."

Together, chapters 8–12 evoke a somewhat blurred picture of the early stages of hazard management. The problems that stymie the processes of assessing hazards and determining levels of tolerable risk highlight the need for better analysis. A partial list of unresolved questions might include the following:

- How can the assessment of chronic consequences and associated exposure standards be improved?

- How can conflicting definitions of risk, which underlie much controversy by scientists and the public, be reconciled?

- What are the alternatives to "muddling through" in judging risk tolerability? Can ethical and value analysis improve the situation?

- How can assessors' biases, described here as ideologies, be made more explicit in the assessment process?

None of these questions has easy or apparent answers and by themselves merely symbolize (and do not encompass) the difficulties encountered in the early stages of hazard management. Yet compared to the later stages of the process, discussed in some length in Part 4, these problems are fairly clear-cut and represent issues that may yield to a concerted attack by a variety of analysts.

8
Automobile Accidents

Christoph Hohenemser and Thomas Bick

Over the last decade, motor vehicle accidents have exacted an average annual toll of 50,000 fatalities, two million injuries, and $12 billion (1977 dollars) in property damage (National Safety Council 1982). When the total cost of fatalities and injuries, measured in lost wages, is added to property damage, the sum amounts to between $25 and $43 billion (1977 dollars), depending on the assumptions made (NHTSA 1978;Faigin 1976). Since the introduction of the motor car at the turn of the century, there have been 2.1 million highway fatalities. This is more than three times the number of all U.S. war dead in this century. Must these costs continue to be paid, or are there solutions that promise major improvement?

Surprisingly, despite the many nonvoluntary aspects of the risks involved in highway driving, much can be accomplished through individual action. This is especially true if individuals are willing to inform themselves in detail about existing risk determinants and then follow up by adjusting their driving habits accordingly. Unfortunately, however, only a few individuals are likely to be effective risk managers. Therefore society must seek to control the conditions that contribute to motor vehicle deaths and injuries.

Whereas the motor vehicle collision hazard is better understood than most other technological hazards, substantial political obstacles impede improved safety. One major difficulty is that consumers generally tend to perceive the problems as less serious than do the experts; therefore major auto makers are wary of government-imposed safety measures that will raise the price of their product.

Variation of Risk

The probability of being involved in a motor vehicle collision varies greatly from place to place, from time to time, and in accordance with other attendant circumstances.

International comparisons. If one consults the data book of the Motor Vehicle Manufacturers Association (1980), one finds the claim that U.S. highways are the "safest" in the world. This statement is supported by Figure 1 (top), which shows the traffic fatalities per 100 million vehicle miles in selected countries. Yet, if one orders these same countries according to fatalities per 100,000 population (National Safety Council 1980), as shown in Figure 1

179

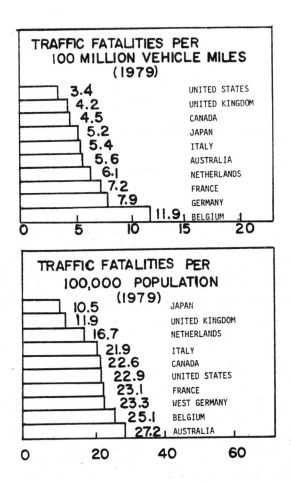

Figure 1. Comparison of auto fatality risks in selected industrialized countries, using two different measures of the risk (National
Safety Council 1980;Motor Vehicle Manufacturers Association 1980).

(bottom), the United States is close to being the most unsafe country. Both of the above observations are, of course, correct. In
the United States and other countries with large land areas, the
advantage in risk per vehicle mile is to a large extent canceled out
by the greater distances driven. Thus, one may drive a mile in the
United States and be four times as safe as in Belgium; but Belgians
drive only a fourth as much as Americans.

    Long-term Trends. In the United States, records of vehicle
fatalities date back to the invention of the automobile. These
statistics reflect the successes and failures of safety management
in a general way. Figure 2 offers an interpretation of these
trends. It shows (top graph) that the risk to the average member of
the public, expressed as the death rate per 100,000 population, has

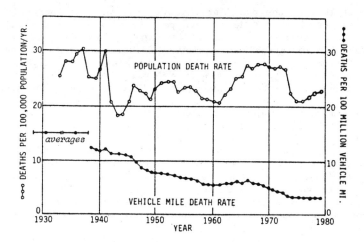

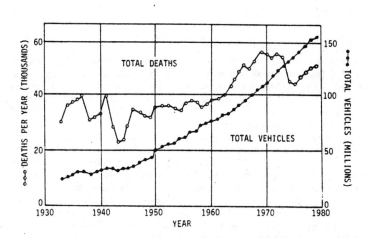

Figure 2.   Four views of motor vehicle risks in terms of historical
trends.   **Top graph:**   The open symbols indicate the death rate
per unit of population.   The closed symbols indicate the death rate
per 100 million vehicle miles.   **Bottom graph:**   The open symbols
indicate the total number of deaths per year.   The closed symbols
indicate the total number of vehicles on the road during each of the
last 45 years. To put it simply: we are dying in larger numbers as
our population increases; and we are driving more but, because of
the increased safety of a unit of driving, the risk to each member
of our population has not changed very much (National Safety Council
1980).

changed little over the past forty years. But the risk per unit of driving, expressed as the death rate per 100 million vehicle miles, is less than a quarter of what it was forty years ago. Underlying these risks (bottom graph) are a slowly rising trend in total deaths (from 40,000 to over 50,000 in forty years) and a rapidly rising trend in number of vehicles and in vehicle miles driven. The increasing trend in total deaths roughly follows population growth except for recent short-term declines. The increasing trend in vehicles is much more rapid than population growth.

State-by-State Variation. Within the United States, fatality rates vary markedly by state. Expressed in fatalities per 100 million vehicle miles, the range covers a factor of three. The highest rates occur in thinly populated Western states such as Wyoming, New Mexico, and Montana; the lowest in densely populated Eastern states such as New Jersey, Rhode Island, and Connecticut. Evidently, high population density makes for safer driving, which is the opposite of what one might naively expect. In fact, urban driving is about twice as safe as rural driving, with rates of approximately two deaths per 100 million vehicle miles for urban driving as compared with about five deaths per 100 million vehicle miles in rural areas. As shown in Figure 3, percent of urban population is a reasonable predictor of risk per 100 million vehicle miles.

When risk is expressed in fatalities per 100,000 population, the effect of urbanization is even greater, as is indicated in Figure 4. This is to be expected if, as is indeed the case, urbanization results not only in lower fatality rates per vehicle mile but in less drivng exposure per person as well.

Regardless of which way the risk is expressed, some states rank anomalously low, whereas others are anomalously high, and there is substantial scatter within a broad "trend band." This emphasizes the fact that there are other causative factors behind traffic mortality rates besides the urban/rural variation, though the latter is clearly one of the most important.

Other Correlates of Highway Mortality. The 45,181 highway fatalities within the United States in 1976 have been subjected to substantial analysis (NHTSA 1977a). This procedure (in conjunction with other sources [Haddon 1972]) yields a number of interesting propositions about how risk varies in relation to a variety of factors. Thus, the fatality rate, expressed as deaths per vehicle mile, varies with time of day or week, type, and vehicle type. It is not uncommon to find the risk changing by as much as two to five times as these variables are altered (see Figures 5 through 9).

Reducing the Risk

One implication of the statistical record is that a large fraction of highway fatalities are, in principle, preventable. To make this clear, it is only necessary to imagine changing the average driving conditions to conditions that are statistically associated with lower-than-average fatality rates. How such a change is to take place poses a serious practical problem for hazard managers. Generally, it is assumed that the solution will involve a variety of legal and regulatory measures that focus on driver behavior, highway safety technology, and vehicle design criteria.

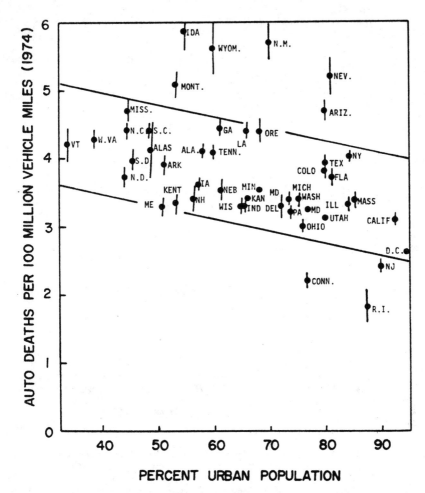

Figure 3. Auto deaths per 100 million vehicle miles plotted as a function of the percent urban population. Though there is considerable scatter in the data, increasing urban population is generally associated with low death rates (Motor Vehicle Manufacturers Association 1975).

Although such an approach will surely lead to improvement in the general situation, the insight that a large fraction of fatalities are preventable can be applied in another way. Thus, someone familiar with the statistical variation in highway fatalities summarized in Figures 5 through 9 could take the viewpoint of a doctor of preventive medicine and write for himself and others a prescription for safer driving. The underlying principle of such a prescription would be conscious individual choice of driving conditions that are statistically associated with low fatality rates. The following is a possible prescription for individual risk reduction (for the sake

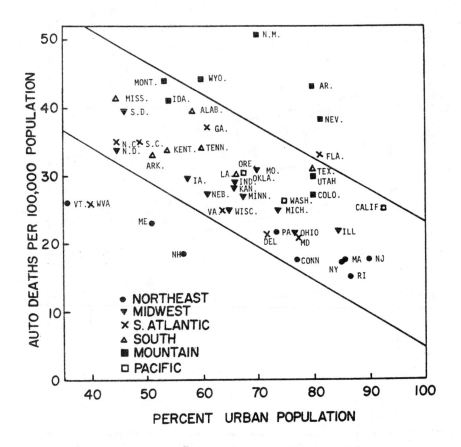

Figure 4. Vehicle deaths per 100,000 population, plotted as a function of percent urban population. The general decrease in death rate with urbanization is considerably more pronounced than shown in Figure 3. Along with a decrease in deaths per vehicle mile, urbanization leads to less driving. This reduces the population death rate in two ways.

of completeness, the use of alcohol and the wearing of seatbelts have been added to the items covered in Figures 5-9):

- Before driving, avoid the use of alcohol or other drugs that reduce alertness. Half of the drivers involved in fatal accidents have positive alcohol blood tests.
- Use seatbelts at all times. If all persons used seatbelts, the annual traffic toll would be reduced by about 20 percent.
- Avoid driving at night (particularly on Friday or Saturdays) when the accident rate is 3.5 times the day-time rate.

- Avoid the use of motorcycles and bicycles on major streets and roads. The fatality rate for motorcycle riders is four times that of automobile users.
- If you can choose the sex or age of the driver and all other things are equal (degree of alertness, adequacy of vision, etc.), let women drive rather than men. On the average, the fatal accident rate for women drivers is less than half that for men. If the choice is between a young or middle-aged man, let the latter drive for the same reason.
- If you have a choice of roadway, opt for turnpikes and interstates. They have less than half the average fatality rate, and less than one-third the rate for rural roads.
- If you have a choice of auto size, and are not concerned about energy saving, choose a large over a small car. In multivehicle accidents, occupant fatality rates in mid-size to full-size cars are less than half those in sub-compacts.

How much will this prescription reduce the individual's risk of death? If all the variables mentioned were independent and within the individual's range of choice, the answer could be obtained by multiplying the various risk-reduction factors together. Dropping out motorcycles, since most people do not use them, this suggests a potential reduction of risk to one-fiftieth of the current level. In fact, of course, the variables that we have considered are not all independent; a risk-reducing change in one area may lead to a risk-increasing change in another. In addition, it is not generally possible to carry out all parts of the prescription. Hence, the actual factor of reduction would be much less than 50. How large it would be is difficult to predict, but we estimate that following our prescription could reduce risk five to ten times.

What is the significance of such an effort? If an individual drives an average amount, then the lifetime probability of dying in a motor vehicle accident is somewhat less than 1.5 in 100. If the individual is a typical suburbanite who drives twice the average amount, the chances increase to 3 in 100. Thus, following the prescription would reduce the lifetime risk well below 1 in 100 for both average and suburban drivers.

Physical Considerations

Seen in the context of all technological hazards, motor vehicle-related deaths and injuries belong to that relatively small class of hazards for which fundamental understanding is fairly advanced. The reasons for this are fairly obvious.

Highway deaths are classically acute—in the sense that their causes are usually close in time to the results produced. Also, they are well identified. The very modifier "highway" suggests this. This sets highway deaths apart from most technological-hazard mortality, which, as chapter 6 notes, is characteristically chronic, with multiple and often indistinct causes. Beyond this, key aspects of highway losses are describable in terms of well-defined and relatively simple physical theory. This again distinguishes this

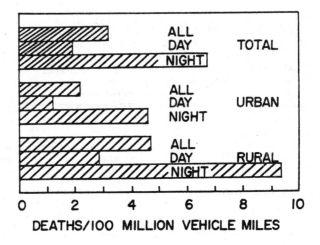

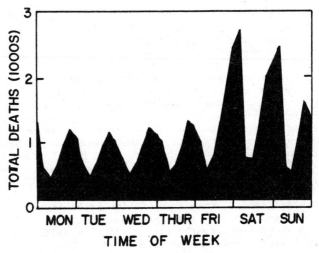

Figure 5. Motor vehicle death rate by time of day, expressed in terms of deaths per 100 million vehicle miles (top) and as total deaths (bottom). The daytime rate is about one-third of the night-time rate (NHTSA 1977a).

hazard from other technological hazards, many of which have no clear theoretical underpinnings. Consideration of some of the more significant aspects of the physical theory underlying highway colli-sion losses can serve as a guide to a variety of control measures.

The Role of Acceleration

Briefly stated, the source of most death and injury is too rapid a change in velocity. This may occur when a moving car crashes, or when a pedestrian or cyclist collides with a moving car.

In some cases the velocity is abruptly increased, in others abruptly decreased. In all cases abrupt changes in velocity produce corresponding forces on affected individuals. If these forces are too large, they lead to injury. It is worth noting in this context that rapid and damaging accelerations are possible for large as well as small velocity changes. Thus, pedestrians may be injured at impact velocities of only 5 to 10 miles an hour. The important thing is not the magnitude of the impact velocity, but the **rate of change of velocity**.

How large must accelerations be before injury results? To provide for a margin of safety, a good rule of thumb is that for properly restrained passengers, the **average** deceleration to which the passenger is exposed should not exceed 30g, or 30 times the acceleration of gravity (Haddon 1975). This is only an approximate rule, however. In most crashes, the acceleration varies widely during the one-tenth second of the crash and exhibits peaks that are two to three times the average (see Figure 10). Nevertheless, the 30g-maximum rule is sufficient here to illustrate the principle features of the physics of crashing.

Frontal Barrier Crashes

Using only the physical definition of velocity and deceleration, it is possible to express the average deceleration in an

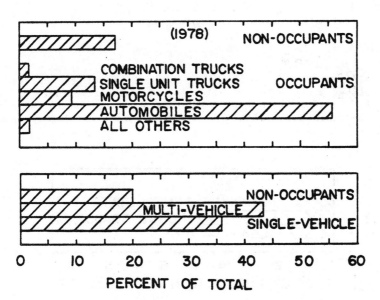

Figure 6. Motor vehicle deaths, as divided among vehicle types (**top**) and collision type (**bottom**). About 20 percent of deaths involve nonoccupants, such as pedestrians (National Safety Council 1980).

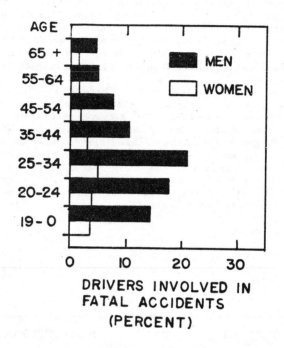

Figure 7. Motor vehicle deaths by age and sex of the driver. Men have higher death rates at all ages, with a particularly dramatic peak in the 15-34 year age group. This phenomenon accounts for the high insurance rate paid by young males (National Safety Council 1980).

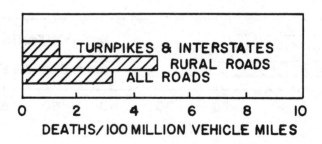

Figure 8. Motor vehicle fatality rates by road type. Turnpikes and interstates have less than half the fatality rate of all roads and about one-third the fatality rate of rural roads (National Safety Council 1980).

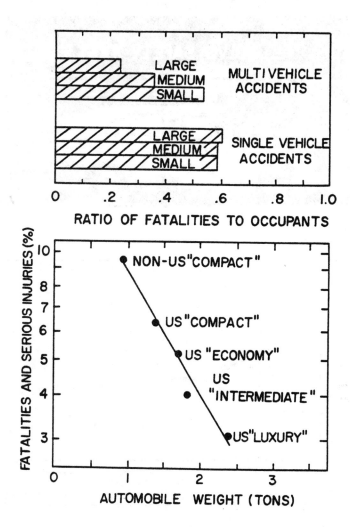

Figure 9. Variation in auto-fatality risk with weight of vehicle. Top: the ratio of fatalities to total occupants by vehicle size and collision type. The mortality ratio of smaller cars is nearly double that of larger cars in multivehicle accidents but varies little with weight in single-vehicle accidents. **Bottom:** Percent fatalities and serious injuries as a function of automobile weight. This shows that when injuries are added to fatalities the relative magnitude of the weight effect is even larger than for fatalities alone. It should be noted, however that the statistics include only a very small number of U.S.-built compacts (NHTSA 1977a; Haddon 1972).

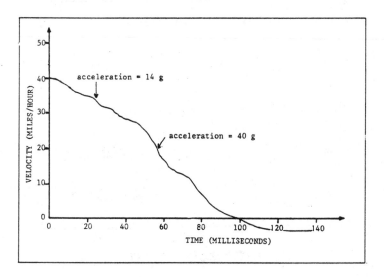

Figure 10. Top: Physical conditions of a Volvo sedan after suffering a 40-mph frontal crash from a barrier. The directors of the test indicate that properly restrained passengers would have survived such a crash because the passenger compartment remained largely intact. Bottom: Graphical display of the velocity and acceleration over the period of the Volvo's crash. The acceleration, measured by the steepness of the plot, reached a maximum of forty times the acceleration of gravity about 60 milliseconds after the beginning of the crash (NHTSA 1977b).

immovable barrier crash as a ratio of the initial velocity squared to twice the stopping distance:

$$\text{average deceleration} = \frac{(\text{initial velocity})^2}{2 \times \text{stopping distance}}$$

To keep the average acceleration in bounds—that is, below 30g—it is thus necessary to make the velocity sufficiently small and/or the stopping distance sufficiently large.

The stopping distance itself is dependent on the velocity and contains all relevant information related to the structure of the vehicle. Since the velocity is more or less fixed by drivng habits and travel-time expectations, the stopping distance is in fact the key variable around which most auto-safety design considerations revolve. For a given vehicle and crash velocity, stopping distance in a head-on collision will depend on the degree to which the front end of the vehicle may be crushed without seriously impacting the passenger compartment. The energy absorbed in crushing the car, though costly in property loss, has the beneficial effect of dispersing the energy of motion of the car without transferring it too rapidly to the passengers.

Stated simply, for frontal barrier crashes, it is the goal of auto-safety engineering to provide enough crushable structure so that crash accelerations are kept below the maximum acceptable level, and to do this without compromising the structural integrity of the passenger compartment. Unfortunately, there are severe limits to this activity. What is sufficient protection may for practical reasons be impossible as speeds approach a certain level. To illustrate, we show in Table 1 the minimum stopping distance needed at each velocity to meet the 30g standard. From this it is clear that for normal highway speeds of 50 to 60 miles an hour, the minimum required stopping distance is approaching dimensions greater than can be reasonably incorporated into the front end of an automobile. Table 1 also makes clear why smaller cars are intrinsically less safe in barrier or comparable crashes; because of their smaller dimensions, they provide less crushable structure and this leads to correspondingly shorter stopping distances. It should also be noted that, whereas crushable structure is desirable in relatively high-speed crashes, it is both unnecessary and expensive for front ends to crush at impact speeds below about 12 miles an hour.

TABLE 1
Minimum acceptable stopping distance in a frontal barrier crash

INITIAL VELOCITY (miles/hour)	MINIMUM STOPPING DISTANCE (feet)
10	0.1
20	0.5
30	1.0
40	1.8
50	2.8
60	4.0

Multivehicle Crashes

When two or more motor vehicles are involved in a crash, the simple description given above for barrier crashes does not apply. In addition to the inherent crushability of the vehicles, an added factor of importance is the mass of the colliding vehicles. For example, when a heavy vehicle strikes a light one, the acceleration of the light one is much greater than that of the heavy one and will in general involve a reversal of direction. This, in effect, places small, light vehicles in double jeopardy. Not only do they have less crushable structure, but they must also bear intrinsically higher accelerations in multicar collisions with heavier vehicles (see Figure 11).

A further interesting example is provided by bus accident statistics (Federal Highway Administration 1977). Buses are much heavier than most other vehicles and will therefore have much smaller accelerations in multivehicle collisions. One should therefore expect greatly reduced passenger fatality rates for buses. This expectation is borne out by Table 2, which indicates that, for intercity bus travel, passenger fatality rates are about one percent of the corresponding auto fatality rates. In 1976 only two intercity bus passengers died in accidents! The usually quoted bus travel fatality rate involves a higher number because the victims of bus crashes are usually not passengers but die outside the bus as pedestrians or occupants of struck automobiles.

Passenger Restraints

The importance of passenger restraints is evident from the physical principles discussed above. Consider what happens if passengers are not restrained. At the moment a vehicle hits a barrier or other vehicle—let us assume head on—the front begins to crumple and the passenger compartment as a whole begins to slow down. An unrestrained passenger, however, continues on at full speed and soon collides with the interior front (the dash) of the slowing passenger compartment. The passenger's acceleration is zero until this collision takes place but subsequently it may be much higher than the acceleration of the passenger compartment as a whole. For example, according to Table 1, in a 30-mile-an-hour frontal barrier crash, the minimum acceptable stopping distance is

Figure 11. Illustration of a multicar crash involving vehicles of considerably different weight. The cars sketched were crashed with a closing speed of 40 miles per hour. The lighter vehicle, weighing approximately 1000 kilograms, is much more severely damaged than the heavier one, weighing 2000 kilograms. This is what must be expected from physical principles, and accounts in large part for the much higher fatality rates for small cars in multicar crashes.

TABLE 2

Comparison of bus and automobile passenger fatality rates (1976)

VEHICLE TYPE	PASSENGER DEATHS PER 100 MILLION PASSENGER MILES	
	all roads	intercity
Automobile	1.4a	0.70b
Intercity bus	---	0.01c

aBased on the 1976 total fatality rate of 1.75 per 100 million passenger miles, reduced by 20 percent to account for non-occupants.
bBased on the 1976 interstate fatality rate of 1.4 per 100 million vehicle miles, an occupancy factor of 2, and no reduction for
non-occupants.
cBased on Federal Highway Administration (1977).

one foot. At that speed, cars will crush within approximately two
feet and thus provide a safe average acceleration for the passenger
compartment. An unrestrained passenger, however, will travel freely
through the passenger compartment, and, upon hitting the dash, experience a stopping distance of 3 to 4 inches, obtained by crushing
the resting, or nearly resting, dash or windshield. This leads to
unacceptable accelerations, to say nothing of possible lacerations.
The safety inherent in a crushable front end is thus available only
if passengers are securely tied to the seat or otherwise restrained
from traveling freely through the passenger compartment. The airbag, one of the most effective methods of passenger restraint, is
schematically illustrated in Figure 12.

Engineering Criteria

What lessons can be drawn from the physics of crashing? Since
the engineering literature on auto safety is extensive and detailed,
(NHTSA 1977b), it is possible to extract several simply stated guid-
ing principles. Following the divisions used by the U.S. Department
of Transportation, these may be stated in nontechnical language as
follows:

Vehicle design criteria

• Increase energy-absorbing capacity of vehicle exteriors
• Increase strength of passenger compartment structures
• Decrease vehicle mass without decreasing energy-absorb-
  ing capacity
• Provide for greater uniformity of vehicle mass
• Provide for adequate and assured passenger restraint

194

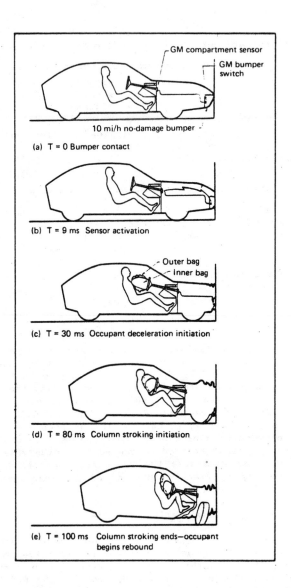

Figure 12. Schematic diagram illustrating the operation of an airbag during a frontal barrier crash. The top picture illustrates that nothing happens when the collision speed is 10 miles per hour: the car front end is not significantly crushed, nor is the airbag deployed. The next pictures show successive "snapshots" of the vehicle as it is crumpled in a 50-mph frontal crash. The time of each picture is given in milliseconds. The airbag is inflated by a pressure sensitive device mounted in the bumper. The essential function of the airbag is to prevent the driver from traveling forward as the passenger compartment slows down (NHTSA 1977b).

Highway design criteria

- Remove roadside obstacles where possible
- Provide "soft" roadside obstacles where removal is not possible
- Where possible, install traffic patterns that prevent contact between multidirectional flows
- Regulate traffic speed and provide for uniform speed
- Design highway to reduce exposure of pedestrians, cyclists, and other similar populations to motor vehicles

These criteria are, of course, not complete and, in any case, are based on the insights of physical theory alone. Other criteria could be suggested based on principles of human perception and behavior.

Although the criteria derived from physical theory are analogous and complementary to our previous prescription for risk reduction, for the most part, they cannot practically be applied by individuals. In the United States, they form the basis of the hazard management carried out by the Department of Transportation. Stated in more technical language, they have been used to set more than 50 vehicle standards since 1966 and an equivalent number of highway standards.

Without going into the detailed debates that have accompanied the setting of vehicle and highway standards, it is evident that the guidelines "derived" from physical principles may be in conflict with one another. For example, adding structural strength to the passenger compartment protects the occupants of the vehicle but may increase the potential for inflicting damage on pedestrians or occupants of other cars; in the jargon of the designers, this is a conflict between protection and aggressivity.

Adding structural strength and energy-absorbing capacity also generally increases the weight of the vehicle, with corresponding negative effects on fuel efficiency; using exotic, high-strength, low-density materials could counter these weight gains but in reality might simply transfer energy demand from vehicle operation to vehicle manufacture.

Though some aspects of such conflicts may be resolved by clever engineering, it is unlikely that all will yield to this approach—a situation similar to that met in other areas of societal hazard management. The usual solution is to approach the problem of standard setting and enforcement from a systems point of view and to study the cost-effectiveness of various measures. Such studies abound in the literature (U.S. Dept. of Transportation 1976) and are used to help the hazard manager choose among conflicting standards and designs. In plain language, cost-effectiveness studies have the function of making explicit the values society places on each of several goals and then using this information to determine which policy provides the greatest safety for the least cost.

Society's Response

Attempts to modify behavior in order to reduce risk have long been components of society's effort to lessen the risks associated with driving motor vehicles. Particular emphasis has been placed on alcohol and speeding, and much less on the factors identified in our prescription for risk reduction. It is hard to say how successful this effort has been. An optimist might attribute much of the historical reduction in deaths per vehicle mile to these efforts. Yet these gains can be attributed equally well to increased maturity of drivers and improved physical environments for driving.

Efforts to avoid and mitigate crashes through highway design have also had a history that dates back to the earliest roads. Essentially all the guidelines listed under highway design criteria have been or are being vigorously, albeit selectively, pursued, usually at enormous societal cost. These costs are often justified by anticipated reduction in congestion and travel time as well as by safety considerations and are financed by gasoline taxes, which are specifically earmarked for the use of the Highway Trust Fund. Some highway design modifications are estimated to cost over $1 million per life saved and are thus among the most expensive hazard control measures that can be envisioned (U.S. Dept. of Transportation 1976). Overall, the gains from improvement in highway design are substantial; for example, the record indicates that the interstate highways have fatality rates less than half the average and less than one-fourth those of the rural highways, which many of them replace (see Figure 9).

In contrast, efforts to require manufacturers to modify vehicle design in order to mitigate or avoid injuries are of more recent origin. Without going into detail, it is clear from a review of current standards that only modest progress has been made in the direction of incorporating the guidelines derived from physical considerations. Thus, while a fair amount has been done to protect occupants through collapsible steering columns, padded dashes, the installation of seatbelts, and the like, little or nothing has been done to reduce the ability of automobile exteriors to inflict damage. Furthermore, the most effective vehicle modification--airbags--has been stalled for more than ten years by the opposition of the automobile industry (see chapter 14). This continuing saga of delay has seen the Carter administration order airbags on all 1982 cars only to have the Reagan administration rescind the order. Subsequent reinstatement by the court has given way to reconsideration and yet another postponement.

The lack of significant progress in vehicle design cannot be explained by the conflicts inherent in the physical-design guidelines discussed above. Results from the Department of Transportation's Experimental Safety Vehicle (ESV) Program indicate that engineers have solved many, if not all, of the problems raised by conflicting design criteria (NHTSA 1977b). For example, cars have been designed with sheet metal-coated foam in critical frontal areas. Such cars, when equipped with airbags, can survive 40- to 50-mph frontal crashes without at the same time doing a great deal of damage to the objects or cars which they collide.

The reasons for the curious imbalance in the response of hazard managers to the available insights of physical theory must thus be sought elsewhere. The slaughter on the highways continues, not for lack of scientific knowledge but because of the multiple ways the auto is embedded in our society, economy, bureaucracy, and in our own activities and attitudes.

REFERENCES

Faigin, Barbara M. 1976. 1975 societal cost of motor vehicle acci-
     dents. DOT HS 802-119 Washington: Dept. of Transportation,
     National Highway Traffic Safety Administration.
Federal Highway Administration. 1977. 1976 accidents of motor
     carriers of passengers. Washington: Dept. of Transportation.
Haddon, William, Jr. 1972. A logical framework for categorizing
     highway safety phenomena and activity. Journal of Trauma
     12:193-207.
Haddon, William, Jr. 1975. Reducing the damage of motor vehicle
     use. Technology Review 77 (July/August):52-59.
Motor Vehicle Manufacturers Association. 1975. MVMA Motor Vehicle
     Facts and Figures '75. Detroit: Motor Vehicle Manufacturers
     Association.
Motor Vehicle Manufacturers Association. 1980. MVMA Motor Vehicle
     Facts and Figures '80. Detroit: Motor Vehicle Manufacturers
     Association.
NHTSA (National Highway Traffic Safety Administration). 1977a.
     Fatal accident reporting system: 1976 annual report. Washing-
     ton: Dept. of Transportation.
NHTSA (National Highway Traffic Safety Administration). 1977b.
     Report on the Sixth International Conference on Experimental
     Vehicles.
NHTSA (National Highway Traffic Safety Administration). 1978.
     Motor vehicle safety 1977. Washington: Dept. of Transporta-
     tion.
National Safety Council. 1980. Accident facts 1980. Chicago:
     National Safety Council.
National Safety Council. 1982. Accident facts 1982. Chicago:
     National Safety Council.
U.S. Department of Transportation. 1976. The national highway
     safety needs report. Washington: Dept. of Transportation.

9
Airborne Mercury

Robert C. Harriss and Christoph Hohenemser

Acute mercury poisoning made headlines around the world in the late sixties and early seventies, and the term **Minamata disease**, referring to death or illness from eating mercury-contaminated seafood, gained wide currency. Minamata, the small Japanese fishing village which gave its name to mercury-related disease, was the site of a factory belonging to Chisso Corporation, one of Japan's major chemical manufacturers. In 1932 the factory was using a process that employed mercury compounds as catalysts. Mercury-contaminated effluent was dumped by the factory into Minamata Bay and absorbed by the fish on which the local residents depended for their diet and their livelihood. A mysterious neurological disorder was recorded in the area in 1953; by 1956 it had reached epidemic proportions. In October 1959 Hajime Hosokowa, a Chisso chemist, determined that the disease was caused by the factory's dumping of mercury into the bay; but the company hid the evidence until the late 1960s. It was not until 1971 that Chisso was forced to stop polluting the bay with mercury. In 1972 a Japanese judge ruled that Chisso was legally responsible for the tragedy and must offer monetary compensation to certifiable victims. By April 1977, 1,046 residents of Minamata had been certified as victims and 3,000 others were awaiting examinations.

Other instances of mercury poisoning began to surface as a result of the publicity given to the Minamata trials. In May 1970 the provincial government in Ontario, Canada, reported the discovery of mercury-contaminated fish in Ontario's rivers and streams; during the summer of 1975 a Japanese medical team examined 89 people who relied heavily on fish in their diet and discovered that 37 had signs of methyl mercury poisoning. Similar studies in Quebec and other parts of Canada revealed even higher percentages of people with Minamata symptoms. In Iraq during 1971-1972 some 500 people died and thousands more were poisoned after eating homemade bread made with wheat contaminated by a methyl mercury fungicide. The studies that followed this epidemic were the first to document the fact that mercury poisoning can be prenatally transmitted. These developments prompted steps to reduce mercury contamination of soils, sediments, and water. A variety of sources provide a comprehensive review of the toxicological literature on mercury (Bidstrup 1964;Grant 1971;NIOSH 1973;OECD 1974;WHO 1976).

In retrospect, it is somewhat puzzling that the acute episodes of a decade ago caught us by surprise. The toxic nature of mercury had been known for more than 400 years: occupational mercury poisoning was reported by Jean Fernel as early as 1579. But what is even more surprising is that roughly 10 years after the recent acute episodes, it is the view of major scientific and regulatory bodies that the mercury problem is largely resolved. Consider the following quotations:

> In light of . . . the fact that nearly all drinking water supplies in the United States are already in compliance with current interim regulations, there is serious question as to whether a standard is needed or serves any useful purpose. (National Research Council 1977a,279)

> Most of us are only exposed to mercury levels well below what is considered acceptable by international experts. In recent years the use of mercury in certain applications has become rigorously controlled. . . . Monitoring programmes have also been extended and coordinated. (Great Britain 1976,iii)

Such sanguine views of the mercury problem are far from justified.

Although the hazard of acute toxicity is not as great as it was a decade ago, it has certainly not been brought fully under control. Major river systems have had to be closed to fishing because of mercury contamination (Environmental Science and Technology 1977); industry has not always complied with the requirements for self-reporting (McFadden 1978); and terrorists have even used mercury to contaminate food supplies (New York Times 1978). Perhaps most disturbing, the acute hazard, as exemplified by the consumption of contaminated food-fish, is being transferred to developing countries, where manufacturers are able to avoid controls imposed in industrialized nations. For example, a recent report from Thailand notes the existence of cases of mercury poisoning in the vicinity of a chlor-alkali plant which was recently relocated from Japan, where mercury emissions are strictly regulated (Suckcharoen, Nuorteva, and Häsänen 1978,113).

Another problem--that of chronic exposure to airborne mercury--remains insufficiently evaluated and may pose a major long-term threat. In fact, it is becoming clear that present strategies for controlling mercury pollution do not prevent the release of mercury into the environment. Instead, they tend to modify the discharge pathways so that what was formerly an acute hazard is often transformed into a chronic one.

Viewed in the general context of technological hazards, chronic exposure to airborne mercury can be seen as one of a class of cases for which scientific information that would be sufficient to lead to clear action is simply not available. Management of this hazard must therefore proceed in an atmosphere of uncertainty. Yet, as we shall show, there is no reason why this should result in paralysis. There is, first of all, sufficient information to permit an estimate of the magnitude of the problem. In addition, there are several options that would reduce chronic exposure without blocking the benefits of mercury use in any significant way.

Perhaps the best statement of the present situation has been made by the Panel on Mercury of the National Research Council (1977b,6):

However, the data are still incomplete and the medical and technical proficiency are too limited to assess conclusively the effect of chronic exposure to low levels of mercury pollution on human beings as well as lower orders of the food chain and the environment. The mercury problem, therefore, clearly needs to be reassessed periodically to take advantage of new data and of the continuing refinement of chemical and analytical techniques.

What follows is such a reassessment. It offers a survey of sources and trends, three methods for evaluating the chronic hazard created by airborne mercury, and an inventory of ways of dealing with the present situation.

Sources of Mercury

Dispersal of mercury as a result of human activity takes place via two principal pathways: release from consumer products and manufacturing, and release from fuels and other minerals as they are burned or processed at high temperature (see Table 1). The two pathways are fundamentally different in the degree of control that can in principle be exercised. The first pathway is capable of being substantially blocked by greater attention to the way in which mercury is handled in manufacturing and in consumer products. In order to block the second pathway it would be necessary to extract mercury vapor at concentrations of less than one part per million from huge volumes of combustion gases in fuel burning and ore processing—a problem that is unlikely to yield to an easy technological fix.

TABLE 1
Mercury mobilization by human activity in the U.S.

METHOD OF MOBILIZATION	AMOUNT MOBILIZED (tons/year)
Manufacturing and consumer products[a]	1310[b]
Fuels and minerals[b]	350

[a]Estimate made for the U.S. Environmental Protection Agency (Van Horn 1975) and includes solid, liquid, and vapor waste emissions.
[b]Our estimate includes 310 tons of vapor, 920 tons deposited on land with potential for vaporization, and 80 tons deposited in water.

Manufacturing and Consumer Products

Mercury is used in electrical equipment, in chlorine and caustic soda manufacture, in scientific instruments, paints, dental supplies, agricultural fungicides, and a number of miscellaneous special uses. Many of these applications require a continuing supply of mercury-bearing rock and, with few exceptions, a continuing release of mercury to the environment. The quantities involved can be estimated by considering mining trends, the amounts included in manufactured products, and waste disposal practices.

Mined quantities are roughly equal to consumed quantities and thus they set an outer limit to the potential quantities of mercury released into the environment. Worldwide production of mercury has been relatively stable over the past decade at about 9,000 tons/year. As shown in Figure 1, U.S. production has fluctuated, averaging between 5 and 10 percent of world production. Unlike certain other minerals, mercury is not being mined at highly accelerated rates.

What happens to the mercury mined in the United States and the impact of regulation on releases of mercury to the environment can

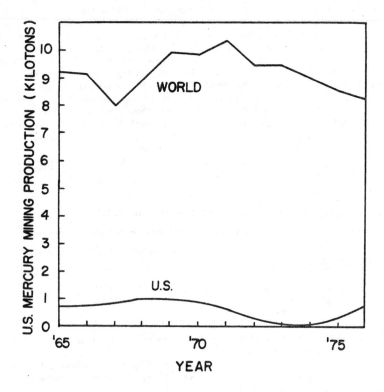

Figure 1. Ten-year trend of world and U.S. mining of mercury. Note that neither the world nor the U.S. mining effort has grown (Data from U.S. Bureau of Mines 1977).

be partially inferred from the data in Figure 2 on industrial demand patterns. With the possible exception of a slight decline in the demand for mercury for use in agriculture, synthetic chemical manufacturing, and general laboratory use, the demand patterns show no evidence of consistent trends. The annual demand for the period 1964-1974 exhibits a low in 1971, followed by a rising trend. Interestingly, the numerous legal and regulatory actions designed to control mercury pollution in the United States since 1970 seem to have had little, or at best a temporary, influence on industrial demand for mercury as a raw material. The demand patterns clearly show that there have been no major substitutions for mercury in its principal uses. In the future it is likely that substitutes for mercury may include nickel-cadmium or other battery systems for electrical apparatus, diaphragm cells for mercury cells in the chlor-alkali industry, organotin compounds in paints, and solid state devices for industrial and control instruments that currently use mercury (U.S. Bureau of Mines 1975; U.S. Bureau of Mines 1977).

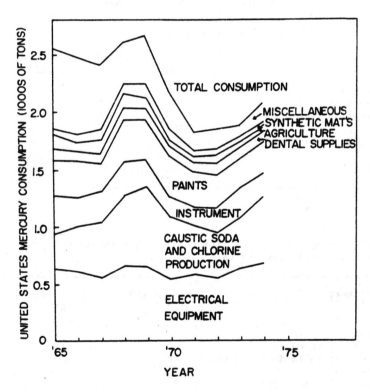

Figure 2. U.S. mercury consumption, by major end uses, 1966-174. Note that total consumption declined about 25 percent, mostly at the expense of miscellaneous uses. At the same time, the dominant four end uses—electrical equipment, caustic soda and chlorine production, instruments, and paints—have experienced an overall increase in consumption (Data from U.S. Bureau of Mines 1975).

How much of the mercury used in manufacturing and consumer products escapes to the environment? The estimates made for the U.S. Environmental Protection Agency (Van Horn 1975) are summarized in Table 2, which indicates that losses of mercury each year represent over 70 percent of consumption, with most losses going into the air or being deposited on land. The British and U.S. experiences with the chlorine industry (Great Britain 1977;Lindberg and Turner 1977), the second largest user of mercury, confirm this pattern (Table 3). Particularly for the United States, this pinpoints large losses associated with land-fill sites and holding ponds.

Mineral Processing and Combustion

Mercury is also dispersed into the air when certain minerals and fuels are processed or burned at high temperature. Included are

TABLE 2
Estimated losses of mercury from U.S manufacturing and consumer products for 1973

PATHWAY	LOSSES tons/year	% of total
Escaped to air	310	17
Deposited on land	920	49
Deposited in water	80	4
Recycled	150	8
Otherwise accounted for	410	22
TOTAL	1870	100

Source: Van Horn (1975).

TABLE 3
Estimated losses of mercury from the chlorine industry in the United States and Britain for the year 1973

PATHWAY	U.S. LOSSES[a] (%)	U.K. LOSSES[b] (%)
Escaped to air	5	9
Deposited in water	1	14
Deposited in land-fill sites and holding ponds	60	27
Unaccounted for, probably located in the plant	34	50

[a]Based on a total annual consumption of 450 tons, as described in Van Horn (1975).
[b]Based on a total annual consumption of 220 tons (Great Britain 1977).

fossil fuels, such as coal, oil, and gas; sulfides and oxides used as ores of iron, copper, and other minerals; and shale used in the manufacture of cement. All of these materials are enriched with mercury through natural geochemical processes (see Table 4). Mercury is released from these materials because mercury boils at 350°C, while combustion, extraction of metals from ore, and the production of cement require temperatures of 1400-2500°C.

To obtain estimates of mercury vapor losses to the environment through these processes, we multiplied the total mass of each material consumed by our best estimate of the mean concentration of mercury in each. The results, which assume that all mercury in the materials is vaporized, are summarized in Table 5. Some observations having policy implications that can be drawn from this table include:

- The 350 tons of mercury that escape into the air each year in the United States as a result of mineral and fuel processing are about equal to the total mercury released into the air from manufacturing and consumer products (see Table 2).
- The combustion of fossil fuels contributes somewhat more than half of the mercury emissions; the processing of ores accounts for most of the remainder.
- Although previous studies of fossil fuels have emphasized the contribution of coal combustion, our calculations clearly identify petroleum combustion as an important source of mercury emissions. This means that changes in the mix of fossil fuel sources will have relatively little effect on the problem.

The problem of mercury pollution is thus the result not only of essential manufacturing processes and consumer goods but of much of our mineral and fuel conversion technology. On the average, humanly produced mobilization of mercury in the United States is more than 1,600 tons per year. Preventive actions focused on acute hazards arising from mercury-contaminated foods have not changed the overall input of mercury into the environment. Therefore, mercury in the air may well pose an unrecognized chronic hazard.

Further complicating the problem of controlling mercury pollution is the transnational nature of its distribution and especially its transport across national boundaries. Thus, as in the well-documented case of acid rain in Scandinavia, the source of mercury pollution is beyond the legal or political control of the affected nations.

Measuring the Mercury Hazard

The effects of mercury on human health have been the subject of a number of comprehensive reviews (Bidstrup 1964;Grant 1971;NIOSH 1973;OECD 1974;WHO 1976;National Research Council 1977a;Great Britain 1976;National Research Council 1977b). Much of the available material deals with the problem of mercury in water, the concentration of toxic levels in fish, and the subsequent acute poisoning of the human populations that eat these fish. Here we are concerned with the hazard of mercury vapor and mercury particulates in air, a

TABLE 4
Mercury enrichment of important minerals and fuels[a]

MINERAL OR FUEL	MERCURY CONTENT (PARTS PER BILLION)
Average rock	~ 70
Coal	200– 600
Oil	70– 130
Shale	~ 400
Ferrous ore	~ 500
Nonferrous ore	400–47,000

[a]Geochemical enrichment is a result of the selective interaction of mercury with both organic and sulfide molecules. Data on the chemical composition of fossil fuels are scattered throughout a large geochemical literature, and usually apply to specific locations and rock types. Thus, reliable estimates of a regional or global average value for the mercury content of minerals and fuels are extremely difficult to obtain. The data in this table derives from a number of sources (Joensuu 1971;Magee, Hall, and Varga 1973;Van Horn 1975;Wedge, Bhatia and Rueff 1976;Block and Dams 1975;Goldberg 1976;Great Britain 1976;Billings and Matson 1977;Kaakinen et al. 1975;Klein, Andren and Bolton 1975).

TABLE 5
Mercury losses to the atmosphere due to mineral processing and fuel combustion

SOURCE	UNITED STATES		WORLD	
	tons/year	% of total	tons/year	% of total
Fossil fuels				
coal	100	29	620	35
petroleum	80	22	270	15
natural gas	20	6	30	2
Iron refining	70	20	450	25
Nonferrous refining	70	20	280	16
Cement production	10	3	130	7
TOTAL	350	100	1780	100

subject which has so far received much less attention. Atmospheric emissions exceed emissions into water by about a factor of 15 (Table 2), yet any effects will be largely chronic and therefore difficult to observe. Perhaps this is why airborne emissions have not been subjected to the stringent regulations applied to waterborne mercury.

The most fundamental tenet of our existing knowledge is that mercury compounds have no known metabolic function. Research to date has not identified any threshold below which there are no toxic effects. Hence, if we are going to be conservative, it is necessary to regard all mercury in the environment as undesirable and potentially hazardous. To be meaningful in the context of hazard management, such a statement must be further elaborated. Below we consider three possible ways of dealing with this problem:

1. One could compare humanly produced emissions to natural background emissions and then permit no more than a modest excess above background levels.

2. One could obtain a measure of the mercury hazard by comparing it, using some suitable index, to other toxic substances and thus arrive at an evaluation of the mercury hazard relative to that of other materials.

3. Depending on the degree of conservatism desired, one could set the permissible level of mercury at or just below the point where well-defined clinical effects become observable.

Natural and Manmade Emissions

When toxic materials occur naturally as well as through human activity, it is customary to compare the magnitude of the humanly produced flux to the natural "background." If the quantity that is humanly produced flux deviates only modestly from the background amount, it is considered acceptable. This principle, which can be roughly justified in terms of the theory of evolution, can serve as a useful qualitative guide in evaluating a hazard of otherwise unknown consequences.

As noted above, for the United States, the emission of mercury vapor due to human activity is estimated at 660 tons per year. Mercury vapor due to natural causes in the United States averages about 1,000 tons per year (Van Horn 1975). Therefore, the humanly produced emissions are, on the average, two-thirds as great as natural emissions into the air. This in itself constitutes a significant deviation from background, but it understates the full extent of the problem.

Mercury emission is in fact highly localized and may exhibit "hot spots" of either human or natural origin. In Figure 3 total mercury emissions are given for each state, with humanly produced and natural emissions indicated separately. It can be seen from the figure that, in a number of states containing heavy industry, humanly produced emissions are four to five times greater than the natural background. In other states which contain large sources of geothermal and/or volcanic emissions, the natural flux far outweighs the natural produced flux.

Application of the principle of "modest deviation from background" is thus not a simple matter. Not only is the humanly produced fraction highly variable, but the background itself fluctuates over a range of about 4. This in turn makes it likely that even more variability would be found if statewide figures were further broken down into county values, or even city and town values. On a local scale, humanly produced mercury vapor levels 100 times greater

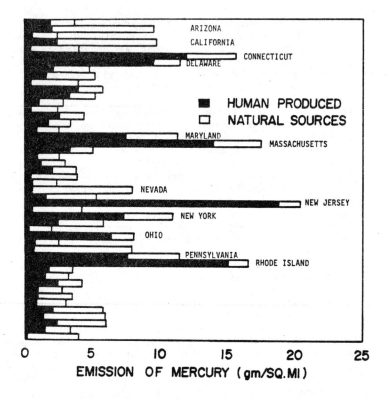

Figure 3. Humanly produced and natural mercury emissions, expressed in grams per square mile, for the 48 contiguous states of the United States. High humanly produced levels occur in heavily industrialized states. High natural levels occur in states with substantial geothermal or volcanic activity (Data from Van Horn 1975).

than background levels or backgrounds 100 times humanly produced levels should not surprise us.

The ratio of humanly produced to natural emission ratios thus gives us cause to worry. Translated into policy, our evaluation should lead to concentrating our efforts on existing "hot spots" rather than on broad-scale attempts to reduce the average levels of mercury in the environment. For the problem is not so much the average amount of mercury in the atmosphere as the degree of fluctuation about the average.

A Comparative Hazard Index

A second way of evaluating the hazard posed by mercury is to compare mercury pollution of the atmosphere to that caused by other metals. Researchers have calculated the ratios of humanly produced emissions into the atmosphere to natural emissions for twelve toxic metals (Garrels, Mackenzie, and Hunt 1975). These ratios, known as **global atmospheric interference factors**, are summarized in Table 6 and indicate that mercury, with an interference factor of 80, has the highest value of the twelve metals.

Since mercury vapor in the air has a ratio of humanly produced to natural emissions of 2 to 3 for the United States, and at most 1 to 5 for certain "hot" states, the result of 80 seems quite remarkable. The apparent discrepancy is partly explained by two factors: first, the calculations on which Table 6 is based include particulate emissions as well as vapor in the humanly produced flux; and, secondly, the calculations neglect low-temperature volatilization of mercury from soil as a source of natural emissions. Both these differences would tend to increase the magnitude of the human interference factor and thus would make the ratios in Table 6 not directly comparable to our results. Yet, because it is difficult to see how the indicated differences could reduce the interference factor to the range—between 1 and 5—that we believe correct for the United States, we introduce a range of 2 to 80 for the interference factor in the following hazard analysis.

As a second step in comparing the twelve similar elements, we have combined the interference factor data with toxicity data compiled by the Environmental Protection Agency (EPA 1976). Specifically, we have divided the interference factor by concentrations considered to be toxic to fresh-water organisms. This yields an index that reflects both the degree of mobilization of the metals in the environment and their relative degrees of toxicity. (In the case of elements where recommended water-quality criteria are not available, these concentrations were derived from lethal concentrations divided by a safety factor of 100.) The results of our calculations, shown in Table 7, indicate a relative hazard ranking that places mercury at the top of the list, even when the modest interference factor of 2 is utilized. Similar results, using somewhat different methods, have been obtained by other investigators (Ketchum 1975).

TABLE 6
Ratio of technological to natural emissions for metals

ELEMENT	TECHNOLOGICAL EMISSIONS / NATURAL FLUX
Nickel	0.9
Vanadium	1.3
Copper	2.3
Arsenic	3.3
Tin	3.5
Zinc	4.6
Cadmium	5.2
Selenium	14
Antimony	28
Molybdenum	29
Lead	70
Mercury	80

Source: Adapted from Garrels, Mackenzie, and Hunt (1975).

TABLE 7
Relative hazard index for some heavy metals

ELEMENT	RELATIVE HAZARD INDEX*
Mercury	40-1600
Cadmium	13
Copper	9
Lead	7
Zinc	4.6
Selenium	0.7
Arsenic	0.7

*The relative hazard index is obtained by dividing ratio of techno-
logical to natural flux shown in Table 6 by the relative toxicity
of each of the metals. Values of relative toxicity used are 0.05
for mercury, 0.4 for cadmium, 0.25 for copper, 10 for lead, 1 for
zinc, 20 for selenium, and 50 for arsenic. The range in hazard
index for mercury results from a corresponding range of 2-80 for
the atmospheric interference factor, as described in text.

Although we acknowledge that obtaining a relative hazard index
in this way relies on limited data that may change over time, we do
not believe that any changes will be sufficiently large to alter the
situation qualitatively. The hazard index we have derived confirms
and expands the analysis of the previous section which was based on
a simple comparison of humanly produced to natural vapor emissions.
It indicated that mercury pollution is, relatively speaking, a
serious hazard. From a practical point of view, it should increase
our motivations for dealing with present "hot spots" of both human
and natural origin.

An Acceptable Exposure Standard

A third approach to dealing with the chronic effects of airborne mercury would be to define the level of harm directly and to
manage the problem by setting a standard defining the level of
acceptable risk. For some toxic materials, such as radioactive
materials, this effort is rationalized by exposure-response curves
which relate human exposure of a given magnitude to the probability
of morbidity or mortality. With an exposure-response curve in hand
and a good estimate of exposure patterns, it becomes possible to
measure the cost incurred at a given exposure or higher and to
compare it to the cost of preventing this exposure. These comparisons can lead to a definition of acceptable risk and hence a standard for acceptable exposure. For example, one could insist that
the cost of mercury-use consequences not exceed the value of the
benefits derived.

A characteristic problem of exposure-response relations is that
effects at low exposure are often ill-defined. Thus, for the classic case of ionizing radiation, human exposure-response data are
limited to levels that are generally higher than those experienced

by the general population, or even by radiation workers. It is therefore necessary to extrapolate high-exposure behavior to low levels of exposure, with all the attendant difficulties involved in such a procedure (Fischhoff et al. 1978).

For airborne mercury the situation is even less well defined than it is for ionizing radiation. Science offers no real quantitative information on the relation between exposure and response for humans, even at high levels of exposure. Instead, clinical observations lead to a series of qualitative and sometimes ambiguous descriptions of experienced morbidity at various levels of ambient air concentrations (NIOSH 1973). These indicate (Table 8, top) discernible effects down to levels of about 10,000 nanograms per cubic meter of air (one nanogram is one billionth of a gram). On an atomic level, this is a concentration that corresponds to 12 mercury atoms for every one billion atoms of air. There is some ambiguity in the apparent lowest level of discernible effects. For example,

TABLE 8
Effects, standards, and environmental levels for mercury

EFFECTS/STANDARDS/LEVELS	CONCENTRATION (nanograms/cubic meter)	REFERENCE
EFFECTS		
Appetite loss, insomnia	10,000-270,000	18
Hyperthyroidism	10,000- 50,000	18
Abnormal reflexes	100,000	18
Chronic neural disorder	800,000	2
Clinically negative results	10,000-120,000	
STANDARDS		
USA-OSHA occupational standard (not to be exceeded in an 8-hour word day)	50,000	1
USA-OSHA occupational standard (ceiling value)	100,000	19
Sweden occupational standard	50,000	19
USSR occupational standard	10,000	19
EPA proposed ambient air permissable maximum	1,000	35
ENVIRONMENTAL LEVELS		
Oceanic air	0.7	8
Continental air, rural	1-10	8
Continental air, urban	1-50	8
Geothermal areas	0.1-0.8	8
Chlor-alkali waste deposits[a]	12-990	12
Air over mercury ore deposits[b]	7-20,000	8

[a] Approaches or exceeds proposed EPA ambient air standard.
[b] Approaches or exceeds OSHA occupational standard and proposed EPA ambient air standard.

the World Health Organization concludes that effects at levels below 50,000 nanograms per cubic meter of air have not been unequivocally established (WHO 1976).

Given this information, standards are set by government agencies at or below the level at which discernible clinical effects are first observed. Depending on the degree of conservatism desired, standards will vary from agency to agency (Table 8, middle). For example, the U.S. Occupational Safety and Health Administration (OSHA) sets the eight-hour average maximum level of mercury in the air at about the level at which health effects may just be detected, thus leaving no appreciable "margin of safety." In contrast, the U.S. Environmental Protection Agency (EPA) recommends a standard for ambient air of 1,000 nanograms per cubic meter, a factor of 50 below the level of discernible clinical effects.

Vapor-phase mercury levels in the air are summarized in Table 8, bottom. In several cases the recommended standards are approached or exceeded. In particular, the EPA ambient air standard is approached or exceeded in the vicinity of chlor-alkali waste deposits, mercury ore deposits, and in industrial air. On the other hand, the OSHA occupational standard is exceeded only in some industrial air.

Taken at face value, the information in the three parts of Table 8 solves the problem of the hazard manager, at least in the conceptual sense. By combining the information on effects, standards, and present ambient levels, he or she can act to reduce ambient levels in excess of given standards. Thus, the OSHA standard should lead to action in noncomplying industrial environments. Correspondingly, the EPA standard should lead to action focused on existing humanly produced and natural sources, such as chlor-alkali waste dumps and mercury mines, which presently approach or exceed the standard.

Yet, do such actions in fact solve the problem? We are skeptical. The fundamental difficulty of setting standards in the absence of adequate exposure response data has not been solved. This is clearly indicated by the variability of the standards that have been set by various agencies. The underlying difficulty is that none of the standards has a clear-cut justification. Any or all may be substantially too low or too high in terms of an acceptable risk criterion that mandates that the cost of mercury-use consequences not exceed the benefits derived.

The setting of standards has, in fact, a rather troubling effect on insufficiently informed members of the public. Standards are apt to be interpreted as defining a threshold below which the effects are negligible or nonexistent, in effect creating a region in which there are only benefits and no costs. Nothing could be further from the truth. The level at which clinical effects just become discernible is not evidence for a threshold of harm but is simply an indication of the present limits of scientific observation techniques. Particularly for chronic effects, these techniques are not nearly as powerful as one might wish. The concrete question for OSHA, which sets its occupational standard just at the level of discernible clinical effects, is: At that level what social costs—measured in terms of morbidity, mortality, and the cost of controls—are being traded for what level of benefits? Although economists may be able to define the cost of controls and the cost of derived benefits, medical scientists cannot yet define the degree of morbidity and mortality. As in the case of many toxic chemicals, therefore, standard-setting for mercury, when examined closely, is in fact a most unscientific matter.

Ways of Coping

Given the information available on chronic exposure to airborne mercury, what realistic ways of coping suggest themselves? Specifically, what actions can reasonably be taken even in the absence of reliable exposure-response information and corresponding standards? We recommend three general approaches to the problem: reduction of the general level of airborne mercury, management of mercury hot spots, and control of mercury levels in industrial air.

Reducing the Mercury Level

Each year in the United States losses of mercury to the environment from manufacturing and consumer products constitute 70 percent of annual consumption of mercury. This is an extraordinarily high rate of loss that offers a number of straightforward opportunities for reduction. It should be possible to double and triple present levels of recycling by the same kind of effort that has been successful with other products, such as returnable beverage containers. For example, as Table 9 indicates, discarded electrical equipment is the source of 95 percent of all mercury entering the environment as solid waste (Crossland 1977). A workable method of controlling this loss through recycling might involve the refund of a "mercury deposit" made on mercury-bearing electrical equipment at the time of purchase.

TABLE 9
Source and forms of mercury entering the environment as solid waste

SOURCE	FORM	AMOUNT (tons/year)	%
Chlor-alkali plants	Brine sludges containing mercury chlorides	8	1.7
Paint manufacture	Phenyl mercury mildewcides in waste sludge	0.8	0.2
	Paint containers with residues of 0.02 to 0.10 percent of mercury	15	3.1
Electrical equipment	Batteries, fluorescent lights, switches	454	95.0

Source: Crossland (1977).

It is also true that a number of substitutes exist for mercury in manufacturing and consumer products. These are already having an impact on mercury utilization (Figure 2) but, so far, without financial incentive the reduction in mercury use is small. The question that needs to be asked is: What reduction in mercury pollution would be achieved if there were a modest incentive for substitution? In the absence of any clear indication of the social costs of present mercury pollution, it is, of course, difficult to know what should be regarded as "reasonable." Conservatively, we therefore propose an incentive that is negligibly small on the scale of the value of the total product of the industry (say less than 1 percent), yet large enough to cause prudent manufacturers to respond.

Finally, as indicated earlier, an appreciable part of humanly produced mercury emissions is the result of fuel combustion and high-temperature mineral processing. These mercury emissions are by far the most difficult to control and may have to remain as a largely uncontrolled, hard-core remnant, a price that has to be paid for our energy and mineral technology. Yet even here there is some hope. It has been shown that the installation of sulfur dioxide scrubbing equipment in fossil fuel-burning power plants can reduce mercury emissions by 33 percent (Kelly 1973). Control of this form of airborne mercury may thus be achieved as a byproduct of a control effort for which the costs and benefits are better known.

Managing "Hot Spots"

One of our most striking findings is that mercury emissions, both humanly produced and natural, are highly variable. Hence, it is quite possible for either natural background or human sources to dominate at a given site by a large margin. This variability places severe limits on an approach such as that above, which is designed to reduce the average level of mercury emissions. This finding leads us to focus on the direct control of existing natural and humanly produced hot spots.

A necessary first step to this end is the collection of better data as a basis for determining the extent of the problem. If and when it is discovered that mercury levels approach or exceed the point at which discernible clinical effects exist, actions to restrict human exposure should be taken. In certain cases, it may be prudent to adopt the principle of distributing exposure over many individuals--a practice that is widely used in the nuclear industry for spreading the burden of radiation exposure. For example, the park ranger living near mercury-emitting geysers or the city-dump operator driving the trash-compacting bulldozer may from time to time have to be replaced by another short-term person.

Managing Industrial Air

Although the extent of the social costs being averted by the present OSHA occupational standard is unknown, the range of experienced ambient air levels in industrial sites may exceed this standard in certain cases. The standard is not conservative since it is at the level where clinical effects just become discernible. We therefore recommend, at minimum, the taking of whatever actions are necessary to meet the standard. If further examination discovered places in industry where so doing would be enormously costly, an alternative approach would be to use temporary workers who receive only limited exposure, as is suggested in the case of hot spots.

A Continuing Problem

Is the hazard of chronic mercury pollution resolved? As we indicated in the beginning of this review, we believe it is not. Though society has, by and large, controlled acute mercury poisoning episodes caused by contaminated food, the hazard of chronic airborne mercury pollution remains. Our discussion has noted at least three broad reasons for this evaluation: (1) control of acute exposures has not reduced mercury mining, mercury use, or mercury mobilization; (2) a study of humanly produced airborne mercury levels and a comparison of these with natural flux and with other toxic metals serve as a warning that mercury is intrinsically a high-hazard material; (3) inadequate exposure-response information prevents a clearcut evaluation of the hazard involved in present exposures.

Even given this situation, however, there are a number of highly plausible actions that can be taken to cope with the problem of airborne mercury pollution. These include reduction of humanly produced mercury vapor levels in the air through recycling and substitution, restriction of the exposure of affected individuals in localized hot spots, and further efforts to reduce mercury levels in industrial air and, failing this, restricting exposure of affected individuals. With these actions it should be possible to reduce the chronic hazard from airborne mercury to about one-tenth its present level without doing obvious violence to associated benefits.

ACKNOWLEDGMENTS

This research was supported in part by the National Science Foundation under grant ENV-77-15334 and the Environmental Protection Agency under grant 801512. Opinions, findings, conclusions, or recommendations herein are those of the authors and do not reflect the views of the National Science Foundation or the Environmental Protection Agency.

REFERENCES

Bidstrup, P. Lesley. 1964. Toxicity of mercury and its compounds. New York: Elsevier.

Billings, Charles E., and Wayne R. Matson. 1972. Mercury emissions from coal combustion. Science 176:1232-1233.

Block, C., and R. Dams. 1975. Inorganic composition of Belgian coals and coal ashes. Environmental Science and Technology 9 no. 2 (February):146-150.

Crossland, Janice. 1977. The wastes endure. Environment 19 no. 5 (June/July):6-13.

Ehrlich, Paul R., Anne H. Ehrlich, and John P. Holdren. 1977. Ecoscience. San Francisco: W.H. Freeman.

Environmental Science and Technology. 1977. Currents. 11 no. 9 (September):844.

EPA (Environmental Protection Agency). 1976. Quality criteria for water. Washington: EPA.

Fischhoff, Baruch, Christoph Hohenemser, Roger E. Kasperson, and Robert W. Kates. 1978. Handling hazards. Environment 20 no. 7 (September):16-20,32-37.

Garrels, Robert M., Fred T. Mackenzie, and Cynthia Hunt. 1975. Chemical cycles and the global environment. Los Altos, California: William Kaufmann.

Goldberg, Edward E. 1976. The health of the oceans. Paris: UNESCO Press.

Grant, Neville 1971. Mercury in man. Environment 13 no. 4 (May):2-15.

Great Britain. 1976. Interdepartmental Working Group on Heavy Metals. Environmental mercury and man. Pollution Paper No. 10. London: HMSO.

Great Britain. 1977. Department of the Environment, Wastes Division. Mercury-bearing wastes. Waste Management Paper No. 12. London: HMSO.

Joensuu, O.I. 1971. Fossil fuels as a source of mercury pollution. Science 172:1027-1028.

Kaakinen, J.W., R.M. Jorden, M.H. Lawasani, and R.E. West. 1975. Trace element behavior in coal-fired power plant. Environmental Science and Technology 9 no. 9 (September):862-869.

Kelly, W.F. 1973. SO_2 scrubber pilot plant: Tennessee Valley Authority Colbert steam plant. EPA Report R2-73-249. Washington: Environment Protection Agency.

Ketchum, Bostwick H. 1975. Biological implications of global marine pollution. In The changing global environment, ed. S.F. Singer,311-328. Dordrecht: Reidel.

Klein, D.H., A.W. Andren, and N.E. Bolton. 1975. Trace element discharges from coal combustion for power production. Water, Air and Soil Pollution 5:71-77.

Lindberg, S.E., and R.R. Turner. 1977. Mercury emission from chlorine production solid waste deposits. Nature 268:133-136.

McFadden, Robert D. 1978. Olin and three ex-aides indicted on dumping mercury in Niagara. New York Times, 24 March, pp. A1,A8.

Magee, E., H.J. Hall, and G. Varga. 1973. Potential pollutants in fossil fuels. NTIS PB-225039. Linden, New Jersey: Esso Research and Engineering Company.

National Research Council. 1977a. Safe Drinking Water Committee. Drinking water and health. Vol. 1. Washington: National Academy of Sciences.

National Research Council. 1977b. Panel on Mercury. An assessment of mercury in the environment. Washington: National Academy of Sciences.

New York Times. 1978. Spanish imports cut by fears of poison. 5 February, p. 9.

NIOSH (National Institute for Occupational Safety and Health). 1973. Criteria for a recommended standard: Occupational exposure to inorganic mercury. NIOSH 73-11024. Cincinnati, Ohio: NIOSH.

OECD (Organization for Economic Cooperation and Development). 1974. Mercury and the environment. Paris: OECD.

Suckcharoen, S., P. Nuorteva, and E. Hasanen. 1978. Alarming signs of mercury pollution in a freshwater area in Thailand. Ambio 7:113-116.

United States Bureau of Mines. 1975. Minerals in the U.S. economy: Ten year supply-demand profiles for mineral and fuel commodities. (1965-74). Washington.

United States Bureau of Mines. 1977. Commodity data series: 1977. Washington.

Van Horn, William. 1975. Materials balance and technology assessment of mercury and its compounds on national and regional bases. EPA 560/3-75-007. Washington: Environmental Protection Agency.

Wedge, W.K., D. Bhatia, and A. Rueff. 1976. Chemical analysis of selected Missouri coals, and some statistical implications. Missouri Department of Natural Resources Report of Investigations Geological Survey No. 60. Rolla: Missouri Department of Natural Resources, Division of Geology and Land Survey.

WHO (World Health Organization). 1976. Mercury. Environmental Health Criteria, 1. Geneva: WHO.

10
Nuclear Power[1]

Christoph Hohenemser,
Roger E. Kasperson, and Robert W. Kates

Nuclear power is in trouble. Despite the results of polls, which have shown repeatedly that the majority (about 60 percent) of the public views nuclear power favorably and thinks it safe, there is a sizable and growing opposition to nuclear technology. Public initiatives for a moratorium on nuclear development were recently defeated in California, Arizona, Colorado, Montana, Ohio, Oregon, and Washington. Nevertheless, similar initiatives are being prepared in another 19 states. Within the industry and in government regulatory agencies, there has been a significant defection of middle-level technologists (Burnham 1976a). Many plants have been delayed or canceled, and capital costs will have risen from $300 per installed kilowatt in 1972 to an estimated $1120 by 1985 (Business Week 1975,100). The price of uranium tripled between 1974 and 1976, and the adequacy of the uranium supply after 1985 is in question (Lieberman 1976;Day 1975).

All this is happening at a time when many features of nuclear technology--low average pollution, cost advantages over coal-and oil-fueled plants in many areas, and replacement of foreign oil resources in electric power generation--should encourage rapid adoption of the technology. What causes the malaise?

Delays, cancellations, and rapidly increasing capital costs are not likely to be decisive in the long run. Recent delays and cancellations have been strongly affected by the decreased demand following the sudden doubling in electric energy prices in 1973 and 1974. Rapidly increasing capital costs are a function of the availability of capital, increases in labor costs, and the recent period of high inflation. These problems are shared by large new fossil-fired plants; solar plants would presumably have similar difficulties if they were available.

We attribute most of nuclear power's problems, therefore, to the issue of safety. For the last two years our interdisciplinary group has studied the safety issue, particularly to see how the risk of rare events enters into the energy policy decisions of our society. At first sight, the case for the safety of nuclear power reactors appears impressive. Some frequently cited statistics and examples are as follows.

1. The maximum permitted annual radiation exposure for persons living at the boundary of a nuclear power plant is 5 millirem. Routine population exposure from all nuclear power plants averages

0.003 millirem per person per year (National Research Council 1972). By comparison, natural and medical sources contribute average exposures of 100 and 70 millirem per person per year (National Research Council 1972;UNSCEAR 1972), and individuals living in buildings constructed of volcanic rock (for example, in Rome) may be exposed to twice the natural background, or about 200 millirem per person per year (UNSCEAR 1972).

2. When coal plants are located in large cities, the population exposure from radioactinides in fly ash is 500 manrem per year.[2] This exceeds permitted radiation exposures from reactors of equivalent power (Wilson and Jones 1974).

3. The most complete study to date of catastrophic reactor risk places the probability of a major radioactive release (release of an appreciable fraction of the volatile fission products found in the reactor) at 1 in 100,000 reactor-years;[3] of core meltdown at 1 in 20,000 reactor-years; and of a loss-of-cooling accident at 1 in 2000 reactor-years (Nuclear Regulatory Commission 1975). These probabilities are given credence by the fact that to date, after 300 reactor-years of commercial reactor operation, there has never been a loss-of-cooling accident (Nuclear Regulatory Commission 1975). With these probabilities, the expected number of prompt and delayed fatalities due to 100 reactors in the United States is only four per year; and the population exposed in the unlikely event of a major reactor accident would have a cancer risk only 1 percent greater than its preexposure risk (Nuclear Regulatory Commission 1975).

4. Although plutonium is a potent carcinogen, substantial quantities ($\simeq 10^5$ kilograms) of it have been handled in the past 30 years with no apparent ill effects: there have been no cancers that can definitely be attributed to plutonium in the several thousand workers who have handled the material (Bair, Richmond, and Wachholz 1974).

In early 1976 a committee of the National Academy of Sciences (NAS) began a study of the risks of various electric power technologies. Although a detailed comparison is an extensive task and must await the NAS report, it is not difficult to characterize and compare the risks of the hydroelectric, coal, and nuclear technologies, the three present options for new baseline electric power. We have done this (Table 1) for four classes of hazards: (1) routine occupational hazards, such as those of mining; (2) routine population hazards, such as the inhalation of pollutants; (3) general environmental degradation, such as destruction of cropland, and (4) catastrophic hazards, such as massive release of radioactivity and dam failures. We conclude from Table 1 that the quantified risks, based on available information, are much larger for coal and hydroelectric than for nuclear power.

Considering Table 1, how do we explain the distrust of nuclear power and the continuing doubts about its safety? We submit that the distrust of nuclear power rests in part on its social history; in part on its unique combination of hazards; and in part on the special way it has been managed and regulated. Furthermore, the public distrust of nuclear power is significantly amplified by the rancorous debate in a polarized expert community.

An Intermingling of Issues

Throughout its 30-year history, nuclear power has inspired some of the major hopes and fears of mankind. Although it is difficult to describe this relationship except in terms of influence or anecdote, to ignore the social history of nuclear power is to misunderstand its present predicament. Many new technologies are born in wartime efforts. None have come to symbolize the destructiveness of war as has the atomic bomb. For better or worse, nuclear power was for many years tied to and overshadowed by the course of military developments. To see this, consider the first 20 years of the nuclear age (Baker 1958).

Immediately after World War II, the United States had a monopoly on nuclear technology. All significant U.S. development efforts were in a military direction. Reactors were built to breed weapons materials and to propel submarines and aircraft carriers, and uranium-235 was isotopically separated for military purposes. Commercial nuclear power was seen as something for the distant future and regarded as highly uneconomical (Weinberg 1972). At the same time, the atomic scientists who had built the bomb persuaded the U.S. government to argue at the United Nations that the nuclear enterprise was so dangerous that nothing short of international ownership would suffice to contain it (Baker 1958). They also exerted considerable influence to establish the Atomic Energy Commission (AEC) as a "civilian" umbrella agency to oversee the nuclear enterprise, with the particular charge to promote and develop commercial as well as military aspects of the technology (Rabinowitch 1950).

The idealism implicit in the U.N. efforts and the establishment of the AEC was short-lived, however, and with the first Soviet atomic tests in 1949 faded quickly into the cold war, the McCarthy period, and the arms race. By 1952 this had culminated in the testing of multimegaton thermonuclear devices by both sides in what was later called by AEC commissioner Thomas Murray "a vacuum of military strategy" (Murray 1957). By 1954 most of the public viewed atomic energy as synonymous with military terrorism in a situation in which the "enemy" was seen as a force of unmitigated evil in the world (Rosenberg 1966;Holsti 1967). Public discussion of alternatives uses of nuclear power was almost nonexistent.

The frozen silence finally thawed when the accidental severe exposure of a Japanese fishing vessel to fallout from the 1954 U.S. Bikini atoll test (Lapp 1958) focused public attention on the worldwide hazard of fallout from nuclear weapons testing. In response to this realization, Adlai Stevenson suggested in the 1956 election campaign that atomic testing be halted. President Eisenhower, while against a halt to testing, countered by proposing Atoms for Peace, a program of international sharing of nuclear technology for peaceful purposes (Baker 1958).

In 1956 there was not a single commercial nuclear power plant in the United States. Development efforts had been limited to experiments with alternative, reactor design concepts, and much of this work had, in fact, been cut back by Eisenhower when he took office in 1953 (Mullenbach 1963). At the same time, notable success had been achieved by the AEC-Westinghouse collaboration on submarine and ship propulsion reactors. To launch Atoms for Peace, the United States thus chose a modified naval reactor for a first demonstration

TABLE 1

Risks from three electric-power technologies. Deaths are the number expected per year for a 100-Mwe power plant. In all cases, man-days lost (MDL) are converted to deaths by 6000 MDL/death (AEC 1974a).

HAZARD TYPE	HYDROELECTRIC	COAL	NUCLEAR
Routine occupational hazard	Construction accidents are significant but the risks are not as large as for coal mining	Coal mining accidents and black lung disease constitute a uniquely high risk	Risks from sources not involving radioactivity dominate. Aggregate risks from all stages of the fuel cycle are less than for coal
Deaths	0.1 to 1.0a	2.7b	0.3 to 0.6c
Routine population hazard	Thought to be benign, although specific cases (for example, the Aswan dam) have produced new health hazards	Air pollution produces relatively high, although uncertain risk of respiratory injury. Significant transportation risks	Low-level radioactive emissions are more benign than corresponding risks from coal. Significant transportation risks remain incompletely evaluated
Deaths		1.2 to 50d	0.03e
General environmental degradation	Permanent loss of free-running streams, agricultural lands, wilderness	Strip mining and acid runoff; acid rainfall with possible effect on nitrogen cycle, atmospheric ozone; eventual need for strip mining on a large scale	Long-term contamination with radioactivity; eventual need for strip mining on a large scale
Catastrophic hazards (excluding occupational)	Major dam failures have occurred, but rarely in modern structures	Acute air pollution episodes with hundreds of deaths are not uncommon. Long-term climatic change induced by CO_2 is conceivable	Risks of reactor accidents are small compared to other quantified catastrophic risks. The problem lies in as yet unquantified risks for the reactors and the remainder of the fuel cycle.
Deaths	1f	0.5g	0.04h

aThis estimate is based on (1) 10,000 man-years to construct a 1000-Mwe hydroelectric dam and generating station; (2) a heavy construction occupational hazard of 0.34 fatality and 1.34 permanently disabling injuries per 1000 man-years, or about 1 fatality equivalent per 1000 man-years (National Safety Council 1973); (3) distribution of construction fatalities over an assumed 100-year useful life of the project; and (4) hydroelectric generation availability of 10 to 100 percent.

bData are from the Atomic Energy Commission (AEC 1974a). Of the 2.7 deaths, 1.1 are due to mining accidents of all kinds, including major mine disasters, and 1.6 are due to black lung disease and other injuries.

cThe lower figure is from the AEC (1974a), the higher figure from David Rose (1974).

dThe lower figure represents transportation accidents only, as given by the AEC (1974a). The higher figure includes an interpretation (National Research Council 1975) of the rather uncertain air pollution epidemiology.

eSee AEC 1974a. The result is consistent with an average annual exposure of 0.035 millirem per individual per reactor, using a cancer risk of 2 x 10-6 cancers per man-rem (National Research Council 1972). The average exposure of 0.035 millirem applies to reactors only. It must therefore be considered as a lower bound for the fuel cycle risk.

fThe figure represents an estimate for dam failure risk based on all historical incidents, as summarized in the **Reactor Safety Study** (Nuclear Regulatory Commission 1975). The number must be taken as an upper bound since many dam failures will not be connected with hydroelectric generation.

gThis is based on the occurrence of one 500-death air pollution episode per year, with one-fifth of the pollution attributable to coal power plants.

hThis estimate is based directly on the **Reactor Safety Study** (Nuclear Regulatory Commission 1975), as discussed in the text, without correction for the incompleteness of the methodology, and must be regarded as a lower bound.

plant. Located in Shippingport, Pennsylvania, this plant went on line in 1958, with a rating of 90 megawatts electric (Mwe). While neither big nor economically competitive, the plant became an important symbol to balance the destructiveness of nuclear weapons in the public's eye.

Meanwhile, the test-ban issue remained the most important public nuclear concern. For seven years (1956 to 1963) it was argued in a context of national security, clean bombs, and dirty bombs, until finally, with the signing of the Moscow treaty, it was literally "driven underground" (Hohenemser and Leitenberg 1967). During this period, despite real doubts about nuclear power economics, extensive plans for commercial nuclear power were developed on a worldwide basis. These plans proved far from realistic and served largely to trigger a new fear that reemphasized the military aspects of nuclear power; that the spread of nuclear power would lead to proliferation of the nuclear weapons capability by making plutonium widely available (see Table 2). Known at the time as the Nth country problem (Davidon, Kalkstein, and Hohenemser 1960;Hohenemser 1962), this fear motivated substantial safeguards in nuclear sharing agreements between the United States and the International Atomic Energy Agency and eventually led to the nuclear nonproliferation treaty in 1968. Proliferation is still feared today and is regarded by some long-time observers, such as Feld (1975), as the single most important hazard of nuclear power.

By 1965, 20 years after the first bombs were dropped, public concern with nuclear policies had subsided to an all-time low. In rejecting President Kennedy's fallout shelter proposals in 1962, the public had shown itself distinctly fatalistic about the prospects for and value of surviving a nuclear war. The first 20 years of the nuclear age thus closed with the balance of terror and nuclear overkill established facts (Lapp 1968). Commercial nuclear power, which had with the start-up of the 500-Mwe plant at Indian Point reached near economic parity with other power sources in 1962 (Mullenbach 1963), made no major impact on a public that now faced news of guerilla war in Vietnam and watched as the number of intercontinental ballistic missiles on both sides increased from the tens to the hundreds to the thousands. In addition, there was no real concern with reactor safety at the time, even though a number of accidents had occurred in experimental reactors (AEC 1971;Zimmerman 1975), and the AEC had outlined rather disturbing conceivable consequences of commercial reactor failure as early as one year before the opening of the Shippingport demonstration plant (AEC 1957).

Since 1965, the public view of nuclear energy has undergone a dramatic and unexpected metamorphosis. Nuclear weapons and nuclear war have disappeared as major issues; the cold war has slowly waned; and although warheads now number in the tens of thousands (SIPRI 1976), threats to the natural environment and a general distrust of high technology have replaced earlier fears. Nuclear power has become controversial, to the bewilderment of nuclear power technologists who for two decades or more have worked on the "peaceful atom" with little doubt about the virtue of the task.

A first attack on commercial nuclear power came late in the 1960s when, as a logical extension of concerns about fallout, the question of routine radioactive emissions from power plants was raised by Sternglass (Boffey 1969), Tamplin (1971), Gofman (1981,

TABLE 2
Plutonium production from civilian nuclear power: projection compared to reality. Data are from Davidon, Kalkstein, and Hohenemser (1960) and Willrich (1973). Plutonium production values are estimates for 1975 from Davidon, Kalkstein, and Hohenemser (1960); to obtain the maximum number of nominal weapons, divide these values by 5 (the critical mass of Pu is $\simeq 5$ kg).

COUNTRY	PLUTONIUM PRODUCTION FOR 1975 (KG/YEAR) ESTIMATED	ACTUAL	PERCENTAGE OF PROJECTION ACHIEVED
Belgium	1500	200	13
Canada	?	600	
China*	?	?	
Czechoslovakia	5000	200	4
France*	8000	600	7
Germany, East	3000+	100	3
Germany, West	6000	1000	18
India*	?	200	
Italy	500+	200	40
Japan	7000	1000	14
Netherlands	3000	100	3
Norway	?	?	
Poland	1800	?	
Rumania	500	?	
Spain	1800	400	22
Sweden	2000	500	25
Switzerland	?	200	
Soviet Union	?	1000	
United Kingdom*	6000+	2000	33
United States*	?	5000	

*These are countries with nuclear weapons. The United States, Soviet Union, France, and United Kingdom have all produced additional plutonium in military reactors. China and India may have done so.

870-871), and others. This issue fit well with growing environmental concerns, which came to a crescendo with Earth Day in 1970. As it turned out, routine emissions were easily shown to be of minor significance compared to other pollutants (see Table 1), and the issue died out soon after Earth Day. But nuclear power had taken on a special status within the environmental movement, and this led in rapid sequence to a whole range of new issues.

During 1971 and 1972, the first large environmental coalition, the Consolidated National Intervenors, assembled around the AEC rule-making hearings on emergency core cooling. These hearings exposed serious inadequacies in AEC safety research and regulation. Questions about AEC safety measures had first been raised by the Union of Concerned Scientists (Forbes et al. 1972), a collaboration of scientists from the Massachusetts Institute of Technology, and were reminiscent of earlier public information efforts in the seven-year debate on fallout. In 1973 Ralph Nader and the Sierra Club took up opposition to nuclear power on a variety of grounds, ranging from safety to economics to unsolved problems of waste disposal. Most recently, nuclear power has become, in the view of the environmental movement, a symbol of high technology, unbridled growth, and centralization—all trends that are being increasingly questioned by activists. Thus, Friends of the Earth argues that "U.S. reliance on fission nuclear power to fill the energy needs of an economy characterized by extravagance and waste needlessly mortgages the peace, welfare, and freedom of future generations" (Friends of the Earth 1976). In contrast, the development of various alternative power sources such as the sun and the wind would "counteract the increased concentration of economic and political power in a few giant energy corporations" and "encourage essentially grass roots efforts involving individual and community action and small businesses" (Friends of the Earth 1976).

The critique of nuclear power is today well advanced. A 1975 Harris poll (Louis Harris and Associates 1975) showed the public strongly divided, with environmentalists leading the way (see Table 3). It is doubtful that a consensus of people would agree today that nuclear power is sufficiently safe. Another perspective on the present appears in Figure 1, where media interest, as a surrogate of public concern, is plotted over three decades. Figure 1 clearly shows the two major periods we have sketched. The first upsurge of interest was during the seven-year debate on nuclear weapons testing; the second reflects the environmental and safety concerns about nuclear power that occupy the present.

It is very likely that the link in the public's mind between nuclear power and weapons testing is more deep-seated than is suggested by the correlations given in Figure 1. For example, Pahner (1975), citing a psychoanalytic study of Hiroshima survivors, argues that a substantial part of the public's concern over nuclear power is displaced anxiety rooted in a fear of nuclear war (Lifton 1967). The fading of the ban-the-bomb marches, then, was not a coming to terms with nuclear weapons, but a repression of fear that is destined to resurface elsewhere. In support of this view, Harris poll findings (Louis Harris and Associates 1975) and opinion surveys that we conducted[4] reveal a widespread public concern that "nuclear power plants may explode."

TABLE 3
Perceived safety of nuclear power plants (Louis Harris and Associates 1975). Abbreviations: VS, very safe; SS, somewhat safe; NSS, not so safe; D, dangerous; and NS, not sure.

GROUP	PERCENTAGE OF PLANTS				
	VS	SS	NSS	D	NS
Public	26	38	13	5	18
Environmentalists	10	25	44	19	2

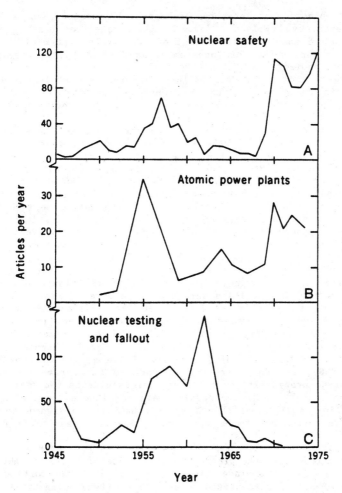

Figure 1. The social history of nuclear power and weapons testing
is illustrated by using media interest as a measure of public interest. (A) nuclear safety in the New York Times, (B) atomic power
plants in Readers' Guide, and (C) nuclear testing and fallout. Data
for (A) were compiled by the authors; the data for (B) and (C) are
from Mazur (1975).

The distrust of nuclear power is thus rooted in the fear of
nuclear weapons and is augmented by concern about pollution and
opposition to high technology and centralization. Is this sufficient to explain all of the distrust? We think not.

Reactor Safety

With the maturing of commercial reactors in the late 1960s, it
became clear that nuclear power poses threats that may be unique in

their combination of catastrophic potential, duration, and scientific uncertainty[5] (Wilson 1973). To illustrate, we begin with the most studied case: the assessment of catastrophic risk in the light water reactor. Using this assessment as a standard, we examine the state of knowledge for the entire fuel cycle.

The hazards of reactor failure were foreseen at least 19 years ago, when the AEC outlined the consequences of conceivable catastrophic accidents for a 150-Mwe reactor in its report WASH 740 (AEC 1957).

The study was updated in 1965 for the 1000-Mwe plants that were then being planned. The WASH 740 report projected as many as 3,400 deaths and 43,000 injuries; the updated version of the report (Mulvihill et al. 1965) showed as many as 45,000 deaths and a disaster area the size of Pennsylvania. Neither WASH 740 nor its updated version had a major public impact at the time, the former because it was overshadowed by the test-ban debate, the latter because it was suppressed for eight years to "avoid great difficulties in obtaining public acceptance of nuclear energy" (Mulvihill et al. 1965).

As the questions raised in the early 1970s about catastrophic reactor failure escalated, the absence of failure probabilities in WASH 740 and its updated version made for a volatile situation. While the AEC argued that the probability of catastrophic occurrences is very low, critics were free to assume or imply the worst, especially since 300 reactor-years of catastrophe-free commercial reactor operation provided no empirical support for the AEC's low core-meltdown probability (AEC 1973). The AEC therefore commissioned a new study under the direction of Norman Rasmussen of the Massachusetts Institute of Technology. Known as the **Reactor Safety Study (RSS)**, it took into account for the first time both consequences and probabilities of catastrophic accidents (Nuclear Regulatory Commission 1975). The results were not inconsistent with those of earlier studies, although the probability assigned to major accidents turned out to be very small. Specific results of RSS may be summarized as follows:

1. The core meltdown probability is 5×10^{-5} per reactor-year. This is larger than the previous AEC estimate of 1×10^{-6} (AEC 1973) and represents an average for the type of 1000-Mwe boiling water reactor (BWR) and pressurized water reactor (PWR) being built in the United States at present.

2. For each reactor type several categories of radioactive releases following core meltdown are identified, and for each of these a probability is found. This analysis makes clear that core meltdown does not necessarily lead to large releases, although it may do so.

3. For each release class, expected consequences are calculated in six categories: prompt fatalities, prompt injuries, delayed cancers, delayed thyroid nodules, genetic effects, and property damage. Employed in obtaining these results were models of weather patterns population densities, as well as the radiation dose-response methodology discussed in the so-called BEIR Report (National Research Council 1972).

4. The separate results from the BWR and PWR were averaged and presented as risk spectra for the six consequences mentioned above. The uncertainty in these spectra ranges from one-fifth to five times the expected risk, as shown in Figure 2.

5. The risk spectrum for prompt fatalities was compared to the spectra for manmade and natural hazards, as shown in Figure 2A. Delayed deaths due to radiation-induced cancer[6] (American Physical Society 1975) were omitted from this comparison, on the grounds that "predictions of this type are not available for non-nuclear events, and so comparisons cannot easily be made" (Nuclear Regulatory Commission 1975).

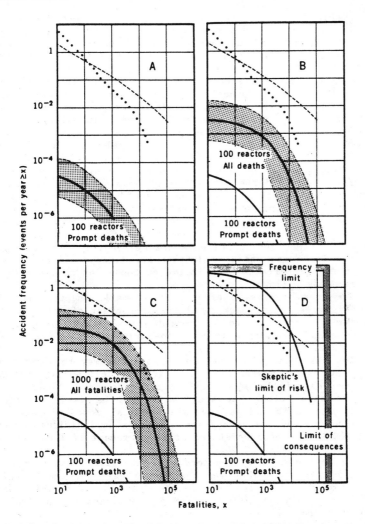

Figure 2. Comparison of the results of the RSS with other catastrophic risk estimates. Natural and nonnuclear manmade catastrophic risks are indicated by dashed and dotted lines, respectively. The RSS results with error bands (shading) are shown as dark, heavy lines. The results in (A) and (B) are from the RSS: those in (C) and (D) are based on increasingly pessimistic interpretations of the RSS, as discussed in the text.

The widely publicized comparison given in Figure 2A appears to settle the nuclear debate as far as accidents are concerned. Yet we know that this is an illusion. Below we explore several issues that transcend the RSS.

Delayed Cancer Deaths. Excluding delayed cancer deaths from the comparison in Figure 2A on the grounds stated above begs an important value question. Alternative characterizations of delayed cancer deaths are as follows:

1. Delayed cancers appear as a $\approx$ 1 percent annual contribution to a preexisting nonnuclear cancer risk. This is statistically an undetectable effect; thus for one rather high-consequence nuclear accident, 170 additional cancers are expected, for an annual total incidence of 17,000.

2. Delayed cancers exceed prompt fatalities in number by a factor of 100 or more. Since there is no acceptable way of discounting future deaths—as we discount future income in economic analysis—we must attribute all deaths, prompt or delayed, to the accident frequency in question, as shown in Figure 2B.

Both of these interpretations are technically correct, although they are based on different crucial value judgments. The dilemma is reminiscent of the fallout debate, in which one side quoted "small" percentage effects and the other "large" absolute numbers. In that case, the perception of large eventual fatalities evidently won out and led the politicians to sign a test-ban treaty.

Genetic Effects. From a value point of view, the treatment of genetic effects is even more problematic than that of cancer deaths. Like cancer deaths, genetic effects are delayed, but unlike cancer deaths, the delay may extend indefinitely. Alternative characterizations of genetic effects are similar to those for cancer risk and may be stated as follows.

1. The risk of genetic effects is a small, undetectable percentage of a preexisting background for nonnuclear effects. Even if all genetic effects lead to death—and they certainly do not—the calculated annual incidence is only one-sixth of the increment of cancer deaths.

2. The absolute number of genetic defects may be larger than the number of cancer deaths since genetic defects propagate for many generations, especially since modern medicine makes possible the survival of those with what would otherwise be fatal mutations. Under future, possibly less favorable medical conditions, an increased genetic load may have drastic effects on individual chances of survival. This cannot be stated as an increment of risk per year (Muller 1955).

Extrapolation to 1000 Reactors. The RSS gives results for 100 light water reactors. Plans for the nation call for the installation of as many as 1000 reactors in the next 30 years. This raises the question of extrapolation. There are at least two alternative views of this problem.

1. It is improper to extrapolate linearly from 100 to 1000 reactors since this does not take into account probable improvements in management and technology with increasing experience (Nuclear Regulatory Commission 1975).

2. One might as well extrapolate, since learning may in whole or in part be canceled by increasing human carelessness as nuclear power proliferates. In addition, learning is strongly attenuated by

present lack of standardization and the fact that an appreciable number of future reactors will be breeders, for which the RSS is irrelevant.

Neither view 1 nor view 2 can be supported or refuted by any available quantitative analysis. Therefore, it is reasonable to plot conservatively both prompt and delayed consequences, as in Figure 2B, and then extrapolate linearly to 1000 reactors, as in Figure 2C. Thus, a reinterpretation of the RSS results (without challenging the methodology of the study) indicates that the risks of catastrophic reactor failure approach the risks of a variety of manmade and natural catastrophic hazards.

A Skeptic's View. A final set of issues deals with a challenge to the methodology of the RSS. A number of critics have stated that the RSS analysis leading to the core meltdown probability is inadequate on the following grounds.

1. Completeness. It is impossible to know whether fault-tree analysis has identified all failure modes, particularly of the common mode variety (Kendall et al. 1974). The RSS agrees, but it is argued that the most important modes, including common ones, have been included.

2. Design Adequacy. Probability and fault-tree analysis cannot deal with inadequacy in reactor design as distinct from statistical failure of components (Kendall et al. 1974;Weatherwax 1975). Experience in the aircraft industry shows that unsuspected design inadequacy is responsible for most early crashes (Hohenemser 1975). The same may be true of reactors.

3. Human failure. As used in the RSS, probability and fault-tree analysis do not deal with certain types of human error, such as willful acts and sabotage. The RSS is, in effect, a statistical study of a perfectly designed machine, with the only sources of failure lying in the statistical malfunction of components and statistically quantifiable operator errors.

None of these criticisms is directed at the quality of analysis done in the RSS within the framework of probability and fault-tree analysis. They are warnings that a skeptical view of the methodology demands that the results be viewed as reasonable lower bounds on accident risk. One may, as some have suggested (Kendall et al. 1974), patch up the methodology or introduce more conservative error limits. Alternatively, one may bypass the RSS analysis for defining risk absolutely and rely instead on bounds defined in part by experience.[7] Two such bounds are (1) the empirical upper bound on accident risk arising from the current 300 reactor-years of catastrophe-free commercial reactor operation and (2) the high-consequence asymptote of the RSS risk spectra, which coincides more or less with the results obtained in the updated version of WASH 740. These bounds of skepticism appear as shaded bands in Figure 2D. Also shown is a risk spectrum of the RSS shape that conforms to these bounds and shows how a rational skeptic might assess reactor risk. The space between this curve and the RSS curve for prompt deaths from 100 reactors is a measure of the gap that currently exists between the strongly skeptical view of nuclear power and the views of a nuclear proponent who accepts the RSS executive summary at face value.

Individual risks calculated from Figure 1 are presented in Table 4, where they may be compared to individual risks from other

TABLE 4
Individual fatality risks (from Nuclear Regulatory Commission 1975)

CAUSE OF ACCIDENT	ACCIDENT RISK PER YEAR (DEATHS PER MILLION)
Principal noncatastrophic risks	
Motor vehicle	300
Falls	90
Fires and hot substances	40
Drowning	30
Poison	20
Firearms	10
Machinery	10
Water transport	9
Falling objects	6
Electrocution	6
Railway	4
Lightning	0.5
Principal nonnuclear catastrophic risks	
Air travel	9
Tornadoes	0.4
Hurricanes	0.4
Fires	0.5
Nuclear reactor risks	
100 reactors, prompt deaths (Fig. 2A)	0.0002*
100 reactors, all deaths except genetic (Fig. 2B)	0.02
1000 reactors, all deaths except genetic (Fig. 2C)	0.2
1000 reactors, rational skeptic's limit (Fig. 2D)	20

*Risks are based on the 15 million people who live within 25 miles of 100 nuclear plants and are candidates for prompt death. All other nuclear risks are based on 200 million people.

hazards, both catastrophic and noncatastrophic. (The individual risk for Figure 2A has been used to characterize catastrophic nuclear risk in Table 1.)

Conclusion. Whether seen through Figure 2 or Table 4, the assessment of catastrophic reactor risk can vary widely, from a point far below other risks to a point that exceeds a number of risks that many consider significant. The assessment of reactor accident risk depends on how we value the future, including the next generation; how we project the future safety of an evolving technology; and how much confidence we have in risk estimation that is based on no direct experience with the event for which risk is assessed. In the end, our answer will depend on whether we are technological optimists or pessimists.

The Rest of the Fuel Cycle

Aside from reactor failure, the light water reactor fuel cycle, shown in Figure 3, is susceptible to several other catastrophic risks. As a first step in characterizing them, we have constructed an exhaustive typology of risks, shown in Table 5. Here, conceivable catastrophic risks are symbolized by initiating events for each hazard and fuel-cycle stage. Below we discuss briefly the present state of knowledge about each of the columns in Table 5.

Nuclear Explosions. The risk of nuclear explosions derives from the possibility that weapons-grade material is illegally diverted from various stages of the fuel cycle. In an international context, this risk was widely discussed 20 years ago and was the principal motivation for the nuclear nonproliferation treaty. More recently, Willrich and Taylor (1974) have emphasized the relative ease of bomb construction and the lack of security against theft from domestic fuel-enrichment and reprocessing plants, as well as plutonium storage facilities. No attempt has, to our knowledge, been made, in a manner that is compatible with the units of Table 1 (expected deaths per reactor year), to evaluate the risk of theft. The prospect of the plutonium economy with annual inventories of 30,000 to 200,000 kilograms (Willrich and Taylor 1974) makes the diversion of a critical mass of ~ 5 kg plausible. At the same time, as far as we know, the military have successfully guarded for 30 years a stockpile of $\sim 100,000$ kg of weapons-grade material, much of it in the form of weapons.

Massive Fission-Product Release. After the reactor, fission product hazards occur in the "back end" of the fuel cycle: in reprocessing, waste disposal, and transport to and from these facilities. These processes are not currently operational in the commercial U.S. fuel cycle. The only commercial reprocessing plant, in

TABLE 5
Typology of catastrophic nuclear risks.

FUEL CYCLE STAGE	HAZARD TYPE		
	NUCLEAR EXPLOSION	FISSION PRODUCT RELEASE	PU DISPERSAL
Mining, milling, and refining			
Enrichment and fuel fabrication	T		S, A
Light water reactor		S, A	S, A
Fuel reprocessing	T	S, A	S, A
Plutonium storage	T		S, A
Waste disposal		S, A	S, A

Key: T, theft; S, Sabotage; and A, accident.

234

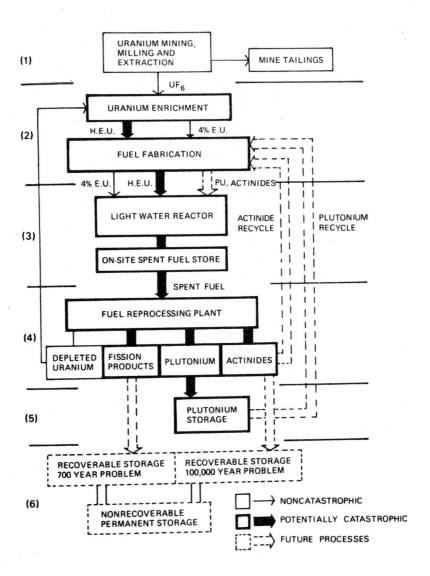

Figure 3. The light water fuel cycle, showing portions currently operational and nonoperational in the United States. The abbreviations E.U. and H.E.U. stand for enriched and highly enriched uranium, respectively.

West Valley, New York, closed in 1972 after six years of operation and is being redesigned and enlarged. A second plant, at Morris, Illinois, has been scrapped because of technical problems that would, among other things, have led to unacceptable occupational exposures. A third plant, at Barnwell, South Carolina, is under construction and is scheduled to open soon. As to the disposal of solid waste, it is still not clear what the product, and therefore the process, will be and where it will be stored (Colby 1976). Most spent fuel is now stored at reactor sites in cooling ponds. Failure to resolve the waste-disposal questions may delay opening and operation of reprocessing plants under construction, even if they are otherwise functional.

In view of the immature state of the back end of the fuel cycle, it is perhaps not surprising that little effort has been spent on risk assessment. What has been done may be summarized as follows.

1. Cohen (1976), in an effort to show that wastes do not pose a significant threat, obtained an upper limit of 0.01 death per reactor-year for random deep burial of solidified wastes. His result depends on treating as equivalent the risk from naturally occurring uranium in rock and the solid waste at the bottom of a deep disposal hole.

2. The AEC (1972) analyzed risks associated with the transportation of spent fuel and waste and estimated accidental fatalities from nonradiological and radiological causes as 0.01 and 10^{-7} per reactor-year, respectively. Ross (1975) challenged this on the grounds that not only volatile fission products (as assumed by the AEC) but also semivolatile fission products can be released in truck accidents accompanied by fires. Our interpretation of his analysis leads to a fatality rate of 0.01 per reactor-year.

Risks not assessed may be more important: consider two cases.

1. If present plans materialize, by the year 2000 there will be 50,000 annual shipments of spent fuel and waste, containing 2 to 3 megacuries each and covering a total of 50 million truck miles (AEC 1972). This would appear to pose a significant sabotage risk.

2. If and when operational, fuel reprocessing plants will handle the fission-product inventories of several reactors. They are potential sources of radiological risk an order of magnitude greater than the risk from reactor accidents. Considering the lack of experience with commercial reprocessing, it is doubtful that even if a study like the RSS were undertaken, meaningful results could be obtained.

Plutonium. A third category of risk involves dispersal of plutonium. In a fully developed fuel cycle this can occur nearly anywhere after the mining of uranium (see Table 5). Plutonium toxicity is based on its alpha activity and, like fission product toxicity, at low doses manifests itself through cancers with a latent period of 15 to 45 years. Although fission-product effects are fairly well defined, however, numerical estimates of plutonium toxicity vary and are controversial.

On some things, however, there is general agreement (Bair and Thompson 1974). Ingested plutonium is almost entirely excreted, and the dominant risk to humans is from inhaled particles. If insoluble, the particles stay in the lung with a half-life of 1000 days; if soluble, they are transported by the blood to the bone and liver and cause cancer at these locations. Plutonium accidents are likely to release insoluble PuO_2; therefore, lung cancer is the dominant risk to humans, and it is reasonable to characterize the toxicity of plutonium by the lung cancer dose. (On the assumption of the linear hypothesis, this is the population dose capable of producing one lung cancer on the average.)

On the basis of a few accidental exposures (Cohen 1975), it is possible to express the lung cancer dose in micrograms of inhaled or deposited plutonium. Table 6 shows a variety of estimates for lung cancer doses. In regard to Table 6, we make the following observations.

1. There have been no cases of lung cancer in 26 plutonium workers who received serious lung doses in 1954 and another 25 who were exposed in 1965 (Bair, Richmond, and Wachholz 1974). Therefore, animal experiments with PuO_2 inhalation (Bair and Thompson 1974) and the experience of underground miners with dust containing natural alpha emitters (National Research Council 1972) constitute the only direct link between lung cancer and internal alpha activity. The nonoccurrence of human lung cancers in the 50 serious plutonium exposures is nevertheless helpful in setting a lower bound on the lung cancer dose.

2. In consensus documents on the biological effects of radiation (National Research Council 1972;UNSCEAR 1972), it is assumed that the effect of internal alpha activity may be predicted on average doses to affected organs. The first four lines of Table 6 are based on this assumption. Either an absolute or relative risk model may be employed. In the first, the expected number of cancers is proportional to the dose; in the second, it is also proportional to the spontaneous cancer rate. This leads to a smaller estimated plutonium lung cancer dose for smokers than for nonsmokers (Gofman 1975a).

3. Despite average dose assumptions made in consensus documents, it is widely agreed that internal alpha doses are almost never evenly distributed (Bair, Richmond, and Wachholz 1974;Bair and Thompson 1974). The effect of dose localization on particle size is illustrated in Table 7. Consequently, Geesaman (1968), Martell (1975), Morgan (1975), and others expect that toxicity depends on particle activity. With large particles of PuO_2 very few cells are exposed, most receive lethal doses, and little if any dose is effective in cancer induction; with small particles the dose structure becomes indistinguishable from an average dose; with intermediate particles, high but nonlethal doses may produce a "resonant" cancer response in a relatively small number of cells. While this model is consistent with available experimental information, no clear-cut evidence of resonant response has yet been found.

It is clear, therefore, that plutonium toxicity poses problems significantly more intractable than those addressed by the RSS. To reach a useful conclusion, it is necessary not only to calculate dispersal probabilities but also to consider the large uncertainties in toxicity. We are therefore unable to report on an assessment of plutonium dispersal that represents a degree of scientific consensus.

A possible useful perspective has been suggested by Gofman (1975b). The amount of ^{239}Pu deposited in the lungs of humans in the United States totals 0.034 gram and results from the dispersal

TABLE 6
Plutonium lung cancer doses as estimated by various authors.

SOURCE	DEPOSITED CANCER DOSE (μg)	
	239PU	REACTOR GRADE PU
Cohen-BEIR absolute risk model[a]	204	38
Gofman-BEIR relative risk model[b]	43	8
Gofman relative risk model		
Smokers[c]	0.058	0.011
Nonsmokers[c]	7.3	1.4
Tamplin-Cochran hot-particle model[d]	0.002	0.0004
Bair-Thompson beagle dog experiments[e]	27	

[a]Estimates based on calculations by Cohen (1975), using the BEIR absolute risk model (National Research Council 1972). The result applies to adults 20 to 30 years of age.

[b]Estimates given by Gofman (1975a), using the BEIR relative risk model (National Research Council 1972) with a lung cancer risk of 0.5 percent of the spontaneous lung cancer risk per man-rem of exposure. The results differ from the preceding ones because current spontaneous rates are used instead of 1945 rates, on which the absolute risk model is based.

[c]Estimates by Gofman (1975a), using the BEIR relative risk model (National Research Council 1972) with modified assumptions: (i) a relative risk of 2 percent of the spontaneous risk per man-rem is used, and (ii) a distinction between smokers and nonsmokers is made, and the much higher "spontaneous" cancer risk of smokers is used. The higher relative risk conversion is justified by previous work of Gofman and Tamplin (see Gofman 1981, 870-871 for a list of references). Although the estimated lung cancer dose for smokers is very small, Gofman argues that it is not consistent with the nonoccurrence of lung cancers in 25 Los Alamos and 25 Dow Chemical workers accidentally exposed in 1944 and 1965, respectively.

[d]Estimates based on the work of Tamplin and Cochran (1974). These authors have considered 1- to 10-μm "hot" particles and have argued that locally high doses must be used in calculating the cancer risk. The results quoted here are based on the Tamplin-Cochran "dose distribution factor" of 10^5 (average dose multiplied by 10^5 to estimate locally high doses near hot particles) and the BEIR absolute risk model. Lung cancer doses as small as those given here are inconsistent with the nonoccurrence of human cancers in the Los Alamos and Dow Chemical exposures, and also with recent hot-particle experiments on animals (Bair, Richmond, and Wachholz 1974).

[e]Estimates based on the work of Bair and Thompson (1974) with beagle dogs (Bair, Richmond, and Wachholz 1974), as suggested by Gofman (1975a). The lowest dose at which all dogs die of lung cancer has been taken as the upper limit for the dog lung cancer dose. The human lung cancer dose was obtained by multiplying by the ratio of the lung mass in humans to that in dogs.

TABLE 7
Relationship of particle size to number of cells at risk for a static lung burden of 0.016 μc of ^{239}PuO$_2$ (AEC 1974a). Static particles are assumed in a structureless human lung of uniform density 0.2gcm^{-3} with an average cell volume of 10^3 μm^3. Cells at risk are taken to be those in a sphere of radius equal to the alpha-particle range (200 μm at the assumed density).

PARTICLE DIAMETER (m)	NUMBER OF PARTICLES	ACTIVITY PER PARTICLE (PC)	CELLS AT RISK	FRACTION OF LUNG (%)
0.1	5.4×10^7	3×10^{-4}	3×10^{11}	30
0.3	2.0×10^6	0.01	1.3×10^{10}	1
0.7	1.8×10^5	0.08	1.2×10^9	0.1
1.0	5.4×10^4	0.3	3.6×10^8	0.03

of ≈400 kg through weapons testing (Bennett 1974). If uptake of accidentally dispersed reactor grade plutonium is not to exceed the effects of fallout, dispersal in a future plutonium economy must be limited to ≈80 kg, assuming equal uptake fractions for the two cases. (Reactor-grade plutonium is about five times more toxic than weapons-grade plutonium.) Cumulative production by the year 2000 may be 10^7 kg (Willrich and Taylor 1974); hence, independent of toxicity, containment will have to be at the 99.999 percent level. The social cost of the ≈0.001 percent escaped plutonium will be 160 to 116,000 lung cancers, depending on which toxicity estimate in Table 6 is used.

Management of Safety

The properties of nuclear power--high technology, large capital investment, rapid growth, abbreviated experience, and low probability-high consequence risks--pose unprecedented regulatory problems. Until recently, these have been compounded by an unhappy marriage between development and regulation in the AEC (Metzger 1972;Primack and von Hippel 1974), an arrangement that dates back to the struggle for civilian control of atomic energy at the end of World War II. In this situation, the overriding priorities for development gave short shrift to pressing safety needs (Gillette 1974a-d). Thus safety research funds have been diverted to support the development of the breeder; quality assurance objectives replaced safety research objectives in the Loss of Fluid Test (LOFT) Program; the safety research budget of the regulatory staff before 1970 remained quite small (see Figure 4); and the regulatory staff was denied access to research findings from national laboratories. Because of the increasing public criticism of nuclear safety in the 1970s and continuing underestimation of the regulatory task, AEC regulatory managers became crisis managers. The recent establishment of the Nuclear Regulatory Commission (NRC) as a regulatory agency and

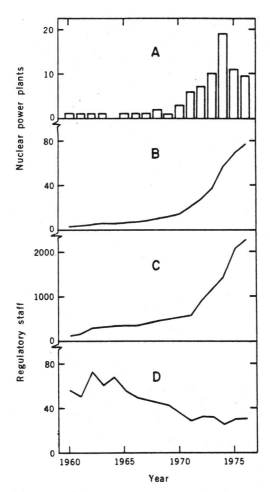

Figure 4. Growth of nuclear power and its regulatory staff. (A)
Number of nuclear power plants achieving commercial operation each
year. (B) Total number of nuclear power plants in commercial operation. (C) Total regulatory staff per nuclear plant in commercial
operation. (D) Regulatory staff per nuclear power plant in commercial operation. Note that as nuclear power grew rapidly in the
1970s, the regulatory staff per power plant declined. The rulemaking hearings on emergency core-cooling systems (ECCS) did not
have an effect on the regulatory staff per plant.

increased funding of regulatory work may eventually solve this
problem. Yet change will be slow, since below the commissioner
level the NRC is staffed largely by former AEC personnel carrying
with them a bureaucratic ethos built over a period of 30 years.
     In the present furor over safety, it is well to remember that
for years the AEC and the Joint Congressional Committee on Atomic
Energy agreed that regulation was in the public domain, while

responsibility for safety lay primarily with private industry. Quite independent of the confusion between development and safety goals, this model of regulation and safety appears, in retrospect, inadequate. The unhappy history of emergency core-cooling systems (ECCS) serves as an apt example (Cottrell 1974).

In 1966, when the AEC identified the loss-of-coolant accident leading to core meltdown as its highest safety priority, it initiated a major research effort and instituted a series of regulatory changes that were designed to ensure the safety of the larger reactors then being developed (Kouts 1975a;Ford and Kendall 1972). Yet in 1969, there were still only three members of the regulatory staff working directly on evaluation of ECCS. In the 1971 ECCS rulemaking hearings, Morris Rosen, then chief of the Systems Performance Branch of the AEC's Division of Reactor Standards, testified that the regulatory staff simply did not have adequate knowledge to make licensing decisions on 100 reactors then pending (Ford and Kendall 1972). It was clear by then that the ECCS problem transcended the capability of any single industry and must, contrary to earlier expectations, be taken over by the government. By 1975, at least 10 years after the initial recognition of the ECCS problem and at a time when the number of commercial reactors stood at about 50, no solution appeared in sight, 90 percent of all current light water research funding was committed to the problem, and in the words of the new director of safety research (Kouts 1975b): "the future program in reactor safety research is largely . . . the future of the ECCS program."

Yet, as we have tried to make clear, there are other significant safety issues in the nuclear fuel cycle. Some, such as waste disposal, may be moving to an early solution. Others such as plutonium dispersal hazards, may never be fully understood, since they involve issues that have been called trans-scientific (Weinberg 1972)--they can be stated in the language of science, but appear for practical purposes to be unanswerable by science.

Finally, it is now becoming clear that the regulation of nuclear safety is impeded by the large capital investments required. These investments go beyond the initial capital (which approaches $1 billion per plant). For example, the official investigative report on a fire in the plant at Browns Ferry, Alabama, called for improved fire prevention designs and noted that retrofitting would cost between $100 and $300 million per plant, with another $500 to $1300 million needed to buy coal for lost electric generating capability (Burnham 1976b). The Indian Point 1 plant on the Hudson River stands idle because of the costs involved in the NRC decision to require ECCS retrofitting. The safety problems involved in "grandfathering" (exempting from retrofitting) can be significant, but because they pose major cost implications, they have emerged as an element in the heated debate among experts.

Rancorous Conflict

Resolution of regulatory problems becomes doubly difficult in a polarized environment. Doubts about credibility and accusations are quick to surface when regulators, by force of circumstance, must obfuscate or risk exposing ignorance. Evidence of the escalating conflict over nuclear energy policy is particularly abundant in the scientific community. In 1975, the Ford Foundation funded a "blue-ribbon panel" to study nuclear energy in the United States. A prime consideration in choosing the panelists was lack of a strong previous position on the problem. A similar NAS study of nuclear risk ran into recruitment difficulties because of the lack of highly qualified "disinterested scholars." A leading journal recently rejected an article by nuclear critics because of its advocacy tone and later accepted one by a proponent of nuclear power, which provoked a stinging rebuttal by the rejected authors (Boffey 1976a). Meanwhile, both sides compete in the number of Nobel laureates and other scientists they can enlist (Boffey 1976b;Walske 1976).

Supporters of nuclear power tend to perceive its opponents as an undifferentiated mass, somewhat irrational and hysterical, committed to the destruction of a technology that is often the lifework of its supporters. The rancorous conflict promotes a "besieged camp" mentality. In the view of some proponents, new issues arise not because genuine new problems have been found, but because outstanding questions have been put to rest and the critics are forced to shift ground.

For critics of nuclear power, the enemy includes the regulators, industry representatives, and supporting scientists who combine in foisting an unsafe technology onto an unknowing and trusting public. Influenced by past cases of censorship and cover-up (Gillette 1972a-d), opponents take at face value no one who speaks in favor of nuclear energy but look immediately for hidden motivations.

The rancorous conflict that feeds on the inadequacies of the regulatory process in turn undermines this process. On strictly human terms, the U.S. regulatory official has a nearly impossible task. Thanks to the Freedom of Information Act, memoranda, letters, and reports are under continuing public scrutiny, and decisions must be made in a "goldfish bowl." The effect is to discourage candor, and when candor survives to blunt its positive impact. For example, when a regulatory task force reported critically on the performance of safety systems during the fire at the Browns Ferry plant (Burnham 1976b), instead of lauding the frankness and openness of the report, nuclear opponents such as Ralph Nader have used it as evidence of everything that is wrong with nuclear reactors and the regulatory process (Walske 1976).

Perhaps the most striking products of the rancorous conflict among experts are current voter referenda that attempt to force a decision despite an apparent lack of public information and understanding on the technical issues that warring factions of scientists and regulators have been unable to resolve (Atomic Industrial Forum 1976). Thus, the California initiative called for a public decison on the effectiveness of all safety systems, the adequacy of waste disposal and storage systems, and improved nuclear accident and liability insurance protection (California 1976).

Conclusion

Weinberg (1972) described the adoption of nuclear power in the following terms: "We nuclear people have made a Faustian bargain with society. On the one hand we offer--in the catalytic nuclear burner [breeder]--an inexhaustible source of energy. . . . But the price we demand of society for this magical energy source is both a vigilance and a longevity of our social institutions that we are quite unaccustomed to." We see the issue of nuclear safety as a Hydra, or many-headed monster—no sooner is one head severed than two others spring up to take its place, and the central head is immortal or nearly so.

Our immediate prognosis is for extension rather than diminution of the opposition to nuclear technology. Public opinion, which has consistently supported nuclear power, is nonetheless deeply divided, much as it was during the war in Vietnam. There is some evidence that wider public exposure to rancorous debate on nuclear power may well stiffen the opposition, as in the Swedish experiment in mass education (Grafström 1975) or in the persistence of opposition despite the initiative defeats.

Our own bias is to keep the nuclear option open, but to proceed cautiously; to press vigorously for solutions to immediate problems; but to forego at this time the implementation of plutonium recycle and the breeder. Time is needed to complete the risk assessment of the light water reactor fuel cycle, to validate experimentally computer codes that serve as substitutes for experience, to resolve such problems as spent fuel transport and waste disposal. Time is also needed to learn to live with or avoid trans-scientific issues such as plutonium toxicity, and intractable social risks such as sabotage, theft, and nuclear weapons proliferation. Finally, time is needed to evaluate long-term energy alternatives not described in Table 1, alternatives that may yet prove to have more favorable characteristics than currently available energy technologies.

Summary

Society seems content to strike a more moderate or uncertain balance with other technologies than with nuclear power. This attitude is traced to the social history of nuclear power, the genuine uncertainty and complexity of safety issues, underestimation of the regulatory task, and the rancorous nature of the debate. Nuclear power is not just another problem of technology, of environment, or of health. It is unique in our time. To be more demanding of nuclear safety may be to apply a double standard, but not necessarily an irrational one.

Our best course appears to be to keep the nuclear option open, work toward the rapid resolution of problems such as waste disposal, but postpone recycling and the breeder reactor. Time is needed to resolve immediate problems such as transport and disposal of nuclear wastes; to come to terms with trans-scientific issues such as plutonium toxicity, sabotage, and weapons proliferation; and to evaluate long-term energy alternatives.

EPILOGUE (JANUARY, 1985)

Nuclear power is still in trouble. More than seven years after the foregoing paper was written the **de facto** moratorium on nuclear power persists and will almost certainly continue to do so, at least for the near term, despite resuscitation efforts by the Reagan administration. A new chapter entitled "Three Mile Island" and an important addition to the chapter on nuclear proliferation have appeared in the social history of nuclear power. Genuine uncertainty over safety issues continues in their wake; an antiregulatory administration threatens to undermine the massive regulatory effort sparked by the accident at Three Mile Island (TMI); a searching review of the Reactor Safety Study found an underestimate of the error bars on accident probabilities and led the Nuclear Regulatory Commission to withdraw its endorsement of the study's estimate of the overall risk of reactor accidents; assumptions concerning the source term for radioactive releases in a major accident are under reconsideration (Payne 1985); emergency response has become a new source of contention, and the rancorous scientific debate has not diminished. Meanwhile, the public distrust of nuclear power appears to have diminished somewhat in intensity but not in substance. Added to these factors of distrust in nuclear power is a powerful economic reality, the deterioration of this energy source's market viability in the face of declining energy consumption and high interest rates.

Continuing scientific debate over nuclear power is quite evident in the response to the accident at Three Mile Island. Since nearly all the radioactivity was contained within the plant, proponents of nuclear power claimed that the accident sequence fell well within the predictions of existing risk assessments (specifically the Reactor Safety Study) and that the success in containing radioactivity illustrated the inherent resiliency of "defense-in-depth" principles of reactor safety assurance. Opponents, pointing to the interaction between technical and human error, see the accident as an indictment of the technlogy and the industry and a harbinger of greater accidents to come. Meanwhile, the Presidential commission appointed to investigate the accident concluded that "fundamental changes" in safety management were required if "risks are to be kept within tolerable limits" (U.S. President's Commission on the Accident at Three Mile Island 1979,7-8).

The many recommendations emerging from at least nine major post-Three Mile Island appraisals have produced an extensive effort to upgrade the safety of nuclear power plants. The most important changes include (1) greater regulatory attention to operating reactors and an increased willingness to order shutdowns and levy fines, (2) a more balanced approach to risk management, with increased attention to slowly developing accident events, to consequence mitigation and emergency response, and to human factors, (3) greatly increased use of probabilistic risk assessment in designing and operating plants, in licensing, and in regulating plants (Nuclear Regulatory Commission 1984), (4) a substantially increased industry effort to upgrade safety assurance at individual plants and to improve its own technical capabilities and training programs, particularly through the work of the Institute for Nuclear Power Operations (INPO), and (5) internal changes within the Nuclear Regulatory Commission aimed at enhancing its inspection and regulatory functions and emergency-response capabilities.

Despite general agreement that safety has been upgraded, the extent of improvement remains uncertain. The spate of new regulations mandated by the NRC, though clearly upgrading safety in some areas (for example, emergency response), may have resulted in a formalism in industry response and a diversion from more imaginative approaches to risk control. On the other hand, the extensive analysis of plants by probabilistic risk assessment (Nuclear Regulatory Commission 1984), intensive effort on improved training of operators, and increased security measures have probably served to lower (perhaps significantly) the risk of major reactor accidents. Open to question is the extent to which either industry or the regulators have achieved the fundamental changes in attitudes and "mindsets" that many post-accident assessments believed central to substantial improvement in safety performance.

Unsettling also is the prospect of future Three Mile Island accidents. If one accepts the recurrence probability of such accidents as falling between one in several hundred thousand to one in several thousand reactor years (as indicated by a probabilistic risk analysis of a particular plant), then one should expect such an accident every five to ten years (assuming 125-150 Gwe installed capacity) and every two to five years somewhere in the world. With current costs for the TMI cleanup projected at over $1 billion, it is increasingly evident that even near misses of major reactor accidents are socially unacceptable.

The past five years have also witnessed continued scientific dispute over the health effects of low level radiation, despite the fact that it is perhaps the best understood of all carcinogenic hazards. Consider, for example, that

> the majority of a scientific committee of the National Academy of Sciences accepted a linear-quadratic extrapolation as the best expression of predicted effects of low-level radiation, with dissenting reports arguing for the linear extrapolation on the one hand and the quadratic extrapolation on the other,

while

> a series of studies have emerged purporting to find higher occupational or therapeutic incidence of cancer than would be predicted by either the linear or the linear-quadratic extrapolation.

Meanwhile, the only significant additional data are those emerging from the new studies of neutron and gamma ray doses and cancer mortality at Hiroshima and Nagasaki, yet these results have been subjected to varying interpretation and appear only to have fueled the debate. The controversy highlights the continuing absence of data that could provide confirmation or rejection of the various interpretations and undoubtedly contributes to a public perception that science is really deeply divided.

Underlying the public distrust of nuclear power has been the continuing concern over waste disposal, an issue that now consistently tops the list of worries over the technology. The public response reflects, at least in part, the inadequate historical performance on radioactive waste disposal. Significant legislative progress, however, has been made on radioactive wastes, as embodied in the Low-Level Waste Policy Act of 1980 and the Nuclear Waste Policy Act of 1982. The extent to which the acts will be implemented in a timely manner remains to be seen; site selection remains a problem for both high-level and low-level wastes and the schedules demarcated for the former appear overly optimistic. Although the Reagan administration has lifted the indefinite deferral of reprocessing, subsidies to support private initiatives have not been forthcoming. Allied Chemical has written off its half share in the Barnwell reprocessing plant, and increasing public concern over nuclear weapons (as well as economic realities), and particularly the plentiful supply of uranium, may reverberate on the long-term prospects for reprocessing.

In regard to this last issue, a significant event for public concern was the Israeli strike in 1981, allegedly for "preemptive" objectives, on the experimental Iraqi reactor. Although subsequent discussion suggests that political considerations may have provided the motivation, the action mobilized the latent public fears over the links between nuclear weapons and nuclear power plants. Similarly, the statement by President Reagan that accumulating spent fuel may be used in the planned expansion of nuclear weapons provoked concern in industry no less than the public (Graham 1981).

As important as these safety issues is the deterioration of nuclear power in the marketplace. The worldwide decline in electric power demand and higher interest rates have resulted in no new domestic orders for nuclear power plants (while dozens of coal plants have been ordered) over the past six years and an overall cumulative cancellation or postponement of over 80 earlier nuclear plant orders. In 1977, our prognosis for nuclear power development was that "increasing capital costs would not be decisive in the long run." A combination of factors—high interest rates, substantial regulatory uncertainty, and saturation of electricity demand—interacted, however, to produce fewer energy plant orders and to deepen utility reluctance to invest in nuclear when plants are added. Given a steady decline over the past decade of projected energy use by the United States, this situation is unlikely to change quickly.

Thus, intentionally or not, time is available to deal with the Hydra-headed issues of nuclear power and to work to regain the public trust in this technology. The mid-1980s offer a clear fork in the road—whether to take advantage of an antiregulatory administration to press ahead in spite of the unresolved safety problems or to pause in order to put in place needed changes, to explore new (and perhaps inherently safer) reactor designs, and to strike the compromises required if nuclear power is again to prosper in the United States.

NOTES

1. This article appeared originally as "The Distrust of Nuclear Power," Science 196 (1 April 1977):25-34 (Copyright 1977 by the American Association for the Advancement of Science) and reflects the authors' views at that point in time. The epilogue reexamines the issue of public attitudes toward nuclear power in 1985.

2. A rem is a unit describing radiation dose in man. It measures the number of ion pairs per unit volume of tissue. A man-rem refers to the accumulation of 1 rem in a single human, or fractions of a rem in several humans such that total dose is 1 rem.

3. A reactor-year is a unit describing the amount of reactor operating experience. For example, 1 reactor-year's experience may be accumulated by one reactor's operating for one year or 12 reactors operating for one month.

4. The surveys involved 100 person-in-the-street interviews in Boston, London (England), and Toronto in 1975, between 200 and 250 similar interviews in each of the three cities in 1976, and 100 lenghthier residential interviews (by random cluster sample) at two reactor sites each in Britain and Canada and at three reactor sites in the United States.

5. R. Wilson, among others, has argued that transportation of liquefied natural gas in ships represents threats comparable to those of nuclear power (see, for example, Wilson 1973). We agree in part, but we would argue that long-term effects of radioactivity set nuclear power apart.

6. The existence of delayed deaths in significant numbers was first recognized by the American Physical Society (1975). Delayed death estimates not very different from these were subsequently incorporated in the final version of the RSS (Nuclear Regulatory Commission 1975). The issue of delayed deaths was not recognized in the draft version (AEC 1974b).

7. For interesting discussion in this connection, the authors are indebted to Jan Beyea.

REFERENCES

AEC (Atomic Energy Commission). 1957. Theoretical possibilities and consequences of major accidents in large nuclear power plants. WASH-740. Washington: AEC.

AEC (Atomic Energy Commission). 1971. Division of Operational Safety. Operational accidents and radiation exposure experience within the Atomic Energy Commission. WASH-1192. Washington: AEC.

AEC (Atomic Energy Commission). 1972. Directorate of Regulatory Standards. Environmental survey of transportation of radioactive materials to and from nuclear power plants. WASH-1238. Washington: AEC.

AEC (Atomic Energy Commission). 1973. The safety of nuclear power reactors (light water cooled) and related facilities. WASH-1250. Washington: AEC.

AEC (Atomic Energy Commission). 1974a. Comparative risk-cost benefit study of alternative sources of electrical energy. USAEC Report WASH-1224. Washington: AEC.

AEC (Atomic Energy Commission). 1974b. Reactor safety study: Draft. WASH-1400. Washington: AEC.

American Physical Society. 1975. Report to the American Physical Society by the study group on reactor safety. Reviews of Modern Physics 47, Supplement no. 1 (Summer).

Atomic Industrial Forum. 1976. Info. no. 94. Washington: Atomic Industrial Forum.

Bair, W.J., C.R. Richmond, and B.W. Wachholz. 1974. A radiobiological assessment of the spatial distribution of radiation dose from inhaled plutonium. WASH-1320. Washington: Atomic Energy Commission.

Bair, W.J., and R.C. Thompson. 1974. Plutonium: Biomedical research. Science 183:715-722.

Baker, P.N. 1958. The arms race. New York: Oceana.

Bennett, Burton G. 1974. Fallout ^{238}Pu dose to man. In Fallout program quarterly summary report. HASL-278. Washington: Health and Safety Laboratory, Atomic Energy Commission.

Boffey, Philip M. 1969. Ernest J. Sternglass: Controversial prophet of doom. Science 166:195-200.

Boffey, Philip M. 1976a. Nuclear foes fault Scientific American's editorial judgment in publishing recent article by Nobel Laureate Hans Bethe. Science 191:1248-1249.

Boffey, Philip M. 1976b. Nuclear power debate: Signing of the pros and cons. Science 192:120-122.

Burnham, David. 1976a. Three engineers quit GE reactor division and volunteer in antinuclear movement. New York Times, 3 February, p. 12.

Burnham, David. 1976b. Inquiry on fire at biggest nuclear plant finds prevention program was essentially zero. New York Times, 29 February, p. 27.

Business Week. 1975. Why atomic power dims today. No. 2407 (17 November):98-106.

California. 1976. Office of the Secretary of State. California voters' pamphlet for the June 8, 1976 primary election. Sacramento, Calif.: Office of the Secretary of State.

Cohen, Bernard L. 1975. The hazards in plutonium dispersal. Pittsburgh: Department of Physics, University of Pittsburgh.

Cohen, Bernard L. 1976. Environmental hazards in radioactive waste disposal. Physics Today 29 no. 1 (January):9-15.

Colby, L.J., Jr. 1976. Fuel reprocessing in the United States: A review of problems and some solutions. Nuclear News 19 no. 1 (January):68-73.

Cottrell, W.B. 1974. The ECCS rule-making hearing. Nuclear Safety 15 no. 1:30-53.

Davidon, William C., Marvin I. Kalkstein, and Christoph Hohenemser. 1960. The nth country problem and arms control. Washington: National Planning Association.

Day, M.C. 1975. Nuclear energy: Second round of questions. Bulletin of the Atomic Scientists 31 no. 10 (December):52-59.

Feld, Bernard T. 1975. Making the world safe for plutonium. Bulletin of the Atomic Scientists 31 no. 5 (May):5-6.

Forbes, Ian A., Daniel F. Ford, Henry W. Kendall, and James J. MacKenzie. 1972. Cooling water. Environment 14 no. 1 (January/February):40-47.

Ford, Daniel F. and Henry W. Kendall. 1972. Nuclear safety. Environment 14 no. 7 (September):2-9.

Friends of the Earth. 1976. Energy and nuclear policy. London: Friends of the Earth.

Geesaman, D.P. 1968. An analysis of the carcinogenic risk from an insoluble alpha-emitting aerosal deposited in deep respiratory tissue. UCRL-50387 and UCRL-50387 Addendum. Washington: Atomic Energy Committee.

Gillette, Robert. 1972a. Nuclear safety (I): The roots of dissent. Science 177:771-775.

Gillette, Robert. 1972b. Nuclear safety (II): The years of delay. Science 177:867-870.

Gillette, Robert. 1972c. Nuclear safety (III): Critics charge conflict of interest. Science 177:970-975.

Gillette, Robert. 1972d. Nuclear safety (IV): Barriers to communication. Science 177:1030-1032.

Gofman, John W. 1975a. The cancer hazard from inhaled plutonium. Yachats, Oregon: Committee for Nuclear Responsibility.

Gofman, John W. 1975b. Estimated production of human cancers by plutonium from world-wide fallout. Yachats, Oregon: Committee for Nuclear Responsibility.

Gofman, John W. 1981. Radiation and human health. San Francisco: Sierra Club Books.

Grafström, E. 1975. Speech at the Workshop on Alternative Energy Strategies, Stockholm, Sweden, June.

Graham, John. 1981. Reagan administration proposes nuclear policy changes. Nuclear News 24 no. 14 (November):30, 32.

Hohenemser, Christoph. 1962. The nth country problem today. In Disarmament, its politics and economics, ed. S. Melman, 238-276. Boston: American Academy of Arts and Sciences.

Hohenemser, Christoph, and Milton Leitenberg. 1967. The nuclear test ban negotiations: 1957-1967. Scientist and Citizen 9:197.

Hohenemser, Kurt H. 1975. The failsafe risk. Environment 17 (January/February):6-10.

Holsti, Ole R. 1967. Cognitive dynamics and images of the enemy. In Image and reality in world politics, ed. John C. Farrell and Asa P. Smith, 16-39. New York: Columbia University Press.

Kendall, Henry W., et al. 1974. Preliminary review of the AEC reactor safety study. Cambridge, MA: Union of Concerned Scientists.

Kouts, Herbert J.C. 1975a. Testimony. In U.S. Congress, House Committee on Interior and Insular Affairs, Subcommittee on Energy and the Environment. Oversight hearings on nuclear energy: Overview of the major issues, 28 April to May 2. 94th Congress, 1st Session. Washington: Government Printing Office.

Kouts, Herbert J.C. 1975b. The future of reactor safety research. Bulletin of the Atomic Scientists 31 no. 7 (September):32-37.

Lapp, Ralph. 1958. The voyage of the lucky dragon. New York: Harper.

Lapp, Ralph. 1968. The weapons culture. New York: Norton.

Lieberman, M.A. 1976. United States uranium resources: An analysis of historical data. Science 192:431-436.

Lifton, Robert J. 1967. Death in life: Survivors of Hiroshima. New York: Random House.

Louis Harris and Associates. 1975. A survey of public leadership attitudes toward nuclear power development in the United States. New York: Ebasco Services Inc.

Martell, Edward A. 1975. Tobacco radioactivity and cancer in smokers. American Scientist 63:404-412.

Mazur, Allan. 1975. Opposition to technological innovation. Minerva 13:58-81.

Metzger, H. Peter. 1972. The atomic establishment. New York: Simon and Schuster.

Morgan, Karl Z. 1975. Suggested reduciton of permissible exposure to plutonium and other transuranium elements. American Industrial Hygiene Association Journal 36 no. 8 (August):567-575.

Mullenbach, Philip. 1963. Civilian nuclear power. New York: Twentieth Century Fund.

Muller, Hermann J. 1955. Radiation and human mutation. Scientific American 193 no. 11 (November):58-65.

Mulvihill, R.J., D.R. Arnold, C.E. Bloomquist, and B. Epstein. 1965. Analysis of United States power reactor accident probability. PRC R-695. Los Angeles: Planning Research Corporation. This is an unpublished draft, from the file "WASH 740 update," Public Documents Room, Nuclear Regulatory Commission, Washington.

Murray, Thomas E. 1957. Reliance on H-bomb and its dangers. Life 42 (6 May): 181-182+.

National Research Council. 1972. Advisory Committee on Biological Effects of Ionizing Radiation. The effects on populations of exposure to low levels of ionizing radiation. Washington: National Academy of Sciences.

National Research Council. 1975. Commission on Natural Resources. Air quality and stationary source emission control: A report of the National Academy of Sciences, National Academy of Engineering and National Research Council to the Committee on Public Works, U. S. Senate 94th Congress, 1st Session. Washington: Government Printing Office.

National Safety Council. 1973. Accident facts 1973. Chicago: National Safety Council.

Nuclear Regulatory Commission. 1975. Reactor safety study. WASH-1400, NUREG-75/014. Washington: Nuclear Regulatory Commission.

Nuclear Regulatory Commission. 1984. Office of Nuclear Regulatory Research. Probabilistic risk assessment (PRA) reference document. NUREG-1050. Washington: Division of Risk Analysis and Operations, Office of Nuclear Regulatory Research, Nuclear Regulatory Commission, September.

Pahner, Philip D. 1975. The psychological displacement of anxiety: application to nuclear energy. In Risk-benefit methodology and application, ed. David Okrent, 557-578. UCLA-ENG 7598. Los Angeles: University of California.

Payne, Jon. 1985. The source term: Phase two. Nuclear News 28 no. 1 (January):29.

Primack, Joel, and Frank von Hippel. 1974. Advice and dissent: Scientists in the public arena. New York: Basic Books.

Rabinowitch, Eugene L., ed. 1950. Minutes to midnight. Chicago: Educational Foundation for Nuclear Science.

Rose, David J. 1974. Nuclear eclectic power. Science 184:351-359.

Rosenberg, Milton J. 1966. Images in relation to the policy process. In International behavior, ed. Herbert C. Kelman, 278-334. New York: Holt, Rinehart and Winston.

Ross, M. 1975. The possibility of release of cesium in a spent fuel transportation accident. Ann Arbor: Department of Physics, University of Michigan.

SIPRI (Stockholm International Peace Research Institute). 1976. Yearbook of world armaments and disarmament: 1976. Stockholm: SIPRI.

Tamplin, Arthur R. 1971. Issues in the radiation controversy. Bulletin of the Atomic Scientists 27 no. 7 (September):25-27.

Tamplin, Arthur R., and Thomas B. Cochran. 1974. A report on the inadequacy of existing radiation protection standards related to internal exposure of man to insoluble particles of plutonium and other alpha-emiting hot particles. Washington: Natural Resources Defense Council.

UNSCEAR (United Nations Scientific Committee on the Effects of Atomic Radiation). 1972. Ionizing radiation: Levels and effects. Report to the General Assembly, United Nations, New York.

U.S. President's Commission on the Accident at Three Mile Island. 1979. The need for change: The legacy of TMI. Washington: Government Printing Office.

Walske, Carl. 1976. Letter. New York Times 15 March, p. 30.

Weatherwax, Robert K. 1975. Virtues and limitations of risk analysis. Bulletin of the Atomic Scientists 31 no. 7 (September):29-32.

Weinberg, Alvin M. 1972. Social institutions and nuclear energy. Science 177:27-34.

Willrich, Mason, ed. 1973. International safeguards and the nuclear industry. Baltimore: The Johns Hopkins Press.

Willrich, Mason, and Theodore B. Taylor. 1974. Nuclear theft: Risks and safeguards. Cambridge, MA: Ballinger.

Wilson, Richard. 1973. Natural gas is a beautiful thing. Bulletin of the Atomic Scientists 29, 7 (September):35-40.

Wilson, Richard, and W.J. Jones. 1974. Energy, ecology and the environment. New York: Academic Press.

Zimmerman, Charles F. 1975. Accidents in the nuclear industry. Ithaca, New York: Department of Agricultural Economics, Cornell University.

11
Hazard Assessment: Art, Science, and Ideology[1]

Robert W. Kates

Human beings appear to become increasingly adept at creating, discovering, or rediscovering threats to themselves and to their environment. A new professional interest, hazard assessment, has developed in assessing these threats. Hazard assessors are becoming more numerous and their products in the form of risk assessments, risk/benefit analyses, environmental impact statements, and technology assessments are widely diffused.

The task is not one for specialists alone; people have always assessed environmental threat: storm, drought, fire, or disease. But for the new and newly discovered hazards, there is strong perception of risk but little experience with consequences. With such uncertainty it is not surprising that hazard-assessment practice is still more art than science and that distinctive, contrasting ideologies flourish.

Hazard Assessment Methods

Hazard assessment is the prime component of the intelligence function of hazard management (chapter 3). For descriptive convenience, Figure 1 separates the overall process into three overlapping elements, but it is important to recognize that in practice the distinctions are blurred. **Hazard identification** is the recognition of a hazard, the answer to the question: what constitutes a threat? Its methods are the methods of research and of screening, monitoring, and diagnosis. **Risk estimation** is the measurement of the threat potential of the hazard, an answer to the questions: how great are the consequences, how often do they occur? Its methods are methods of knowing: revelation, intuition, and extrapolation from experience. **Social evaluation** is the meaning of the measurement of threat potential, an answer to the question: how important is the estimated risk? Its methods are methods of comparison: aversion, balance, and cost/benefit analysis.

Hazard Identification

For much of human history, the identification of environmental hazards arose from the direct human experience of harmful events and consequences or from the application of ritual or magic. Technological hazards too often manifest themselves experientially as

251

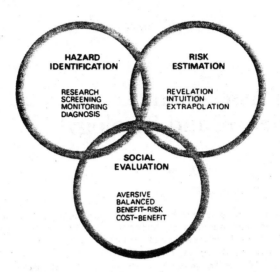

Figure 1. Elements of hazard assessment. Risk assessment is composed of three overlapping elements. Hazard identification is the recognition of risk; estimation is the measurement of the threat potential of the risk, and social evaluation, the appraisal of the meaning and importance of the risk.

surprises, chance discoveries, outbreaks—often disastrous. Thus Minamata disease is a graphic reminder of the hazards of mercury (chapter 9) and toxic shock syndrome dramatizes the hazards of tampons—thereby alerting society to new and rediscovered threats. But beyond such painful experiences, the identification of new and newly discovered hazards will continue to rely for the most part on science.

Basic research or "pure science" is not directed toward hazard assessment; it deals with knowledge for its own sake. Nevertheless, fundamental scientific inquiry discovers threats, albeit somewhat randomly, and provides the basis for directing and interpreting more purposeful search. "Critical" science[2] engages in a directed, intensive search for environmental or technological hazard as part of its effort to redress the perceived imbalance between technology and the human environment. But the institutionalized task of hazard identification falls to "practical" or "applied" science, employing screening, monitoring, and diagnosis.

In **screening**, a standardized procedure is applied to classify products, processes, phenomena, or persons for their hazard potential (see, for example, Committee 17 1975), whereas **monitoring** (in health studies, surveillance) observes, records, and analyzes the same for the recurrence of hazardous events or their consequences (SCOPE 1971;Munn 1973). In **diagnosis**, the identification of hazards takes place through analysis of indicators or symptoms of consequences (World Health Organization 1972, 1976; Engle and Davis 1963). Each of these methods has distinctive historical origins and preferred usage in certain disciplines and professions, and only recently, in the context of such activities as Earthwatch (Jensen, Brown, and Mirabito 1975) is there emerging some searching comparison of these methodologies.

Implicit in these methods of hazard identification is a sequence in the suspectability of hazard potential. Screening procedures are akin to "fishing expeditions." Monitoring implies knowledge of threat potential, where the purpose of monitoring is to measure variation in some critical indicator, the cumulation of a hazardous condition, or the failure of a protective device. Diagnosis implies the ready existence of hazard-indicative "symptoms," some abnormal set of events or consequences—the location, etiology, or treatment of which are in doubt. Any complex socioenvironmental problem may call upon all the methods of hazard identification.

A current example is the recent and growing discovery of threats to the atmospheric ozone column that serves to protect us from ultraviolet radiation and resultant skin cancer. The basic chemistry of ozone formation and its observed concentration in the stratosphere dates back to Chapman's work in 1930. Stratospheric ozone, a molecule of oxygen that contains three atoms rather than the usual two, is formed when ultraviolet solar radiation dissociates O_2 into two atoms, freeing each to recombine with other O_2 molecules to form O_3. But only in the context of the United States debate over development of supersonic transport (SST) did the hazard potential emerge. James McDonald (1971), an atmospheric scientist with an interest in public policy issues, connected the distribution of skin cancer with latitudinal variations in ozone concentration. His favored mechanism for ozone destruction was water vapor injection from the SST. In this he erred, overestimating the effect. Crutzen (1970) and Johnston (1971), drawing on their basic research, proposed NO_x as the major catalyst of ozone destruction. An applied governmental research program, the Climate Impact Assessment Program, validated most of these early hypotheses at a cost of US $20 million (National Research Council 1975).

Once the potential for ozone destruction was recognized, basic knowledge of chemical reactions suggested other catalytic agents. Continuing laboratory experiments revised rate constants for many key reactions. Among the currently recognized agents of ozone destruction are atomic warfare, the space shuttle, nitrogen fertilizer, and chlorofluorocarbons in aerosol cans, solvents, and refrigeration. New efforts monitor ozone, its destructive catalysts, and incidence of skin cancer. Modeling efforts have led to a set of repeated, systematic risk assessments (National Research Council 1975, 1976a, 1976b, 1979a, 1979b, 1982;Great Britain 1976, 1979).

The case of ozone illustrates that all attempts at hazard identification pose problems of reliability (serious hazards do not get identified); of cost (of collecting large amounts of expensive data little used or of little use); and of bias (the data are misleading in some consistent way). The most serious problem, however, is the proliferation of unknown hazards. It seems unlikely that random research thrusts, underfinanced critical science, or massive screening, monitoring, and diagnostic methods can keep pace with the creation of environmental threat. In this context it is sobering to note that just a few years ago, atmospheric scientists had proposed monitoring a commercial chlorofluorocarbon, not because of its effect on ozone but because it was deemed an inert, nonreactive tracer!

Risk Estimation

Revelation, through divine or supernatural inspiration, of the likelihood of threatening events and their consequences is as old as the sacred prophetic religious experiences or as common as the astrology column of the newspaper (Jahoda 1971). Its value clearly depends on the degree of belief and number of believers. Intuition shares some qualities with revelation, but it is internally generated and is employed in both science and everyday experience (Westcott 1968).

Scientific risk estimation for the most part rests on extrapolation: forward into the future from experience; backwards from possible future unknown but imagined events to their known precursors; or sideways by analog and transfer of parallel experience from different but similar places, situations, or things. A great deal of ingenuity has gone into refining methods of extrapolation: extending the underlying data base, clarifying the meaning of probability (Savage 1954), developing more precise and powerful mathematical methods (Green and Bourne 1972), creating tree-like logical sequences of events and consequences (Nuclear Regulatory Commission 1975), modeling systems, quantifying subjective estimates (Keeney and Raiffa 1976;Selvidge 1973), and stretching imagination by scenarios (Erickson 1975).

The ingenuity and limitations of such extrapolative methods are well exemplified in the Reactor Safety Study (Nuclear Regulatory Commission 1975), discussed in detail in chapter 10. A landmark in the art of risk estimation, the study has undergone wide emulation, review, and criticism (American Physical Society 1975; Risk Assessment Review Group 1978). It may command a degree of belief in the hypothetical (Häfele 1974), because, the financial toll of Three Mile Island notwithstanding, more than 300 years of commercial reactor experience have witnessed no catastrophic accidents. Thus the study (and its users) must rely on varied substitutes for experience: logical analysis, understanding of physical laws, frequencies of component failures, and radiation dose-response curves derived from studies of animals and war victims. Using these in combination, it is possible to estimate the risk spectrum for catastrophic events as shown in Figure 2. The very complexity of the process of risk estimation weakens its credibility. And even for those who are not skeptical of its methods, the Reactor Safety Study offers very differing interpretations (see also chapter 10).

All such extrapolative methods, then, are hampered by common and sometimes subtle distortions of assumptions and method[3] and by the limits of human cognitive processes (Tversky and Kahneman 1974; Slovic, Fischhoff, and Lichtenstein 1976,165). But most difficult is the "prison of experience"--humans are at risk from threats greater than or different from individual and collective experience (Kates 1976,133). And extrapolative methods, no matter how ingenious, can only enlarge but not escape such containment. Indeed, the causal structure, as developed in chapter 2, underscores the uncertainty inherent in all risk estimates.

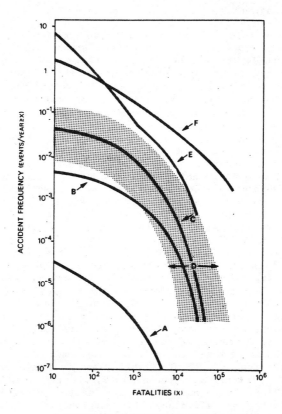

A REACTOR SAFETY STUDY (100 NUCLEAR POWER PLANTS,
    PROMPT DEATH)

B LATENT DEATHS FROM CANCER (100 NUCLEAR POWER PLANTS,
    ALL DEATHS OVER 30 YEARS)

C EXTRAPOLATION FOR 1000 NUCLEAR POWER PLANTS
    BY YEAR 2000 (ALL DEATHS)

D UNCERTAINTY LIMITS

E TOTAL MAN CAUSED HAZARD EVENTS

F TOTAL NATURAL HAZARD EVENTS

Figure 2. Risk of death from U.S. commercial nuclear power reactors. The Risk Spectrum is a graph that relates the frequency and
magnitude of a catastrophic nuclear accident. Spectrum A is the one
displayed in the Executive Summary of the Reactor Safety Study, and
it applies to prompt fatalities only.   In this widely reproduced
graph, the risk of a nuclear power plant is many orders of magnitude
below comparable risks of manmade and natural events.   Using the
data provided in the Reactor Safety Study, it is possible to reduce
this apparent margin of safety by: 1) adding in the latent deaths
that will occur from radiation-induced cancer over a 30-year period
(Spectrum B); 2) extrapolating to 1000 nuclear reactors by the year
2000, a target of U.S. energy policy (Spectrum C); and 3) adding to
that extrapolation the uncertainty limits of the Reactor Safety
Study (Spectrum D).   Extending the spectrum in this manner is a
matter of judgment or bias, not of factual disagreement.

Social Evaluation

In the aversion of hazard or of risk itself, little or no consideration is given to comparison with other risks and benefits. Aversive methods lie embedded in culture as taboos, in society as absolute standards or regulations, and in individuals as avoidance preferences. Aversion as taboo may be considered "primitive," whereas as a regulatory standard (e.g., zero tolerance for carcinogens in food) it may be considered modern and indeed scientific (Douglas 1978). In contrast to the absolutes and imperatives of aversion, balanced risk methods--described in chapter 3 and discussed in chapter 12--seek to compare and equalize consequences. Comparisons of specific hazards with natural background levels (National Research Council 1980,66-67) and with other hazards prevalent in society (Cohen and Lee 1979;Wilson 1979) serve to encourage or inform some action or to reveal some inconsistency. Some studies compare risks in terms of the cost-effectiveness of control (Sinclair, Marstrand, and Newick 1972;U.S. Department of Transportation 1976;Siddall 1981). Other studies compare risks and benefits (Crouch and Wilson 1982) as in risk/benefit analyses or in some overall cost/benefit analysis. Again, much ingenious effort has gone to improving the data base for comparisons (Rowe 1977), to seeking revealed societal preferences for acceptable levels of risk (Starr 1972,17;Otway and Cohen 1975), to illuminating inconsistencies between different accepted risks and between different communities, cultures, and nations (Roschin and Timofeevskaya 1975;Winell 1975;Whyte and Burton 1980;Derr et al. 1981;Douglas and Wildavsky 1982), to comparing benefits and costs that have multiple attributes (Gardiner and Edwards 1975;Gros 1975), and to improving the making of judgments (Hammond and Brehmer 1969;Pill 1971;Howard, Matheson, and North 1971). These comparisons are limited by the data base but more importantly by differences in distributions of costs, risks, and benefits (see chapter 7).

Immediate benefits need to be compared to uncertain, amorphous, or long-term costs, or widely diffused benefits need to be compared to risks that fall heavily on a specific population or place. And hazards with low probabilities of occurrence but catastrophic consequences need to be compared to hazards of higher probability but less serious consequences.

Thus for the ozone hazard cited above, the social utility of the Concorde and of future SSTs, the convenience of aerosol sprays, and perhaps even the production of food (using nitrogen fertilizer) will have to be weighed against uncertain estimates of increase in skin cancer. And the risks of coal-produced electricity--the exacerbation of respiratory disease and increases in premature deaths for the exposed public, black-lung disease and accidents among miners--need to be compared to nuclear hazards of rare occurrence and latent effect. Consensually accepted methods for making such comparisons are not available.

Risk Assessment Ideology

The perception of hazard is strong, the facts of risk are ambiguous, the methods of analysis are limited and still evolving. It is not surprising, then, that hope, fear, and faith enter the risk-assessment process as overriding views or assumptions that in archetypal expression border on ideology. Each view assumes a fundamental imbalance between prevailing risk assessments and their hazard potential. Each begins with the implicit assumption that the true hazard potential is greater than, less than, or different from, the prevailing risk assessment.

Tip of the Iceberg

For some risk assessors, the hazard is almost always greater than the risks assessed. Since for them the consequences of technology are too recent to be apparent, they assess only the tip of the iceberg:

> The roll of casualties of our time is incomplete. Among those numbered in hundreds every year we have counted invalid survivors of spina bifida, patients accidentally injured during cardiac catheterization, and those disabled by reactions to such drugs as chloramphenicol. Rising casualties numbering thousands annually result from the health environment surrounding certain infants born in our cities, from the vulnerability of young people to head injuries, drug addiction, and crime, and from chronic lung disease associated with air pollution. Increasing numbers, in the tens of thousands every year, suffer or die from arteriosclerotic heart disease or are disabled by the frailties of age. Other casualties may be on the way: additional victims of environmental pollution, more infants surviving with genetic defects, more casualties of affluence, made useless by automation or retirement from boring work, more artificially supported survivors, and more casualties of new drugs. Though these numbers may in a sense be outweighed by a rising standard of living, better education, less work, and less discomfort, they are surely enough to cause concern. (Ford 1970,262)

For these tip-of-the-iceberg assessors, by the time the roll of casualties is complete, it is already too late; such are the latent effects of carcinogens or mutagens.

The basic methods of tip-of-the-iceberg assessors complement their concern. They search for new hazards, try to estimate consequences, particularly from maximum events, and attempt to predict long-term effects. At the same time they avoid estimating the probability of events' leading to harmful consequences, arguing that in the absence of adequate experience these will tend to be underestimates. They favor the use of the scenario that stretches the imagination, renders the incredible more credible, and suggests the greater hazard lurking beneath seas of complacency.

Count the Bodies

For some risk assessors, the hazard is almost always less than the risks assessed. Because of scientific and technical advance, administrative oversight, and the long-term increase in societal ability to cope with threat, people are demonstrably better protected. If the environment appears less secure to many, it is because of changes in social expectations, certain processes of communication, and recurrent waves of public fad or mood.

Social values and expectations of security change, becoming more demanding over time, as evidenced in movements for consumer, environmental, and occupational safety. The dramatic increase in communication makes for exaggerated assessments. Improved reporting of events previously unreported creates an illusion of their increase and of global threat for what may be highly localized problems. And these trends may overlap with secular or cyclical changes in attitudes. Recurrent waves of pessimism are thought to alternate with periods of optimism, especially among intellectuals and elites. The populace, and especially youth, is currently seen as suspicious of authority, hostile to science, and attracted by irrationality. The public is viewed as ill-informed, depersonalized, and frustrated by the bigness, complexity, and remoteness of phenomena that have an impact on its life.

These assessors see themselves as struggling for fact, caution, and rationality to "count the bodies," not the speculations. Thus they tend to limit themselves to short-run consequences, arguing that these are reasonably knowable. In estimation, they favor quantifying the likelihood (usually small) of events and making comparisons with the likelihood (usually higher) of everyday hazards that are seemingly acceptable to society. Their favored method is quantification by reduction, extrapolating from unknown to known events. This fault-tree and event-tree methodology emphasizes the contingent nature of catastrophic hazard and its ensuing low probability.

Worry Beads

Finally, for some risk assessors the major hazards are different from those for which risks have been assessed. They accept the insights of those who assert that the visible risks assessed are but the tip of the iceberg as well as those of the skeptical statistician, technologist, or social commentator who knows that hindsight will show that many perceived risks have been exaggerated. Their concern is that the societal ability to assess risk is limited, expandable but not infinite, and in danger of being squandered on the unimportant while failing to identify the truly perilous.

Proponents of the "worry-bead" hypothesis argue that individuals and societies have a small, relatively fixed stock of worry beads to dispense on the myriad threats of the world. People are not irrational, but they are constrained in their rationality either by human limitations of cognition and judgment; by cultural, ideological, or personal aversions toward certain risks and the discounting of others; by ignorance, misunderstanding, or limited experience; or by the sheer number and complexity of threats confronting them. Societal capacity to worry intelligently exceeds that of individuals, thus it is possible to divide the labor and the anxiety. But even this expanded capacity, in this view, is less than the threats perceived, and to both individuals and societies, where and when to rub one's worry beads is baffling and difficult to rationalize even if desired.

Thus worry-bead assessors strive first to improve overall strategies of hazard identification. In examining evaluation methods, they study empirically the societal response to threat to determine "what is," not simply "what ought to be." Their favored methods are those designed for improving and making easier decision and choice and for allocating the appropriate institutional mechanisms and group processes to the "right" type of hazard.

Ideology in Nuclear Risk Assessment

All of these ideologies of risk assessment surface in the rancorous debate over nuclear power described in chapter 10. "Tip-of-the-iceberg" assessors readily accepted studies of maximum consequences (AEC 1975) even as they discounted studies of their probability (Nuclear Regulatory Commission 1975). Just the reverse seemed to characterize the "count-the-bodies" school, epitomized in the Reactor Safety Study, where the executive summary overdramatized the low probability of major accidents and minimized their consequences. Similarly, the laying to rest by the count-the-bodies school of one risk issue in the nuclear debate, such as environmental leakage under normal operations or functioning of the emergency core cooling system, only serves to encourage the tip-ofthe-iceberg assessors to identify new and troubling issues—such as containment durability, human error, weapons proliferation, radioactive waste, and evacuation plans.

Our own view is that of the "worry beads" assessors. This leads us to consider the catastrophic potential of the entire fuel cycle (as in Table 1, chapter 10), to recognize the validity of both the "double standard" claim of the count-the-bodies assessors and the "understudied and poorly understood" claims of the tip-of-the-iceberg assessors for some areas of the nuclear fuel cycle, and to try to understand the basis for the social distrust of nuclear power (chapter 10). Our fears that the acrimonious debate thwarts needed progress in dealing with perhaps the greater hazard of long-term energy needs lead us and others (Bupp and Derian 1978) to propose a compromise solution that combines limits on nuclear expansion with attention to the most pressing safety and waste issues, while keeping the nuclear power option open (Kasperson et al. 1979).

Living with Ideology

In individual risk assessors, these archetypes of hazard assessment ideology are clearly overdrawn and individuals display a mix of attitudes. Yet the typology can be readily applied, as has been done in the case of nuclear power above. As representative approaches, the archetypes are not easily displaced. Such is the nature of the environmental hazard problem.

The review by Lawless of 45 major public alarms over technology found that in over a fourth of the cases, the threat was not as great as originally described by opponents of the technology, but in over half of the cases, the threat was probably greater than that admitted by the proponents of the technology and the problem was allowed to grow. Early warning signs were present and mostly ignored in 40 percent of the cases, and existing technology assessments (which usually include a risk assessment) were judged by the study team as surely helpful in only about 40 percent of the cases (Lawless 1977).

In the classification of 93 hazards, from which we developed the taxonomy described in chapter 4, the scientific literature serves to define, within a single order of magnitude, the major characteristics of each hazard. These order-of-magnitude differences sufficed to differentiate across the set among hazard risks that differed by up to a million or more. Hence it was possible to bridge differences between conflicting assessments, the ranges of which were much smaller than those between hazards. At the same time, the analysis based on the causal structure of hazards (chapter 2) affirms the transscientific nature of much of hazard assessment. Real limits hamper our ever knowing the certain "true risk"--because we lack the theoretical understanding as to cause (e.g., cancer), because we are unable to conduct experiments (e.g., ethics of human experimentation), or because we cannot achieve consensus on how to weight the attendant value issues (e.g., equity vs. efficiency).

Thus, as the theory and methodology of hazard assessment continue to evolve and improve as they have over the past decade (chapter 1), there is hope for greater scientific consensus on what is known about the hazards assessed, what needs to be known and how to learn it, and what the limits of knowing are. But it is highly improbable that even improved procedures of hazard identification, risk estimation, and social evaluation can cope with the proliferation of threats. The burden of hazard needs reduction, not because many serious risks cannot be assessed and coped with, but because all of them cannot be.

ACKNOWLEDGMENTS

The work on which this paper is based has been supported by a United Nations Environment Programme (UNEP) grant to the International Council of Scientific Unions, Scientific Committee on Problems of the Environment (ICSU/SCOPE), Mid-Term Project No. 7, "Communication of Environmental Information and Societal Assessment and Response," and by the Electric Power Research Institute for the Workshop on Comparative Risk Assessment of Environmental Hazards in an International Context, held in Woods Hole, Massachusetts, from March 31 to April 4, 1975. In its preparation I have been greatly assisted by Mimi Berberian of Clark University.

NOTES

1. This chapter is an updated and revised version of "Assessing the Assessors: The Art and Ideology of Risk Assessment," Ambio 6 no. 5 (1977):247-252. Reproduced by permission.

2. J. Ravetz (1971) describes the emergence of "critical" science, which seems preferable to the somewhat self-righteous "public-interest" variety.

3. It is not unusual to find in complex risk assessments that very diverse data, expressed originally in different measurement

scales, extrapolated by all three methods, with assumptions of process both deterministic and random and relationships both contingent and dependent, are then combined together in a single number or value.

REFERENCES

AEC (Atomic Energy Commission). 1957. Theoretical possibilities and consequences of major accidents in large nuclear power plants. WASH-740. Washington: AEC.

American Physical Society. 1975. Report to the American Physical Society by the study group on reactor safety. Reviews of Modern Physics 47, Supplement no. 1 (Summer).

Bupp, Irvin C., and Jean-Claude Derian, 1978. Light water: How the nuclear dream dissolved. New York: Basic Books.

Chapman, S. 1930. A theory of upper-atmospheric ozone. Memoirs of the Royal Meteorological Society 3(26):103-125.

Cohen, Bernard, and I-Sing Lee, 1979. A catalog of risks. Health Physics 36 (June):707-722.

Committee 17. 1975. Environmental mutagenic hazards. Science 187:503-514.

Crouch, Edward A.C., and Richard Wilson. 1982. Risk/benefit analysis. Cambridge, Mass.: Ballinger.

Crutzen, P.J. 1970. The influence of nitrogen oxides on the atmospheric ozone content. Royal Meteorological Society Journal 96:320-325.

Derr, Patrick, Robert Goble, Roger E. Kasperson, and Robert W. Kates. 1981. Environment 23 no. 7 (September):6-15,31-36.

Douglas, Mary. 1978. Purity and danger. London: Routledge and Kegan Paul.

Douglas, Mary, and Aaron Wildavsky. 1982. Risk and culture: An essay on the selection of technological and environmental dangers. Berkeley and Los Angeles: University of California Press.

Engle, R.L., and B.J. Davis. 1963. Medical diagnosis: Present, past and future. Archives of Internal Medicine 112:512-543.

Erickson, N.J. 1975. Scenario methodology in natural hazards research. Boulder, Colorado: Institute of Behavioral Science, University of Colorado.

Ford, Amasa B. 1970. Casualties of our time. Science 167:256-263.

Gardiner, P.C., and W. Edwards. 1975. Public values: Multiattribute utility measurement for social sciences decisionmaking. Social Science Research Institute Report 75-5. Los Angeles: University of Southern California.

Great Britain. 1976. Central Unit on Environmental Pollution. Chlorofluorocarbons and their effect on stratospheric ozone. Pollution Paper no. 5, London: HMSO.

Great Britain. 1979. Central Unit on Environmental Pollution. Chlorofluorocarbons and their effect on stratospheric ozone: Second report. Pollution Paper no. 15, London: HMSO.

Green, A.E., and A.J. Bourne. 1972. Reliability technology. New York: Wiley-Interscience.

262

Gros, J. 1975. Power-plant siting: A Paretian environmental approach. Research Memorandum RM-75-44. Laxenburg, Austria: International Institute for Applied Systems Analysis.

Häfele, Wolf. 1974. Hypotheticality and the new challenge: The pathfinder role of nuclear energy. Minerva 12 (July):314-315.

Hammond, Kenneth, and Berndt Brehmer. 1969. Cognition, quarrels, and cybernetics. Program on Cognitive Processes Report No. 117. Boulder, Colorado: Institute of Behavioral Science, University of Colorado.

Howard, R.A., J.E. Matheson, and D.W. North. 1971. The decision to seed hurricanes. Menlo Park, Calif.: Stanford Research Institute.

Jahoda, G. 1971. The psychology of superstition. Baltimore: Penguin.

Jensen, Clayton E., Dail W. Brown, and John A. Mirabito. 1975. Earthwatch: Guidelines for implementing this global environmental assessment program are presented. Science 190:432-438.

Johnston, H. 1971. Reduction of stratospheric ozone by nitrogen oxide catalysts from supersonic transport exhaust. Science 173:517-522.

Kasperson, Jeanne X., Roger E. Kasperson, Christoph Hohenemser, and Robert W. Kates. 1979. Institutional responses to Three Mile Island. Bulletin of the Atomic Scientists 35 no. 10 (December):20-24.

Kates, Robert W. 1976. Experiencing the environment of hazard. In Experiencing the Environment, ed. S. Wapner, S. Cohen, and B. Kaplan, 133-156. New York: Plenum Press.

Kates, Robert W. 1978. Risk assessment of environmental hazard. SCOPE 8. New York: John Wiley for the Scientific Committee on Problems of the Environment.

Keeney, Ralph L., and Howard Raiffa. 1976. Decisions with multiple objectives. New York: Wiley.

Lawless, Edward T. 1977. Technology and social shock. New Brunswick, N.J.: Rutgers University Press.

McDonald, James E. 1971. Testimony in Congressional Record: Senate, 19 March.

Munn, R.E. 1973. Global environmental monitoring system: GEMS. SCOPE Report 3. Toronto: International Council of Scientific Unions, Scientific Committee on Problems of the Environment.

Munn, R.E., ed. 1975. Environmental impact assessment: Principles and procedures. SCOPE Report 5. Toronto: International Council of Scientific Unions, Scientific Committee on Problems of the Environment.

National Research Council. 1975. Climatic Impact Committee. Environmental impact of stratospheric flight: Biological and climatic effects of aircraft emissions in the stratosphere. Washington: National Academy of Sciences.

National Research Council. 1976a. Committee on Impacts of Stratospheric Change. Halocarbons: Environmental effects of chlorofluoromethane release. Washington: National Academy of Sciences.

National Research Council. 1976b. Committee on Impacts of Stratospheric Change. Halocarbons: Effects on stratospheric ozone. Washington: National Academy of Sciences.

National Research Council. 1979a. Committee on Impacts of Stratospheric Ozone Change. Protection against depletion of strato- spheric ozone by chlorofluorocarbons. Washington: National Academy of Sciences.

National Research Council. 1979b. Committee on Impacts of Stratospheric Change. Stratospheric ozone depletion by halocarbons: Chemistry and transport. Washington: National Academy of Sciences.

National Research Council. 1980. Committee on the Biological Effects of Ionizing Radiation. The effects on populations of exposure to low levels of ionizing radiation: 1980. Washington: National Academy Press.

National Research Council. 1982. Committee on Chemistry and Physics of Ozone Depletion. Causes and effects of stratospheric ozone depletion: An update. Washington: National Academy Press.

Nuclear Regulatory Commission. 1975. Reactor safety study. WASH-1400, NUREG 75/014. Washington: NRC.

Otway, Harry J., and J.J. Cohen. 1975. Revealed preferences: Comments on the Starr benefit-risk relationship. Research Memorandum 75-5. Laxenburg, Austria: International Institute for Applied Systems Analysis.

Pill, Juri. 1971. The Delphi method: Substance, context, a critique and annotated bibliography. Socio-Economic Planning Sciences 5:57-71.

Risk Assessment Review Group. 1978. Risk assessment review group report to the Nuclear Regulatory Commission (Lewis Report). NUREG CR-0400. Washington: Nuclear Regulatory Commission.

Roschin, Alexander V., and L.A. Timofeevskaya. 1975. Chemical substances in the work environment: Some comparative aspects of USSR and US hygienic standards. Ambio 4 no. 1:30-33.

Rowe, William D. 1977. An anatomy of risk. New York: Wiley.

SCOPE (Scientific Committee on Problems of the Environment). 1971. Global environmental monitoring. SCOPE Report no. 1. Stockholm: SCOPE.

Savage, L.J. 1954. The foundations of statistics. New York: Wiley.

Selvidge, J. 1973. A three-step procedure for assigning priorities to rare events. In Fourth Research Conference on subjective probability, utility and decision-making. Boulder, Colo.: University of Colorado.

Siddall, E. 1981. Risk, fear and public safety. AECL-7404. Mississauga, Ontario: Atomic Energy of Canada Limited.

Sinclair, Craig, Pauline Marstrand, and Pamela Newick. 1972. Innovation and human risk: The evaluation of human life and safety in relation to technical change. London: Centre for the Study of Industrial Innovation).

Slovic, Paul, Baruch Fischhoff, and Sarah Lichtenstein. 1976. Cognitive processes and societal risk taking. In Cognition and social behavior, ed. J.S. Carroll and J.W. Payne. Potomac, Md.: Laurence Erlbaum.

Starr, Chauncey. 1972. Benefit-cost studies in sociotechnical systems. In Perspectives on benefit-risk decision making. Washington: National Academy of Engineering.

Tversky, Amos, and Daniel Kahneman. 1974. Judgment under uncertainty: Heuristics and biases. Science 185:1124-1131.

U.S. Department of Transportation. 1975. The national highway safety needs report. Washington: DOT.

Westcott, M.R. 1968. Toward a contemporary psychology of intuition. New York: Holt, Rinehart and Winston.

Whyte, Anne V., and Ian Burton, eds. 1980. Environmental risk assessment. SCOPE 15. New York: John Wiley for the Scientific Committee on Problems of the Environment.

Wilson, Richard. 1979. Analyzing the daily risks of life. Technology Review 81 (February):41-46.

Winell, Margareta. 1975. An international comparison of hygienic standards for chemicals in the work environment. Ambio 4(1):34-36.

World Health Organization. 1972. Health hazards of the human environment. Geneva: WHO.

World Health Orgainzation. 1976. Health hazards from new environmental pollutants: Report of a WHO study group. WHO Technical Report Series, 586. Geneva: WHO.

12
Weighing the Risks[1]

Baruch Fischhoff, Paul Slovic,
and Sarah Lichtenstein

The bottom line in hazard management is usually some variant of the question, "How safe is safe enough?" It takes such forms as: "Do we need additional containment shells around our nuclear power plants?" "Is the carcinogenicity of saccharin sufficiently low to allow its use?" "Should schools with asbestos ceilings be closed?" Lack of adequate answers to such questions has bedeviled hazard management.

Of late, many hazard management decisions are simply not being made--in part because of vague legislative mandates and cumbersome legal proceedings, in part because there are no clear criteria on the basis of which to decide. As a result, the nuclear industry has ground to a halt while utilities wait to see if the building of new plants will ever be feasible (**Business Week** 1978) the Consumer Product Safety Commission has invested millions of dollars in producing a few puny standards (chapter 16). Observers wonder whether the new Toxic Substances Control Act can be implemented (Culliton 1978), and the Food and Drug Administration is unable to resolve the competing claims that it is taking undue risks and that it is stifling innovation.

The decisions that are made are often inconsistent. Our legal statutes are less tolerant of carcinogens in the food we eat than of those in the water we drink or in the air we breathe. In the United Kingdom, 2,500 times as much money per life saved is spent on safety measures in the pharmaceutical industry as in agriculture (Sinclair, Marstrand, and Newick 1972). U.S. society is apparently willing to spend about $140,000 in highway construction to save one life and $5 million to save a person from death due to radiation exposure (Howard, Matheson, and Owen 1978).

Frustration over this state of affairs has led to a search for clear, implementable rules that will tell us whether or not a given technology is sufficiently safe. Various authors (e.g., Lowrance 1976 and Rowe 1977) discuss criteria for determining acceptable risk. Four approaches are most frequently used in attempting to make this assessment. They are cost/benefit analysis, revealed preferences, expressed preferences, and natural standards. Respectively, they would deem a technology to be safe if its benefits outweigh its cost; if its risks are no greater than those of currently tolerated technologies of equivalent benefit; if people say that its risks are acceptable; if its risks are no greater than

265

those accompanying the development of the human species. Each of
these approaches has its pros and cons, its uses and its limitations.

Cost/Benefit Analysis

Cost/benefit analysis attempts to answer the question of whether the expected benefits from a proposed activity outweigh its
expected costs. The first steps in calculating the expected cost of
a project are: to enumerate all the adverse consequences that might
result from its implementation; to assess the probability of each
such consequence; and to estimate the cost or loss to society whenever the consequence occurs. Next, the expected cost of each possible consequence is calculated by multiplying the cost of the consequence by the probability that it will be incurred. The expected
cost of the entire project is computed by summing the expected
losses associated with the various possible consequences. An analogous procedure produces an estimate of the expected benefits
(Fischhoff 1977;Stokey and Zeckhauser 1978). The most general form
of cost/benefit analysis is decision analysis, in which the role of
uncertainty, the subjective nature of costs and benefits, and the
existence of alternative actions are made explicit (Brown, Kahr, and
Peterson 1974;Howard, Matheson, and Miller 1976).

These procedures, and decision analysis in particular, are
based on appealing premises and are supported by sophisticated
methodology. Furthermore, they permit considerable flexibility;
analyses are readily revised to incorporate new options and new
information. An important advantage of these methods for decision
making in the public sphere is that they are easily scrutinized.
Each quantitative input or qualitative assumption is available for
all to see and evaluate, as are the explicit computational rules
that combine them.

Decision analysis and its variants have a number of potentially
serious limitations, perhaps the most important of which are unreal-
istic assumptions about the availability of the data needed to
complete the analysis. Performing a full-dress analysis assumes,
among other things, that all possible events and all significant
consequences can be enumerated in advance; that meaningful probability, cost, and benefit values can be obtained and assigned to
them; and that the often disparate costs and benefits can somehow be
made comparable to one another.

Unfortunately, it is sometimes impossible to accomplish some
of these tasks, while in the case of others, the results are hardly
to be trusted. Despite the enormous scientific progress of the last
decade or two, we still do not know all or even most of the possible
physical, biological, and social consequences of any large-scale
energy project (Fischhoff et al. 1978). Even when we know what the
consequences are, we often do not, or cannot, know their likelihood.
For example, although we know that a nuclear reactor core melt-down
is unlikely, we will not know quite how unlikely until we accumulate
much more on-line experience. Even then, we will be able to utilize
that knowledge only if we can assume that the reactor and the attendant circumstances remain the same (e.g., no changes in the
incidence of terrorism or the availability of trained personnel).
For many situations, even when a danger is known to be present, its

extent cannot be known. Whenever low-level radiation or exposure to toxic substances is involved, consequences can be assessed only by tenuous extrapolation--either downward from the consequences of high-level exposure to human beings or from observation of exposure in animals (Najarian 1978).

In all these instances, we must rely upon human judgment to guide or supplement our formal methods. Research into the psychological processes involved in producing such judgments offers reason for concern, since this research demonstrates that people (including experts forced to go beyond the available data and rely on their intuitions) have a great deal of difficulty both in comprehending complex and uncertain information and in making valid inferences from such information (Slovic, Fischhoff, and Lichtenstein 1979). Frequently, these problems can be traced to the use of judgmental heuristics--mental strategies whereby people try to reduce the difficult tasks to simpler judgments. These strategies may be useful in some situations but in others they lead to errors that are large, persistent, and serious in their implications. Furthermore, individuals are typically unaware of these deficiencies in the judgments.

Even if all the consequences could be enumerated and their likelihood assessed, placing a price tag on them poses further difficulties. Consider, for example, the problems of placing a value on a human life. Despite our resistance to thinking about life in economic terms, the fact is that, by our actions, we actually do put a finite value on our lives. Decisions about installing safety features, buying life insurance, or accepting a more hazardous job for extra salary all carry implicit judgments about the value we put on a life.

Economists have long debated the question of how best to quantify the value of a life (Linnerooth 1977). The traditional economic approach has been to equate the value of a life with the value of a person's expected future earnings. Many problems with this index are readily apparent. For one, it undervalues those in society who are underpaid and places no value at all on people who are not in income-earning positions. In addition, it ignores the interpersonal effects of a death which may make the loss suffered much greater than any measurable financial loss. A second approach, which equates the value of life with court awards, can hardly be considered more satisfactory (Holmes 1970;Kidner and Richards 1974).

Some have argued that the question, "What is a life worth?" is poorly phrased and what we really want to know is, "What is the value placed on a particular change in survival probability?" (Linnerooth 1977). One approach to answering this second question is by observing the actual market behavior of people trading risks for economic benefits. For example, one study examined salary as a function of occupational risk and claimed to find that a premium of about $200 per year was required to induce workers in risky occupations (coal mining, for example) to accept an increase of .001 in their annual probability of accidental death (Thaler and Rosen 1976). From this finding, it was inferred that society should be willing to pay about $200,000 to prevent a death. A replication of this study by Rappaport (1977) produced a value of $2,000,000; thus, even if one accepts the assumption underlying this approach, a definitive value may still elude us.[2]

Decision analysis attempts to accommodate the uncertainties inherent in the assessment of problems and of the values of the variables involved through the judicious use of **sensitivity analysis**. The calculations of expected costs and benefits are repeated using alternative values of one troublesome probability, cost, or benefit. If each reanalysis produces the same relative preponderance of expected costs or benefits, then it is argued that these particular differences do not matter.

Unfortunately, however, there are no firm guidelines regarding which of the data might be in error or what range of possible values ought to be tested. A further problem with sensitivity analysis is that it typically tells us little about how the uncertainty from different sources of error is compounded or about what happens when different data are subject to a common bias. The untested assumption is that errors in different inputs will cancel one another, rather than compound in some pernicious way (Fischhoff 1980).

In the end, determining the quality of an analysis is a matter of judgment. Someone must use intuition to determine which inputs are of doubtful validity and which alternative values should be incorporated in sensitivity analyses. Essentially, that someone has to decide how good her or his own best judgment is. Unfortunately, an extensive body of research suggests that people tend to overestimate the quality of such judgments (Slovic, Fischhoff, and Lichtenstein 1979).

Revealed Preferences

An alternative approach to determining acceptable risks is the method of **revealed preferences** advocated by Chauncey Starr (1969). This approach is based on the assumption that, by trial and error, society has arrived at an "essentially optimum" balance between the risks and benefits associated with any activity. As a result, it is assumed that economic risk and benefit data from recent years will reveal patterns of acceptable risk/benefit tradeoffs.

Acceptable risk for a new technology is defined as that level of safety associated with ongoing activities having similar benefits to society. Starr argued the potential usefulness of revealed preferences by examining the relationship between risk and benefit across a number of common activities.

From this analysis, Starr derived what might be called laws of acceptable risk:

- The acceptability of risk is roughly proportional to the third power (cube) of the benefits.
- The public seems willing to accept risks from voluntary activities, such as skiing, roughly a thousand times greater than it would tolerate from involuntary activities, such as food preservatives, that provide the same level of benefit.
- The acceptable level of risk is inversely related to the number of persons exposed to that risk.

Figure 1 depicts the results of Starr's analysis, whereas Figure 2 shows our own expanded replication of Starr's study, in which we examine 25 activities and technologies, including the eight

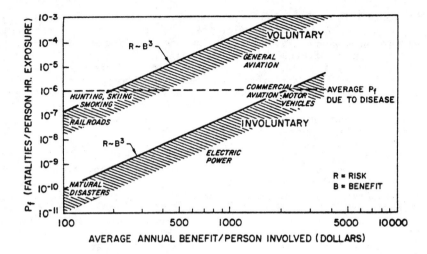

Figure 1. A comparison of risk and benefit to U.S. society from various sources. Risk is measured by fatalities per person per hour of exposure. Benefit reflects either the average amount of money an individual participant spends on a particular activity or the average contribution an activity makes to a participant's annual income. The best-fitting lines were drawn by eye with error bands to indicate their approximate nature. <u>Source</u>: Starr (1972).

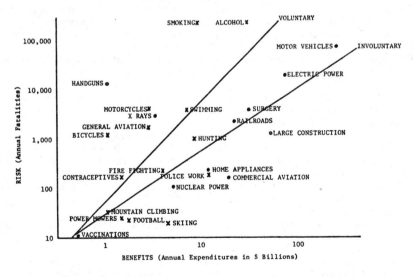

Figure 2. One possible assessment of current risks and benefits from 25 activities and technologies. Items are marked with an X, if voluntary; with a closed circle, if involuntary. Handguns and large construction dams, marked here with open circles, defy classification as primarily voluntary or involuntary and do not enter into the calculation of the two regression lines shown in the figure.

he used. This replication uses somewhat different methods. Whereas Starr estimated risk in terms of fatality rate per hour of exposure, we have used annual fatalities. This change is motivated in part by the greater availability of data for the latter measure and in part because the definition of exposure to some hazards (for instance, handguns, smoking, antibiotics) is elusive. Whereas Starr measured benefit either by the average amount of money spent on an activity by a single participant or the average contribution the activity made to a participant's annual income, we have used the single measure of total annual consumer expenditure.

Like any other economic measure of benefit, expenditure has its limitations. It includes "bad" as well as "good" expenditures; for example, money spent on the abatement of pollution caused by an industry is weighted as heavily as the value of the product it manufactures. A second problem is that this measure ignores distributive considerations (who pays and who profits). A third problem is that the market price may not be responsive to welfare issues that are critical to social planning. Does the price of cigarettes take account of smokers' higher probability of heart disease or cancer? Does the price of pesticides adequately reflect the increased probability of various deleterious effects on the one hand and the increased yield of foodstuffs on the other?

Expenditures for private goods (whose purchase is the result of the decisions of individual consumers) were obtained from trade and manufacturing associations, whereas public services, such as police work or fire fighting, were estimated by using government expenditures on payroll and equipment. No attempt was made to calculate the secondary and tertiary economic benefits of a product or service (for example, the increase in agricultural yield attributable to the use of pesticides), or the present value of past structural investments (for example, airport terminals, acquisition of wilderness areas), or contributions to distributional equity.

Despite the differences in procedure, our analysis produced results similar to Starr's. Overall, there was a positive relation between benefits and risks (slope = .3, correlation = .55). Furthermore, at any given level of benefit, voluntary activities tended to be riskier than involuntary ones (compare alcohol and surgery or swimming and nuclear power).

Although based upon an intuitively compelling logic, the method of revealed preferences has several drawbacks. It assumes that past behavior is a valid predictor of present preferences, perhaps a dubious assumption in a world where values can change quite rapidly. It is politically conservative in that it enshrines current economic and social arrangements. It ignores distributional questions (who assumes what risks and who gets what benefits?). It may underweigh risks to which the market responds sluggishly, such as those involving a long time lag between exposure and consequences (as in the case of carcinogens).

It makes strong (and not always supported) assumptions about the rationality of people's decision making in the marketplace and about the freedom of choice that the marketplace provides. Consider the automobile, for example. Unless the public really knows what safety is possible from a design standpoint and unless the industry provides the public with a set of alternatives from which to choose, market behavior may not indicate what a reflective individual would decide after thoughtful and intensive inquiry.

A revealed-preferences approach assumes not only that people have full information but also that they can use that information optimally, an assumption that seems quite doubtful in the light of much research on the psychology of decision making. Finally, from a technical standpoint, it is no simple matter to develop the measures of risks and benefits needed for the implementation of this approach.

Expressed Preferences

Both cost/benefit analysis and revealed-preferences analysis must infer public values indirectly, using procedures that may be both theoretically and politically untenable. The **expressed preferences** approach tries to circumvent this problem by asking people directly what levels of safety they deem acceptable.

The appeal of this approach is obvious. It elicits current preferences; thus it is responsive to changing values. It also allows for widespread citizen involvement in decision making and thus should be politically acceptable. It allows consideration of all aspects of risks and benefits, including those not readily converted into dollars and body counts. Some ways of obtaining expressed preferences are through referenda, opinion surveys, detailed questioning of selected groups of citizens, interviewing "public interest advocates," and hearings.

Recently, we conducted a series of expressed-preference studies paralleling Starr's revealed-preference study (chapter 5). We asked people to judge the total risk and benefit for each of thirty activities and technologies, including those used by Starr. Contrary to Starr's presumption, our respondents did not believe that society had managed these activities and technologies so as to allow higher risk only when higher benefit is obtained (see Figure 3a). In their view, society currently tolerates a number of activities with very low benefits and very high risks (alcoholic beverages, handguns, motorcycles, and smoking). Some very safe activities were judged to have very great benefits (antibiotics, railroads, vaccinations).

When we asked people what level of safety would be acceptable for each of the thirty activities and technologies, they responded that current levels were too safe 10 percent of the time, about right 40 percent of the time, and too risky about 50 percent of the time ("too risky" was defined as "indicating the need for serious societal action'). Thus for these individuals, the historical record used by the revealed-preferences approach apparently would not be an acceptable guide to future action.

When acceptable levels of safety were compared with perceived benefits, a relationship emerged much like the one obtained by Starr. Participants believed that greater risk should be tolerated for more beneficial activities and that a double standard is appropriate for voluntary and involuntary activities (Figure 3b)[3]

Similar studies were conducted with students, members of the (generally liberal) League of Women Voters, and members of a (generally conservative) community service club. Although the groups disagreed on the evaluation of particular items, their judgments showed the same general pattern of results as shown in Figure 3.

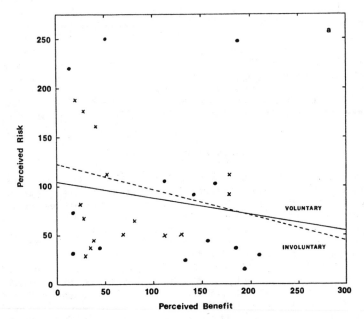

Figure 3a. Average judgments--of 76 members of the League of Women Voters in Eugene, Oregon--of current risks and benefits for 30 activities and technologies. The two best-fit lines indicate that these respondents saw no systematic relationships between current risks and benefits, a sharp contrast to the pattern shown in Figures 1 and 2.

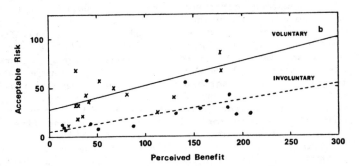

Figure 3b. A comparison between judgments of acceptable levels of risk and judgments of current benefits. Respondents (members of the League of Women Voters) believed higher risks should be tolerated for beneficial activities and for voluntary (X), as opposed to involuntary (dots), activities. Summarizing the contrast among Figures 1, 2, and 3, the individuals we questioned believed that Starr's hypothesized relationships should be obtained in a society in which risk levels are adequately regulated. They also thought, however, that their current world did not achieve that ideal. Source (Figures 3a and 3b): Fischhoff et al. (1978). Reproduced by permission of Elsevier North Holland, Inc.

One frequent criticism of the expressed-preferences approach is that safety issues are too complicated for ordinary citizens to understand. The results just cited suggest, however, that, in some situations at least, motivated lay people can produce orderly, interpretable responses to complex questions.

A related criticism is that, when it comes to new and complex issues, people do not have well-articulated preferences. In some fundamental sense their values may be incoherent—not thought through (Fischhoff, Slovic, and Lichtenstein 1980). In thinking about acceptable risks, people may be unfamiliar with the terms in which the issues are formulated (social discount rates, miniscule probabilities, megadeaths). They may have contradictory values (a strong aversion to catastrophic losses of life and a realization that they are not more moved by a plane crash with 500 fatalities than one with 300). They may occupy different roles in life (parents, workers, children), which produce clear-cut but inconsistent values. They may vacillate between incompatible, but strongly held, positions (freedom of speech is inviolate, but it should be denied to authoritarian movements). They may not even know how to begin thinking about some issues (how to compare the opportunity to dye one's hair with a vague, minute increase in the probability of cancer twenty years from now). Their views may change over time (say, as the hour of decision or the consequence itself draws near), and they may not know which view should form the basis of a decision.

In such situations, where people do not know what they want, the values they express may be highly unstable. Subtle changes in how issues are presented—how questions are phrased and responses are elicited—can have marked effects on their expressed preferences. The particular question posed may evoke a central concern or a peripheral one; it may help to clarify the respondent's opinion or irreversibly shape it; it may even create an opinion where none existed before.

Three features of these shifting judgments are important. First, people are typically unaware of the extent of such shifts in their perspectives. Second, they often have no guidelines as to which perspective is the appropriate one. Finally, even when there are guidelines, people may not want to relinquish their own inconsistency, creating an impasse.

Natural Standards

A shared flaw of the approaches described above is that all of them are subject to the existing limitations of society and its citizens. It might be desirable to have a standard of safety independent of a particular society, especially for risks whose effects are collective, cumulative, or irreversible. One such alternative is to look to "biological wisdom" to insure the physical well-being of the human species—not to mention the well-being of other species (Tribe, Schelling, and Voss 1976). Rather than examining (recent) historical time for guidelines, one might look to geological time, assuming that the optimal level of exposure to pollutants is that characteristic of the conditions in which the species evolved.

Specific proposals derived from this approach might be to set allowable radiation levels from the nuclear fuel cycle according to natural background radiation and to set allowable levels of chemical wastes according to the levels found in archaeological remains (Ericson, Shirahata, and Patterson 1979). These standards would not constitute outright bans, as some level of radiation-induced mutation is apparently good for the species and traces of many chemicals are needed for survival. Since exposure has varied from epoch to epoch and from place to place, one could establish ranges of tolerable exposure.

Perhaps the best-known criteria for risk acceptability based on natural standards are those for ionizing radiation set by the International Commission on Radiological Protection. The standards set by this small, voluntary, international group are subscribed to by most countries in the world. Their underlying assumptions include the following:

> **The maximum permissible dose levels should be set in such a way that, in the light of present knowledge:**
>
> 1. they carry a negligible probability of severe somatic or genetic injuries; for example, leukemia or genetic malformations that result from exposure to individuals at the maximum permissible dose would be limited to an exceedingly small fraction of the exposed group; and
> 2. the effects ensuing more frequently are those of a minor nature that would not be considered unacceptable by the exposed individual and by the society of which he is a part. Such frequently occurring effects might be, for example, modifications in the formed elements of the blood or changes in bone density. Such effects could be detected only by very extensive studies of the exposed individual. Effects such as shortening of life span, which may be proportional to the accumulated dose, would be so small that they would be hidden by normal biological variations and perhaps could be detected only by extensive epidemiological studies. (Morgan 1969)

Figure 4 shows how U.S. Atomic Energy Commission standards compared with natural background levels of radiation in 1976. It also compares current levels of SO_2 and NO_2 with background levels, indicating the implications of invoking natural standards in these contexts.

Natural standards have a variety of attractive features. They avoid converting risks into a common monetary unit (like dollars per life lost). They present issues in a way that is probably quite compatible with people's natural thought processes. Among other things, this approach can avoid any direct numerical reference to very small probabilities, for which people have little or no intuitive feeling (Lichtenstein et al. 1978). Use of natural standards should produce consistent practices when managing the same emission appearing in different sources of hazards.

As a guide to policy, natural standards are flawed by the fact that our natural exposure to many hazards has not diminished. Thus, whatever new exposure is allowed is an addition to what we are already subjected to by nature and thereby constitutes excess

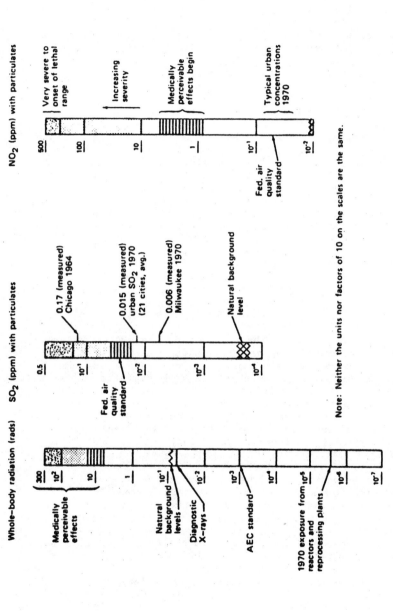

Figure 4. Comparison of pollutant standards, background levels, manmade exposures, and health effects for radiation, SO₂, and NO₂.

"unnatural" exposure (although conceivably within the range of toleration).

A second problem is that most hazards increase some exposures and reduce others. Trading off different exposures brings one back to the realm of cost/benefit analysis.

Another problem arises when one considers completely new substances for which there is no historical tolerance (saccharin, for example). In such cases, a policy based on natural standards would tolerate none of the substance at all, unless it involved no risk. The Delaney Amendment, which outlaws the addition of any known carcinogen to food, is consistent with this approach.

The technical difficulties of performing this type of analysis are formidable. Indeed, while there may be some hope of assessing natural exposure to chemicals and radiation that leave traces in bone or rock, appraising the natural incidence of accidents and infectious disease is probably impossible. Furthermore, should such an analysis be completed, it would quickly become apparent that the ecology of hazard in which humans live has changed drastically over the eons—mostly for the better, as in the case of the reduced incidence of infectious disease (chapter 6). The biological wisdom (or importance) of restoring one componemt of the mix to its prehistoric values would demand careful examination.

In addition to whatever difficulties there may be with their internal logic and implementation, natural standards are likely to fail as a sole guide to policy because they ignore the benefits that accompany hazards and the costs of complying with the standards.

Multiple Hazards

Our discussion so far has focused on the acceptable risk associated with individual hazards. What additional problems are created by considering many hazards at once? There are some 50,000 consumer products and 60,000 chemicals in common use in the United States (chapter 16;Maugh 1978). If even a small fraction of these presented the legal and technical complexities engendered by saccharin or flammable sleepwear (not to mention nuclear power), it would take legions of analysts, lawyers, toxicologists, and regulators to handle the situation. If hazards are dealt with one at a time, many must be neglected. The instinctive response to this problem is to deal with problems in order of importance. Unfortunately, the information needed to establish priorities is not available; the very collection of such data might swamp the system.

Even if legions of hazard managers were available, the wisdom of tackling problems one at a time is questionable. Responsible management must ask not only which dangers are the worst but which are the most amenable to treatment. A safety measure that is reasonable in a **cost/benefit** sense may not seem reasonable in a **cost-effectiveness** sense. That is, if our safety dollars are limited, finding that the benefits of a particular safety measure outweigh its costs does not preclude the possibility that even greater benefits could be reaped with a like expenditure elsewhere. The hazard-by-hazard approach may cause misallocation of resources across activities (for instance, giving greater protection to nuclear plant operators than to coal miners) or even within activities [protecting crop dusters but not those in the fields below (Berman 1978)].

The cumulative danger from a problem that appears in many guises may be hidden from a society that tackles hazards one by one. The current cancer crisis seems to reflect an abrupt realization of a risk distributed in relatively small doses over a very large number of sources. The nuclear industry has only recently been alerted to the possibility that temporary workers who receive their legal limit of radiation exposure in one facility frequently move on unnoticed to another and another (Nuclear Regulatory Commission 1978;Melville 1981).

Proponents of new products or systems can often argue persuasively that the stringent risk standards imposed upon them by the public constitute an irrational resistance to progress. After all, many currently tolerated products have much greater risks with appreciably less benefit. The public may, however, be responding to its overall risk burden, a problem outside the purview of these proponents. From that perspective, one of the obvious ways to reduce a currently intolerable risk level is to forbid even relatively safe new hazards, unless they reduce our dependence on more harmful existing hazards.

Treating hazards individually may obscure solutions as well as problems. Hazard managers must worry not only about how to trade lives and health for dollars but also about how to do so in an equitable fashion. Resolving equity issues in the context of an individual hazard often demands either heroic theoretical assumptions or considerable political muscle. Looking at the whole portfolio of hazards faced by a society may offer some hope of circumventing these problems. No one escapes either the risks or the benefits of all aspects of a society. Indeed, both are often implicitly traded between individuals. I live below the dam that provides you with hydroelectric power in the summer whereas you live near the nuclear power plant that provides me with electricity in the winter. In this example, the participants might view the trade as equitable, without recourse to complex distributional formulas. Although such simple dyads may be rare, looking at the total distribution of risks and benefits in a society may possibly produce clearer, sounder guidelines for resolving equity issues than would solutions generated for individual hazards.

Facing Political Realities

Models that do not capture the critical facts about a hazard will not pass muster before the scientific community. Approaches that fail to represent the political realities of a situation will be rejected by those interests that are underrepresented. No one method can serve the needs of all the environmentalists, industrialists, regulators, lawyers, and politicians involved with a particular hazard. These people appropriately view each specific decision as an arena in which broader political struggles are waged.

In theory, any of the approaches described here should find some support among "public interest" advocates and some resistance among technology proponents since all of them make the decision process more open and explicit than it was in the dark ages of hazard management when matters were decided behind closed doors. However, the enchantment of the public wanes some when closed doors are replaced by opaque analyses that effectively transfer power to the minute technical elite who perform them (McGinty and Atherly 1977). In such cases, "public interest" advocates may resist formal analysis, feeling that avoiding disenfranchisement is more important than determining acceptable levels of risk. The battle brewing in the United States over the use of cost/benefit analysis to regulate toxic substances and other hazards may largely hinge on these concerns (Chemical and Engineering News 1978;Carter 1979).

For other members of the public, the openness itself is a sham, since each of these approaches makes the political-ideological assumption that society is sufficiently cohesive and common-goaled that problems can be resolved by reason and without confrontation. Sitting down to discuss a decision analysis would, in this view, itself constitute the surrender of important principles. Cooperation may even be seen as a scheme to submerge the opposition in paper work and abrogate its right to fight the outcome of an analysis not to its liking (Fairfax 1978). Such suspicions are most easily justified when the workings of the decision-making process are poorly understood. It is not hard to imagine the observers of a decision analysis accepting its premises but balking at its conclusions when the results of the analysis are complex or counterintuitive. At the extreme, this would mean that people will only believe analyses confirming their prior opinions.

Proponents of a technology would probably prefer to have the determination of risk acceptability left to their own corporate consciences. Barring that (or the equivalent captive regulatory system), proponents may find it easier to live with adversity than with uncertainty. As a result, one would expect industry increasingly to advocate routine approaches with rigorous deadlines for making decisions. From this perspective, the zenith of the influence of the Toxic Substances Control Act may have been reached immediately after its enactment. At that moment, industry practice could respond only by making all products as safe as possible, not knowing which substances would actually be dealt with or how stringently. Cynically speaking, the sooner and more precisely the rules are laid down, the more efficacious the search for loopholes can be.

One could draw similar caricatures of the hidden agendas of other (would-be) participants in hazard management. The point of such an assessment is not to argue that reasonable management is impossible but that all approaches must be seen in their political contexts. Such a broadened perspective may help us to understand the motives of the various participants and the legitimacy that should be assigned to their maneuvers.

In so doing, a crucial issue will be deciding whether society should have goals higher than maximizing the safety of particular technologies. Such goals might include developing an informed citizenry and preserving democratic institutions. In this case, the process could be more important than the product, and it would be important for society to provide the resources needed to make meaningful public participation possible (Casper 1976). Such participation would require new tools for communicating with the public—both for presenting technical issues to lay people and for eliciting their values (Slovic, Fischhoff, and Lichtenstein 1979). It might also require new social and legal forms, such as hiring representative citizens to participate in the analytic process, thereby enabling them to acquire the expertise needed by the governed to give their informed consent to whatever decision is eventually reached. Such a procedure might be considered a science court with a lay jury. It would consider any or all of the analytic techniques described here as possible inputs to its proceedings. It might also place the logic of jurisprudence above the logic of analysis, acknowledging that there is no single way to determine what risks are acceptable.

The forums in which safety issues are currently argued were not designed to deal with such problems. H. R. Piehler has, in fact, argued that the legal system could hardly have been designed more poorly for airing and clarifying the technical considerations that arise in product liability suits (Piehler et al. 1974). Much public opinion about hazards derives from the testimony of experts. Often this testimony is offered in rancorous debates between experts trying to cast doubt on the probity of their opponents (Mazur 1973). In addition to creating negative attitudes toward scientists, such spectacles tend to destroy public confidence in the possibility of ever understanding or satisfactorily resolving these issues.

Natural disagreements in areas of incomplete knowledge are aggravated by the feeling that "bad evidence drives out good." A two-handed scientist ("on the other hand . . . while on the other . . .") may be bested by a two-fisted debater intent on acquiring converts. Decisions about controversial technologies might be improved if all participants publicly subscribed to an established code of behavior. Some possible rules might be:

- Never cite a research result without having a complete, accessible reference.
- Never cite as fact a result supported only by tenuous research findings.
- Acknowledge areas in which you are not an expert (but are still entitled to an opinion).

Like rules of parliamentary procedure, this code would formalize values that many people espouse but have difficulty upholding in practice (fairness, mutual respect, etc.).

Muddling Through Intelligently

No approach to acceptable risk is clearly superior to the others. To exploit the contributions each of these methods can make, careful consideration must be given to the social and political world in which they are used and to the natural world in which we all live. Our social world is characterized by its lack of orderliness. Since hazards are not the only consideration in hazard-management decisions, the best we can hope for is some intelligent muddling through. Recognizing this, we should develop and apply the various approaches to hazard management not as inviolate ends in themselves but as servants to that process. The openness of formal analyses must be assured in order to avoid suspicion and rejection of whatever conclusions are finally reached. When the available numbers are not trustworthy, we should content ourselves with digitless structuring of problems. When good numbers are available, but the issues are unfamiliar, great care must be taken in designing suitable presentations. When we do not know what goal we want to reach, value issues should be framed in a variety of ways and their implications carefully explored.

A distinctive characteristic of our natural world is that it typically is both unknown and unknowable to the desired degree of precision. We must not only acknowledge this uncertainty but also devote more of our efforts to determining its extent. The most critical input to many hazard-management decisions may be how good our best guess is. The real alternatives may be: "If we don't understand it, we shouldn't mess with it" and "If we don't experiment, we'll never know what it means" (Goodwin 1978).

Uncertainty about facts and uncertainty about values both imply that determining the acceptability of a hazard must be an iterative process, partly because, as time goes on, we learn more about how a hazard behaves and how much we like or dislike its consequences. In other words, it takes experience that acknowledges the experimental nature of life to teach us what the facts are and what we realy want.

Iteration is essential to any well-done formal analysis. A measure of the success of any analysis is its ability to inform (as well as to reflect) our beliefs and values. Once the analysis is completed, we may then be ready to start over again, incorporating our new and better understandings. In this light, many of the nonpolitical critiques generated by the **Reactor Safety Study** (Nuclear Regulatory Commission 1975) reflect its success in deepening the respondents' perspectives. As an aid to policy, the study's main weakness lay in attempting to close the books prematurely and thereby failing to take adequate account of these criticisms.

Whereas a good analysis should be insightful, it need not be conclusive. At times, it may not be possible to reach any analytic conclusion, for example, when inter- and intra-personal disagreements are too great to be compromised. If people do not know what they want or if a topic is so politicized that no solution will ever be acceptable, analysis should perhaps best be treated as a process for deepening knowledge and clarifying positions. Performing the sort of calculations that lead to a specific recommendation would, in such cases, only create an illusion of analyzability.

A Combined Approach

The disciplinary training of scientists shows them how to get the right answers to a set of specially defined problems. The problems raised by hazard management are too broad to be solved by any one discipline (chapter 3). No one knows how to get the right answer. All we can do is avoid making the particular mistakes to which each of us is prone. The more scientific and lay perspectives applied to a problem the better chance we have of not getting it wrong.

Just as no single discipline has all the answers, no one of the approaches discussd above provides a sufficient basis for determining what levels of safety are acceptable. In attempting to solve the problems inherent in the other methods, each approach engenders problems of its own.

Are better approaches likely to come along? Probably not, for it seems as though all attempts to rule on the safety of particular hazards share common conceptual and operational difficulties whose source lies in the very attempt to reduce the problem to manageable size. What we can hope for is to understand the various approaches well enough to be able to use them in combination so that they complement one another's strengths rather than compound each other's weaknesses.

ACKNOWLEDGMENTS

This research was supported by the National Science Foundation under Grant ENV77-15332 to Perceptronics, Inc. of Eugene, Oregon. Any opinions, findings, and conclusions or recommendations expressed in this publication are those of the authors and do not necessarily reflect the views of the National Science Foundation. We are grateful to Robert Kates and Richard Wilson for comments on an earlier draft of the manuscript and to Michael Enbar for much of the analysis in Figure 2.

NOTES

1. Except for minor revisions appropriate to this volume, this chapter is reproduced with permission from Environment 21 no. 4 (May 1979):17-20, 32-38, a publication of the Helen Dwight Reid Educational Foundation (HELDREF).

2. These assumptions are essentially those underlying the revealed-preferences approach described below.

3. One complication of this latter relationship is that the degree of voluntariness of some activities proved to be rather ambiguous (e.g., handguns). A second is that double standards were also observed with other qualitative aspects of risk such as perceived control, familiarity, knowledge, and immediacy.

REFERENCES

AEC (Atomic Energy Commission). 1974. Comparative risk-cost bene-fit study of alternative sources of electrical energy. WASH-1224. Washington: AEC.

Berman, D.M. 1978. How cheap is a life? International Journal of Health Sciences 8:79-99.

Brown, Rex V., Andrew S. Kahr, and Cameron Peterson. 1974. Decision analysis for the manager. New York: Holt, Rinehart, and Winston.

Business Week. 1978. Nuclear dilemma: The atom's fizzle in an energy-short world. No. 2566 (December 25):54-68.

Carter, Luther J. 1979. An industry study of TSCA: How to achieve credibility. Science 203:247-249.

Casper, Barry N. 1976. Technology policy and democracy. Science 194:29-35.

Chemical and Engineering News. 1979. Cost/benefit analyses sought for regulations. 57 no. 4 (22 January):7.

Culliton, Barbara J. 1978. Toxic substances legislation: How well are laws being implemented? Science 201:1198-1199.

Ericson, Jonathon E., Hiroshi Shirahata, and Clair C. Patterson. 1979. Skeletal concentrations of lead in ancient Peruvians. New England Journal of Medicine 300:946-951.

Fairfax, Sally K. 1978. A disaster in the environmental movement. Science 199:743-748.

Fischhoff, Baruch. 1977. Cost-benefit analysis and the art of motorcycle maintenance. Policy Sciences 8:177-202.

Fischhoff, Baruch. 1980. Clinical decision analysis. Operations Research 28:28-43.

Fischhoff, Baruch, Christoph Hohenemser, Roger E. Kasperson, and Robert W. Kates. 1978. Handling hazards. Environment 20 (September):16,20,32-37.

Fischhoff, Baruch, Paul Slovic, and Sarah Lichtenstein. 1980. Knowing what you want: Measuring labile values. In Cognitive processes in choice and decision behavior, ed. T. Wallsten, 117-141. Hillsdale, N.J.: Erlbaum.

Fischhoff, Baruch, Paul Slovic, Sarah Lichtenstein, Stephen Read, and Barbara Combs. 1978. How safe is safe enough? A psychometric study of attitudes towards technological risks and benefits. Policy Sciences 9:127-152.

Goodwin, R. 1978. Uncertainty as an excuse for cheating our children: The case of nuclear wastes. Policy Sciences 10:25-43.

Holmes, R. A. 1970. On the economic welfare of victims of automobile accidents. American Economic Review 60:143-152.

Howard, Ronald A., James E. Matheson, and Katherine E. Miller, eds. 1976. Readings in decision analysis. Menlo Park, CA: Decision Analysis Group, Stanford Research Institute.

Howard, Ronald A., James E. Matheson, and David L. Owen. 1978. The value of life and nuclear design. In Probabilistic analysis of nuclear reactor safety: Topical Meetings, May 8-10, 1978, Los Angeles, California, eds. D. Okrent and E. Cramer. LaGrange Park, Ill: American Nuclear Society.

Kidner, R., and K. Richards. 1974. Compensation to dependants of accident victims. Economic Journal 84:130-142.

Lichtenstein, Sarah, Paul Slovic, Baruch Fischhoff, Mark Layman, and Barbara Combs. 1978. Judged frequency of lethal events. Journal of Experimental Psychology: Human Learning and Memory. 4:551-578.

Linnerooth, Joanne. 1977. The evaluation of life-saving: A survey. Joint IAEA/IIASA research report. Vienna: International Atomic Energy Agency.

Lowrance, William W. 1976. Of acceptable risk. Los Altos, Calif.: William Kaufmann.

McGinty, Laurence, and Gordon Atherly. 1977. Acceptability versus democracy. New Scientist 74:323-325.

Maugh, Thomas H. 1978. Chemical carcinogens: The scientific basis for regulation. Science 201:1200-1205.

Mazur, Allan. 1973. Disputes between experts. Minerva 11:243-262.

Melville, Mary H. 1981. The temporary worker in the nuclear power industry. CENTED Monograph Series, no. 1. Worcester, Mass.: Center for Technology, Environment, and Development (CENTED).

Morgan, Karl Z. 1969. Present status of recommendations of the International Commission on Radiological Protection, National Council on Radiation Protection and Federal Radiation Council. In Health Physics, ed. A.M.F. Duhamel. Vol. 2. New York: Pergamon Press.

Najarian, Thomas. 1978. The controversy over the health effects of radiation. Technology Review 81 no. 2 (November):74-82.

Nuclear Regulatory Commission. 1975. Reactor safety study: An assessment of accident risks in U.S. commercial power plants. WASH-1400, NUREG-75/014. Washington: Nuclear Regulatory Commission.

Nuclear Regulatory Commission. 1978. Notices, instructions, and reports to workers, inspections; standards for protection against radiation: Proposed rule. Federal Register 43 no. 25 (February 6):1865-1868.

Piehler, H.R., Aaron D. Twerski, Alvin S. Weinstein, and William A. Donaher. 1974. Product liability and the technical expert. Science 186:1089-1093.

Rappaport, Edward B. 1981. The demand for improvements in mortality probabilities. Ph.D. diss. Department of Economics, University of California, Los Angeles.

Rowe, William D. 1977. An anatomy of risk. New York: Wiley.

Sinclair, Craig, Pauline Marstrand, and Pamela Newick. 1972. Innovation and human risk: The evaluation of human life and safety in relation to technical change. London: Centre for the Study of Industrial Innovation.

Slovic, Paul, Baruch Fischhoff, and Sarah Lichtenstein. 1979. Rating the risks. Environment (April):14-20,36-39.

Starr, Chauncey. 1969. Social benefit versus technological risk. Science 165:1232-1238.

Starr, Chauncey. 1972. Benefit-cost analysis in sociotechnical systems. In Perspectives on benefit-risk decision making, ed. Committee on Public Engineering Policy, 17-42. Washington: National Academy of Engineering.

Stokey, Edith, and Richard Zeckhauser. 1978. A primer for policy analysis. New York: Norton.

Thaler, Richard, and Sherwin Rosen. 1976. The value of saving a life: Evidence from the labor market. In Household Production and Consumption, ed. N. Terleckyj, 265-297. New York: Columbia University Press.

Tribe, Laurence H., Corinne S. Schelling, and John Voss, eds. 1976. When values conflict: Essays on environmental analysis, discourse, and decision. Cambridge, Mass: Ballinger.

Part Four

Overview: Managing Hazards

Hazard management is the purposeful activity by which society informs itself about hazards, decides what to do about them, and implements measures to control or mitigate their consequences. Chapter 3 identified four major types of management activity (hazard assessment, control analysis, strategy selection, and implementation and evaluation) and classified the participants as managers (individuals, technology sponsors, policy makers, regulators, and assessors) and monitors (adversarial groups and the media). A comprehensive case study of hazard management should consider the roles of all major participants and analyze the four kinds of management activity. Case studies rarely achieve this level of completeness, but they may nevertheless inform hazard management by contributing to the formulation, testing, and refinement of conceptual structures such as those presented in the first part of this volume.

The seven chapters that follow fall short individually of meeting this test of comprehensiveness. This is scarcely surprising insofar as each inevitably focusses variously on certain major activities or participants (Table 1). Except for the propositional inventory (chapter 13), which derives from a survey of the literature, each chapter constitutes a case study of a specific hazard (automobiles, PCBs, contraceptives, television) or hazard manager (the Consumer Product Safety Commission, the United States Congress). Three of the hazards entail releases of energy or materials, with both acute and chronic consequences. The fourth, television, brands information—not energy or materials—a potential threat to humans and what they value. The two case studies of hazard managers look at a regulatory agency (the Consumer Product Safety Commission) and a policy maker (Congress) as key participants in the management process. Chapters 13-19 comprise specific studies, undertaken by different members of the Hazard Assessment Group (at Clark University) and at different stages of our research, primarily to explore new topics rather than to illustrate or exemplify concepts. Yet taken together, as Table 1 indicates, they provide an overall comprehensiveness not found in any single study.

"Tales of Woe" (chapter 13) comprises an early (1977) propositional inventory of hazard management based upon a literature review embracing 41, mostly book-length, studies. Where multiple books on a specific hazard or hazard manager were available, author Branden

Johnson chose the most scientific or balanced treatment. Even so, however, the sample of 41 betrays a basic flaw in the available literature—its domination by studies of managerial failure. Apparently, few people write of successful hazard management. Johnson concentrates on the characterization of participants, culling from his sample synthetic portraits of "businessmen" (technology sponsors), "legislators" (policy makers), "bureaucrats" (regulators), and "scientists" (assessors) and sketching in lesser detail roles for the media and the public. The preponderantly negative characterizations, attributable in part to the bias of evaluation inherent in the choice of topics, may be somewhat misleading. They are certainly stereotypical: businessmen are greedy, scientists are "two-handed," bureaucrats are indecisive, and legislators are desultory. Perhaps more troubling is the virtual absence of social science theory to "explain" the behavior of these participants.

The next six chapters, though far from unmitigated success stories, paint a less lopsided picture and attempt to identify and analyze both success and failure in hazard management. Again, each chapter achieves a different level of comprehensiveness.

"Regulating Automobile Safety" (chapter 14) approaches comprehensiveness—particularly when read in conjunction with chapter 8. A limitation to a single hazard chain—the acute consequences of automobile accidents—detracts from the completeness of the study, which excludes other hazard chains such as occupational hazards or health and environmental hazards from automobile emissions. The chapter addresses hazard assessment, analysis of actual and potential controls, past and current control strategy, and several examples of success and failure in implementation. Despite a focus on regulation, the discussion of passive restraints considers interactions among regulators, technology sponsors, adversarial groups, and the public. The authors observe grimly that safety gains notwithstanding, society tolerates annually between 20 and 30 deaths per 100,000 population—a threshold that "permits" an annual slaughter of over 50,000 deaths and over two million injuries. In tolerating such statistics, society has let slip the opportunity to implement technological and behavioral measures to control a manageable hazard—automobile accidents.

The study of polychlorinated biphenyls (PCBs) defines four major hazard chains (occupational exposure, food contamination, environmental release, and waste disposal) and provides for each an examination of the range of control actions, an account of the history of implementation, and some evaluation of success as well as failure. Compared to automobile accidents, the hazard consequences of PCBs are delayed and poorly understood. As food contaminants and suspected carcinogens, PCBs may well trigger a high level of perceived risk that might explain why, even in the face of uncertainty, United States society has taken a less permissive stance toward PCBs and gone so far as to cease production and to ban most traditional applications. Yet the PCB hazard persists in ubiquitous products and wastes and in the scurry to find substitutes, themselves potentially hazardous, for its beneficial purposes. In short, the ban has not ended a 50-year history of dealing with PCBs.

The hazards of consumer products date back much further—at least to the lead glazing of the first clay pot. But the elevation of a hazard to a societal "problem" is often signalled by the

TABLE 1
Focus of hazard management studies

HAZARDS	PARTICIPANTS							MANAGERIAL ACTIVITY			
	MANAGERS					MONITORS		HAZARD ASSESSMENT	CONTROL ANALYSIS	STRATEGY SELECTION	IMPLEMENTATION AND EVALUATION
	Individuals	Technology Sponsors	Policy Makers	Regulators	Assessors	Adversarial Groups	Media				
TALES OF WOE • 41 studies of hazard management	+	++	++	++	++		+				
AUTOMOBILES • accidents	+	+	+	++		+		++	++	++	++
POLYCHLORINATED BIPHENYLS (PCBs) • occupational • food contamination • environmental • waste		+		++				++	++	+	++
CPSC • 10,000 consumer products	+		+	++	+	+		++	+	+	++
CONTRACEPTIVES • oral • intrauterine • injectable • postcoital	+	+		++	+	+	+	++	++	++	++
TELEVISION • violence • advertising • stereotyping	++	++	+	++	++	++		++	+	+	++
CONGRESS • 36 hazards 1957-1978			++					+	++	+	+

++ major focus
+ minor focus

creation of a bureaucracy entrusted with its management. In that sense, the establishment of the Consumer Product Safety Commission (CPSC) a mere decade ago marks a new recognition of consumer products as potential hazards. Chapter 16, a case study of the CPSC as hazard manager, describes a new attack on an old problem. In establishing the CPSC, Congress sought to incorporate three major structural innovations: openness and accessibility to consumers, insulation from political interferences, and generic authority over an awesome domain of some 10,000 products. The CPSC has overcome its growing pains to develop an innovative computerized injury-surveillance network for identifying and monitoring hazards. Also, a few false starts notwithstanding, the agency had, by the end of the 1970s, come to terms with setting priorities for its control efforts. Although intended as an in-depth study of a regulatory agency, the case study focusses on a number of value issues in hazard management--particularly those encountered in the setting of priorities and the selection of management strategies. Chapter 16 examines the CPSC at two intervals--at five years in the original paper and at ten years in a brief epilogue--and reports the Commission's mixed performance as an experiment in regulatory reform but increased capability (prior to Reagan cutbacks) in hazard control.

In a sense, the next chapter also deals with an experiment: how an old agency manages a novel class of products. The Food and Drug Administration (FDA) had long regulated the food contaminants and dangerous drugs that society had numbered among the first technological hazards. But four new contraceptive technologies, developed between 1960 and 1980, did not quite fall into either category and thus required special handling. In chapter 17, then, author Mary Lavine examines societal management of the four innovations: oral contraceptives, injectable contraceptives, DES (diethylstilbestrol) as a postcoital drug, and intrauterine devices (IUDs). The study provides exposure-consequence data for each technology and assesses the roles of various participants--individual consumer, technology sponsor, regulator, the media, and adversarial groups. Confounding this simplified roster of participants is the medical practitioner who may function as partly assessor, partly consumer, and partly consumer's counselor. In attempting to manage contraceptives, the FDA has run up against old problems in new guises: problems of identifying hazards, of balancing risks and benefits, and of informing exposed populations. Each of these problems has entailed the use or consideration of new managerial strategies--innovative practices that hold promise for application with other technological hazards.

Nearly all of the technological hazards that make their way into this volume involve releases of hazardous energy or materials. Such releases link hazards to the fundamental flows of the biosphere: the triad of energy, materials, and information. Early in our research we recognized the existence of information hazards--the disjunctive release of toxic information that either overlooks or poisons human relationships. This novel concept of a generic information hazard, related particularly to an electronic revolution that permits the storage, production, and dissemination of enormous amounts of information, does not enjoy wide recognition in the risk assessment literature. Thus it was difficult to study the phenomenon on a par with the more familiar energy and material hazards. Nevertheless, chapter 18 comprises a case study of the most common information hazard--television. In a sense, the hazard is well studied; in the past 25 years over 3,000 books, articles, and reports have examined the effects of television viewing on human behavior--and more than 90 percent of these reports have appeared in the last decade. Even this massive effort has yet to produce a comprehensive risk assessment of the television hazard. Moreover, despite congressional attention and a widespread call for government intervention, the individual, particularly the parent, was and still is the prime hazard manager. The regulatory agency charged with managing television maintains a virtually unblemished record of impotence (the industry is essentially self-regulated). Meanwhile, an imposing array of revolutionary sister technologies--videotaping, computerized video games, and visual display terminals (VDTs)--poses new problems without the existing ones' being understood or controlled. As chapter 18 shows, even the most widely acknowledged hazard of television--the negative effects of televised violence on human behavior--has yet to undergo effective control. On the whole, major participants (individuals, technology sponsors, regulators, and assessors) have failed to engage successful managerial activity (particularly in terms of hazard assessment and control implementation) to control this most pervasive and intrusive hazard.

Whereas the United States Congress must share responsibility for the ineffectual response to the television hazard, it can claim more general success as hazard manager. The final chapter of this volume evaluates congressional performance in managing technological hazards. An examination of legislation enacted between 1957 and 1978 serves to identify a roster of 179 laws and a congressional agenda of 36 hazards. Congressional policy making appears to steer an incrementalist course, adding cautiously to existing legislation rather than enacting new innovative laws. At the same time, author Branden Johnson credits Congress with addressing most of the hazards that have commanded societal concern. The case study focusses on a single participant--Congress as policy maker--engaged primarily in control analysis but also in hazard assessment, strategy selection, and implementation and evaluation. In the face of growing public awareness of technological hazards, legislative activity remains remarkably constant over time and deals on a recurring basis with a well-established list of hazards. Insofar as Congress proceeds in a desultory fashion, its performance as a hazard manager is somewhat mixed, though chapter 19 qualifies more as success story than tale of woe.

Most of the case studies in this volume allow for an evaluative mix that represents a departure from the usual preoccupation with failure, noted in the introduction (chapter 1) and so tellingly documented by Johnson's propositional inventory (chapter 13). Moreover, the case studies in this Part--and, for that matter, in part 3 as well--go some distance toward filling other gaps in knowledge, as identified in chapter 3.

To varying degrees, five of the six case studies deal with some of the successful aspects of hazard management. The study of television fares less well as a success story, but even that tale is not completely negative. The chapters on automobiles, PCBs,

contraceptives, and television reach for comprehensiveness in looking at most types of managerial activity and most relevant participants, but there are still missing pieces (recall Table 1).

The stereotypical portraits of managers and monitors who emerge
from Johnson's propositional inventory drive home the need to scru-
tinize the roles of other participants, especially technology
sponsors. To that end, our Hazard Assessment Group has recently
undertaken a study of hazard management by industry. The studies in
Part 4 persist in focussing on regulation as a form of management,
but chapter 19 turns the spotlight on a policy maker (Congress), and
the television study looks at all major participants except for the
media.

As for managerial activity, all six case studies in this part
succeed remarkably in covering all four types of activity. The
studies of automobiles (chapter 14) and contraceptives (chapter 17)
are virtually comprehensive. Hazard assessment and implementation
and evaluation are subject to heavy focus in five, and control analysis enters into four, of the six case studies. Only strategy
selection receives short shrift in a majority of the studies.

The more comprehensive and balanced the case study, the greater
its utility for informing the management of technological hazards.
As Table 1 indicates, the case studies in part 4 are uneven in
coverage, but this does not discount their value. The intentionally
narrow but in-depth study of a single regulatory agency (the CPSC)
or a specific hazard (PCBs) yields its own particular contributions
to our understanding of hazard management.

Thanks to chapters 18 (television) and 19 (Congress), we know
more about participants other than regulators, although the television study highlights the scantiness of our knowledge of individuals
and technology sponsors as hazard managers. The studies of automobiles and PCBs enhance our understanding of available options for
controlling hazards. A heightened awareness of innovative control
measures--for dispensing user information and for identifying and
monitoring hazardous consumer products--emerges from the studies of
contraceptives and the CPSC. The successful application of the
causal model, introduced in chapter 2, to such diverse hazards as
PCBs, contraceptives, and automobile crashes adds credibility to our
conceptual framework. The television study provides a measure of
empirical grounding for a new concept (information hazards). All
six studies provide evidence of, though not criteria for, managerial
success as well as failure.

Part 4 both identifies and narrows some of the gaps in our
knowledge. In short, although we still know more about assessing
hazards than we do about managing them, we come away from the case
studies with a sense of having gained on the problems delineated in
chapter 3.

13
Tales of Woe:
A Literature Survey

Branden B. Johnson

When it comes to managing technological hazards, gaffes and blunders abound. Most of the 45 incidents detailed by Lawless (1977) are sagas of failures that have prompted one observer to conclude: "Indeed, based on the Lawless study, I estimate that there has been one significant technological blunder each week, week in and week out, for thirty-five years" (Coates 1982,22).

The Lawless casebook is only one of the 41 published studies of hazard management (listed in Table 1) to undergo critical examination by the present author. Authors of these tales of woe, who include scientists, bureaucrats, journalists, citizen activists, and a congressional committee, cast a critical eye at the management of hazards ranging from agricultural technology to television. Taken together, the studies yield critical, unflattering, and stereotypical portraits of all hazard managers--bureaucrats, industrialists, legislators, and scientists alike. The stereotypes may conflict with the views of the managers themselves and to some extent with those of the general public, but they are nonetheless disturbingly consistent. Convincing as they are as individual case studies, the reports are collectively less useful for improving future management of technological hazards.

This chapter describes the methods for selecting the 41 studies, extracts from those studies composite portraits of hazard managers, and considers the influence of those portraits on the hazard management process. The author summarizes here the findings of a lengthier, more detailed study (Johnson 1979).

Selecting and Analyzing the Studies

The selection process entailed four criteria: (1) the sample set must include a wide variety of hazards, (2) the subject of the study must be hazard control and not the hazard itself (for example, one would choose a study of a regulatory agency's monitoring of the testing of a potentially hazardous chemical instead of a study of the potentially hazardous attributes of that chemical), (3) one should select the most scholarly treatment of a particular hazard, and (4) one should give preference to book-length studies of specific hazards (only 4 of the 41 studies are shorter than book-length).

The original literature survey (Johnson 1979), conducted during the early years of Clark University's Hazard Assessment Group, aimed

TABLE 1
Hazard management studies*

===

GENERAL	Lawless, <u>Technology and Social Shock</u> Lowrance, <u>Of Acceptable Risk</u> Rowe, <u>An Anatomy of Risk</u>
ADVISORY GROUPS	Primack and von Hippel, <u>Advice and Dissent</u>
AGRICULTURAL TECHNOLOGY	Hightower, <u>Hard Tomatoes and Hard Times</u>
AIR POLLUTION	Crenson, <u>The Un-Politics of Air Pollution</u> Esposito, <u>Vanishing Air</u> Jones, <u>Clean Air</u>
AIRPORT NOISE	Nelkin, <u>Jetport: The Boston Airport Controversy</u> Stevenson, <u>The Politics of Airport Noise</u>
AUTOMOBILES	Nader, <u>Unsafe at Any Speed</u>
BIOLOGICAL/PSYCHOLOGICAL ENGINEERING	Packard, <u>The People Shapers</u>
CANCER	Rettig, <u>Cancer Crusade</u>
CHEMICALS	Carpenter, "Legislative Approaches... in the Regulation of Chemicals"
COSMETICS	Nader, "The Regulation . . . of Cosmetics"
DAM SAFETY	U.S. Congress, House Committee, <u>Dam Safety</u>
DIOXIN	Fuller, <u>The Poison that Fell from the Sky</u>
ENVIRONMENTAL PROTECTION AGENCY	Environmental Protection Agency <u>Environ- mental Emergency Response</u> National Research Council <u>Decision Making in the Environmental Protection Agency</u> Quarles, <u>Cleaning Up America</u>
FLUORIDATION	Crain, Katz, and Rosenthal, <u>The Politics of Community Conflict</u>
FOOD ADDITIVES	Verrett and Carper, <u>Eating May Be Hazardous to Your Health</u>

TABLE 1 (continued)
Hazard management studies*

FOOD AND DRUG ADMINISTRATION	Turner, The Chemical Feast
FOREST FIRE CONTROL	Schiff, Fire and Water
GENETIC ENGINEERING	Goodfield, Playing God
JUDICIARY	Bazelon, "Coping with Technology Through the Legal Process"
LIGHT-WATER NUCLEAR REACTORS	Bupp and Derian, Light Water
LIQUEFIED NATURAL GAS	van der Linde, Time Bomb
MERCURY	D'Itri and D'Itri Mercury Contamination: A Human Tragedy
MICROWAVES	Brodeur, The Zapping of America
NATIONAL ACADEMY OF SCIENCES	Boffey, The Brain Bank of America
NOISE	Kavaler, Noise: the New Menace
OCCUPATIONAL HEALTH	Ashford, Crisis in the Workplace Brodeur, Expendable Americans
OIL SPILLS	Steinhart and Steinhart, Blowout
OZONE	Dotto and Schiff, The Ozone War
PARTICLE ACCELERATOR	Lowi, et al., Poliscide
PESTICIDES	Blodgett, "Pesticides: Regulation of an Evolving Technology" Graham, Since Silent Spring
SMOKING	Fritschler, Smoking and Politics
TELEVISION	Mankiewicz and Swerdlow, Remote Control

*Full bibliographic citations appear in the References at the end of this chapter.

to derive an overview of hazard management and managers. To that end, the author constructed a propositional inventory, a technique well-known to the social and behavioral sciences. The box on page 295 elaborates on this method of extracting from the literature propositions, or generalizations, about the management of technological hazards.

The inventory process described in the box serves to make two important points. First, the propositional "evidence" by no means constitutes evidence in the usual scientific sense. No attempt is made to "prove" anything with it; instead, raw information is transmuted, with a dash of imagination and license, into generalizations. There is no need for generalizations to be supported conclusively by facts or even to be true; the only requirement is that they be **reasonable** extrapolations of the authors' views **and** that they might be useful "pre-hypotheses" for future studies (in this case, of managing technological hazards). Second, these propositions are not the only ones that could have been derived from the case studies. Persons starting with different assumptions and interests might well have come up with a different inventory, no less valid than this one. The important criterion is a proposition's usefulness, not its truth. Thus the evidence--in the form of supporting quotations from the literature--that accompanies a propositional inventory allows for easy assessment (not proof) of its "reasonableness."

Creating a propositional inventory is **scientific** to the extent that it requires sufficient agreement, among one's peers, as to its validity so that empirical testing of the propositions as hypotheses is undertaken; it is **artistic** in the sense that propositions must be sculpted out of a mass of details and assertions, with the final result determined in large part by the aesthetic sense of the researcher.

The summary that follows derives from a propositional inventory of 41 studies, published in 1978 or earlier, which treat the management of technological hazards. Some of the studies (e.g., Lowrance 1976;Lawless 1977;Rowe 1977) focus on overall hazard management and comparisons of management of various hazards. Most of the remaining reports comprise book-length case studies of specific hazards of hazard managers. Only four of the studies (Blodgett 1974;Carpenter 1974;Nader 1974;Bazelon 1977) constitute article-length accounts or hazard control. The entire inventory yielded 175 propositions, 83 of which define the behavior of hazard managers.

Definitions of Hazard Managers

BUREAUCRAT: A government official who succumbs to industry pressure, unless he is being entirely passive in order to avoid controversy.

The 23 case studies that yield propositions on the bureaucrat as hazard manager draw a composite picture of an official whose most sacred duties are to avoid rocking the boat, to look and not leap, to put off until tomorrow anything that could have been done last week:

HOW TO CREATE A PROPOSITIONAL INVENTORY

A proposition is a generalization about some subject of interest. One does not need to be a scientist to look for propositions, but imagination and patience are prime requirements.

The propositions in this study derive not from detailed descriptions of specific incidents but from more general statements in the studies. The following example may give an inkling of the creative process:

Cancer Crusade (Rettig 1977,64) contains the following paragraph:

> The biomedical research community's consistent view has been that review of contracts should be performed by some peer-review mechanism. The award process, moreover, should sharply constrain administrative discretion and require that all funded contract research be judged meritorious by the scientific community. Finally, the contract instrument should be used sparingly for special purposes like animal procurement and not for imparting direction to the research enterprise. Kenneth Endicott and his staff, on the other hand, thought that the review process should lead to a judgment about program relevance of research as well as an assessment of the quality of work being proposed. Consequently, the award process should place decision authority in the NCI's professional staff, for legal as well as functional reasons. Finally, the contract should be used when the integration of components of a complex program of directed research is desired. The historical unfolding of these issues, however, was a rather complicated experience.

The author found something significant—exactly what, he was not sure on his first reading of the paragraph—about this discussion of the research management process at the National Cancer Institute, so a quotation was extracted that included its major points. After much rearranging and discarding of similar quotations from other studies (four others provide equivalent "evidence"), the author came up with proposition #30 (the second part of which comes from the other studies):

> Scientists do not want research restrictions; they feel that they should do it because they can do it.

The process of creating a propositional inventory thus involves several steps:

(1) selecting an appropriate case study;

(2) reading the case study for apparently significant generalizations or generalizable descriptions;

(3) putting the resultant quotations in rough groupings;

(4) discarding insignificant items and distilling more important quotations down to the minimum necessary to convey information;

(5) defining and refining categories of "proto-propositions"; and

(6) repeating steps 4 and 5 as often as necessary to be able to

(7) list propositions and their supporting evidence in a coherent manner.

> Policy statements are particularly vulnerable to debate and criticism. The incremental changes that result from piecemeal planning are far less controversial and bureaucracies thus tend to avoid circulating long-term plans. (Nelkin 1974,158)

> It is not that the government decision-makers are corrupt, but that their sense of public duty is constantly eroded by industry contacts and the consideration of short-term impacts on industry. . . . (Verrett and Carper 1974,96)

> The pressure for action has emerged as much from publicity and public opinion about egregious instances of chemically induced cancer as from normal decision-making channels. (Rettig 1977,306)

Bureaucratic desire to avoid "loss of face" has resulted in the persistence of outmoded policies, with suppression of unfavorable research and promotion of studies that support the agency's credo:

> Bureaucrats are constantly involved in struggles to expand and defend their empires. This is a dangerous business, and what bureaucrats perhaps most desire from their science advisors is protection: protection against surprise by new technological developments and protection of their policies against political attack (Primack and von Hippel 1974,46). Agencies have sometimes even been able to suppress a report, though [National Academy of Sciences] officials claim that is a thing of the past. (Boffey 1975,62)

> . . . criticism issuing from interested governmental agencies . . . only partly succeeded in curbing the distribution of mis-leading publicity and in effecting a redistribution of the research program . . . . The Service's initial reaction had been to commence or continue research to "prove a theory instead of to find the facts." At first, controversy froze thinking, preventing any infusion of fresh viewpoints; only gradually did attitudes thaw. (Schiff 1962,172-173)

Information problems of agency personnel extend beyond public relations. Bureaucrats do not seek help from other agencies because that would reflect on their own competence; moreover, facts in themselves may be more trouble than they're worth:

> For the centers in Washington to be ignorant, ignorance itself had to be a policy . . . relevant knowledge, although usually easy to get, makes decisions harder to make. This tells us something more about the function of ignorance . . . . (Lowi et al. 1976,284)

This bureaucratic wish to avoid controversy results in a rather curious relationship with the public: officials who assume that their own plans and actions reflect a public consensus view the public as an "ignorant mob" from whom hazard information must be kept at all costs.

> . . . a highly trained technocratic elite has exercised responsibility for deciding technical questions according to its own perception of what was in the interest of French society [and] has only rarely been influenced by public opinion. (Bupp and Derian, 1978,106)

> The "accommodation" which is the usual goal of politics may cause the rejection of cold scientific findings if they are too disruptive to the electorate. (Carpenter 1974,23)

Bureaucrats in these studies come off as being more interested in preserving the power and integrity of their respective agencies than in pursuing the public interest.

> *BUSINESSMAN: A money-grubbing, short-sighted man who has to be forced by regulatory agencies to control hazards, assumes that his products are safe until proven dangerous, and sees hazard crises as public relations problems rather than as scientific challenges.*

Most of these studies paint the industrialist as intentionally villainous, wilfully bent on the pursuit of the dollar to the deliberate exclusion of all other concerns. Of the 17 case studies that yielded propositions about businessmen as hazard managers, only one (Dotto and Schiff 1978,20) betrayed any willingness to characterize them as perhaps themselves victims of inadequate science. The propositions about businessmen offer three major criticisms.

The overwhelming sin of industry is that its primary criterion for technology assessment is the profit motive:

> What contractor would pay these prices when he could be underbid for a job by another using less quiet machines? (Kavaler 1975,121)

> Corporate greed being what it sometimes is, . . . economic survival is accomplished routinely by falsifying or, with a shrug, taking that calculated risk with other people's lives. (van der Linde 1978,136)

Eleven of the books lend support to two other major propositions--that industrialists implement hazard control only under external compulsion and that they perceive crises as the concern of the public relations department:

> By now, the . . . company was desperately committed
> to finding a solution to the problem. It was the
> only way out from the constant pressure of the gov-
> ernment and the news media. (Steinhart and Steinhart
> 1972,68)

> It is easier and cheaper to launch a new public rela-
> tions cover-up than it is to deal directly with the
> environmental problem. (Esposito 1970,74)

> Whenever criticism of smoking grew, the tobacco
> interests responded by more research or more public
> relations expenditures. (Fritschler 1975,23)

Whereas the bulk of the propositions about businessmen lie
among the three foregoing criticisms, several other propositions are
also evident. Noteworthy, for example, are three different treatments of industrial "ignorance":

> Increasingly, communications within a company or
> industry appear to segregate into horizontal layers,
> and the individuals in top management tend to become
> isolated from the technological base. They are thus
> greatly surprised when the public becomes concerned
> over the [hazards] that their companies produce.
> They are, it would seem, often poorly aware of even
> the large technologically related trends in the
> world, or else callous of the public interest . . . .
> (Lawless 1977,527)

> People in the military-electronics industry complex
> don't want to know the extent of the problem. If
> they knew about it they might have to admit they knew
> about it, and then might even have to do something
> about it, which would cost a lot of money both in
> terms of litigation and preventive measures.
> (Brodeur 1977,188)

> . . . the whole thing [the fluorcarbon/ozone theory]
> had come as a complete shock to Du Pont. The shock
> was due not to the fact that industry had not consid-
> ered the environmental impact of fluorocarbons, but
> to the fact that they had. In 1972, Du Pont had
> issued an invitation to fluorocarbon manufacturers
> around the world to attend a seminar on "the ecology
> of fluorocarbons." . . . "It is prudent that we
> investigate any effects which the compounds may
> produce on plants or animals now or in the future."
> The companies funded several research projects, and
> by 1974 [the year of the theory] the results of these
> studies indicated that fluorocarbons posed no major
> environmental problems in the lower atmosphere.
> (Dotto and Schiff 1978,20)

Curiously, in the 41 studies reviewed, these three quotations are the only references to a rather provocative subject: the availability and use of information about hazards. The discussion by Dotto and Schiff is the sole indication that failure of industry to control hazards may in some instances be due to the inadequacy of current scientific theory and data, rather than to malevolence aforethought.

The modern industrialist, if one believes these studies, does not manage hazards so much as he "manages" public concern and regulation.

> LEGISLATOR: *A lawmaker who is roused to simu=lated action only by public pressure, pursues votes and public attention, and considers oversight of the executive agencies unnecessary.*

Some authors perceive legislators as resistant to taking action on technical matters. The case studies attribute resistance to various forces, including defense of a congressional committee's jurisdiction, confidence in the capability of certain scientists, or lack of public support.

> The . . . issue was seriously examined by Congress only after it had become the subject of a full-scale national debate, led by environmental groups and largely informed by independent scientists. (Primack and von Hippel 1974,25)

There is one partial exception to this view of congressional inaction:

> In air pollution policy development . . . members of Congress have been consistently involved in the [proposal] formulation process, often with members of the executive opposing their proposals for extended federal authority. (Jones 1975,54)

Although Jones presents the 1970 policy debate on air pollution control (due to perceived public pressure for action) as an exception to Congress's "usual" passivity, subsequent events suggest that inaction may no longer be so common.

The need to gather votes means that the appearance of action is more valuable to politicians than action itself, which may alienate colleagues and financial supporters (Nader 1965,342;Steinhart and Steinhart 1972,110; Mankiewicz and Swerdlow 1978,59). A legislator who **does** oppose a technology may well be seeking personal publicity rather than expressing genuine concern about hazards:

> Fluoridation offers a great temptation to the politician . . . it can be opposed purely for the sake of attracting headlines; it can be opposed as a symbol of encroaching socialism; it can be opposed to "show a thing or two" to some rising politicians and bureaucrats. (Crain, Katz, and Rosenthal 1969,184)

About half of the thirteen case studies that describe legislators suggest that legislative oversight is lax or nonexistent and that Congress's long-term influence on agencies is weak:

> Hearings on agricultural research budgets . . . are left pretty much to the land grant college community, buttressed by its agribusiness colleagues. The appropriations process . . . is little more than a chance for special interests to press for particular research projects or facilities. (Hightower 1972, 137)

> Specific instances where . . . oversight hearings led to USDA adjustment of policy . . . are infrequent. (Blodgett 1974,106)

One study, though, concludes that an agency that defies its congressional constituency may be harassed or even destroyed (Fritschler 1975,118).
These studies present a less consensual picture of legislators than that of other hazard managers. But the predominant image is one of drowsy solons awakened only by the sound of ballot boxes being dusted.

> *SCIENTIST: A researcher who says he wants no restrictions on his choice and handling of research and avoids considering nontechnical implications of such research, but fails to voice disagreement with erroneous interpretations of his research by the corporations and agencies that fund it.*

Almost half of the twelve studies yielding generalizations about scientists referred to the scientists' inclinations to avoid research restrictions:

> Their position was and still is: no one—neither public agency nor peer group nor society—can say to a scientist, "Thou shalt not do a particular experiment." (Goodfield 1977,100)

> [Scientists] have tended to feel that their mission is to seek the Truth, whatever it may turn up . . . . If something can be done, it should be done. And anyhow it will be done . . . . (Packard 1977,329)

There is some countervailing evidence offered. Occasionally, as in the case of recombinant DNA (Goodfield 1977,10 and 14), scientists will be the first to recognize and act to control the hazards of their research. European scientists are much more amenable than their American counterparts to government intervention in research (Goodfield 1977,107-108). Such exceptions, however, tend to reinforce rather than contradict the view of scientists that prevails in the foregoing quotations.
Scientists are reluctant to "go public" for fear their results will be misunderstood by the public (Brodeur 1977,92):

> . . . they faced—so it seemed to them—a dangerous
> situation of science under strong attack . . . .
> They feel it bitterly unfair that their motives and
> actions are so misunderstood, if not downright mis-
> represented, and some of them clearly interpret the
> whole episode as an awful warning of what happens
> when scientific disagreements are debated out in the
> open . . . . Their instinct now is to "hole up."
> (Goodfield 1979,203)

Insofar as disagreements among experts and cautious, qualified
statements by scientists frequently confuse the public and legislators (D'Itri and D'Itri 1977,261;Lawless 1977,424;Primack and von
Hippel 1974,270), the aversion to speaking out is quite justifi-
able:

> . . . strong polarizations developed among the tech-
> nological experts . . . in most cases, it is probably
> safe to say, they often added unnecessary elements of
> confusion for the decision makers and the public,
> without appreciably helping in the resolution of the
> problem. (Lawless 1977, 498)

Although, or perhaps because, about half of the studies were
written or co-authored by scientists, they contain scarcely a hint
of the usual idealistic picture of researchers. At best, one comes
away with the feeling that scientists are only too human and
fallible.

Bureaucrats, businessmen, legislators, and scientists are the
subjects of 80 percent of the propositions about hazard managers.
They are not the only groups to be stereotyped, however, for the
communications industry, the judiciary, and the public also come in
for their share.

A few propositions concern the communications industry. Media
coverage of hazards and hazard management leans toward sensationalism and oversimplification (Lowrance 1976,113;Steinhart and Steinhart 1972,80;Dotto and Schiff 1978,84). On the other hand, one
study relates how an attempt to **avoid** sensationalism actually
fostered the suppression of a story about a new hazard—namely,
fluorocarbons (Dotto and Schiff 1978,21). The limited role of the
media in managing hazards is attributed to the complexity and speed
of the impact of hazard issues (Rettig 1977,294). Still another
study, taking specialty magazines for auto buffs as a case in point,
argues that the media avoid offending those industries that supply
them with advertising revenue and technical information (Nader
1965,14-15). All in all, the media do not fare too well.

The judiciary receives its share of criticism. Slow and conservative, courts usually act only after injury has occurred—this
dawdling can benefit a beleaguered industry:

> When the FDA wishes to remove a product from the mar-
> ket, it relies heavily on a voluntary recall proce-
> dure, rather than . . . seizure . . . both the FDA
> and the affected company are aware that court action
> can be slow, and that it can provide opportunities

> for delay by a manufacturer. Thus, the threat of a
> legal action is a two-edged sword to the FDA. (Nader
> 1974,116)

Some studies suggest that the adversary environment of the courts is
not conducive to good hazard management, because the issues involved
in such management are complex and indeterminate (Bupp and Derian
1979,166;Steinhart and Steinhart 1972,68;Stevenson 1972,7). Moreover, the perspective of the courts is too narrow:

> Courts see only cases that are brought before them
> and must therefore focus their attention on the
> rights and wrongs of those particular cases. They
> simply cannot engage in the kind of balancing act
> that is required to set priorities and allocate na-
> tional resources. (Bazelon 1977,832)

Given these drawbacks, the judiciary's most important contribution
to hazard management may have been to stir other hazard managers
into motion:

> The courts are ill-equipped in legal precedent, tech-
> nical expertise, and overall perspective to begin
> resolving problems like those caused by jet noise.
> The most significant function of the courts has been
> to stimulate legislators to initiate plans for alle-
> viating the situation by other means. (Stevenson
> 1972,64)

Several studies direct criticism toward the public. Public
demand for hazard control often occurs even when neither the hazard
nor knowledge of its consequences has increased. But acceptance of
this proposition conceals sharply divergent viewpoints on the cause
of its manifestation. For some, public opinion is the culprit:

> The actual public need for pollution control was only
> marginally more serious in the early 1970's than it
> had been in the preceding decades. Yet public policy
> underwent a vast change. What happened? . . . The
> "environmental crisis" was a happening in the area of
> public opinion, . . . testimony to the political
> power of public opinion once its fury is unleashed.
> (Quarles, 1976,171)

For others it is the product of political manipulation:

> The "squawk" potential refers to a condition for an
> issue that is distasteful or unacceptable to a par-
> ticular value group to become a major issue, blown up
> through dire predictions of consequences, based pri-
> marily on half-truths, but flamed by competing com-
> mercial news media. The objective is to . . .
> [generate] enough concern, perhaps hysteria, to
> affect elective governmental bodies, regulatory agen-
> cies, and the courts. On this basis, any issue has

the potential for becoming a major "squawk." (Rowe 1977,63)

This proposition that public demand for hazard control is not the result of objective risk assessment contrasts with statements (Crain, Katz, and Rosenthal 1969,137 and 138) that public concern for hazards stems from doubts about a technology's benefits and sponsors rather than from ignorance about technology. Other studies, noting a lack of public concern about hazards, attribute it variously to (1) the perception of important benefits from a technology (Brodeur 1977,317-318) or (2) to government inaction (Dotto and Schiff 1978,174).

The case studies overwhelmingly portray (with little contradictory evidence) hazard managers of all sorts as short-sighted, narrow-minded, venal, greedy, self-serving, passive, and, ultimately, utterly inadequate. Depending upon one's point of view, the "definitions" of manager-types given above may seem amusing, outrageous, or perhaps even matter-of-fact. These images certainly indicate a pervasive distrust of, and disenchantment with, hazard managers on the part of study authors.

Managers as Captives

One explanation of poor performance is the apparent tendency of various hazard managers to be "captured" by other groups—that is, they come to believe that their interests and goals are synonymous with those of the "captors." A large proportion of the propositions support this explanation.

Despite the voluminous literature generally on congressional lobbyists, only a few propositions deal with outside influence on legislators. Perhaps this is the result of industrialists' perceptions that, for technological hazards, their influence can be most effectively applied outside the legislative process. One study (Blodgett 1974,219) supports this with evidence that industry prefers administrative agreements to inflexible legislative solutions; but another (Fritschler 1975,53) says lobbyists view Congress as being more responsive to them than to the bureaucracy. Although one author (Quarles 1976,153) suggests that the legislature may be hesitant to weaken a hazard control bill for fear of **appearing** to be under the control of special interests, the consensus of these studies is that Congress is generally willing to accommodate business.

Indeed, study authors betray little doubt that public agencies are, willingly or otherwise, in the hip pockets of those whom they are supposed to control. A couple of studies (Crenson 1971,124; Mankiewicz and Swerdlow 1978,60-61) suggest that the mere reputation of industry power prevents regulatory action. Even without such unseen allies, however, business employs a variety of tactics in its fight to remain independent:

- Business uses industry-induced agency weakness as an argument against strengthening its powers (Turner 1976, 46).

- Industry successfully pits agencies against each other and Congress to avoid regulation when there is no public support for government action (Fritschler 1975,9-10).
- Industry invokes the excuse of technological difficulties to disguise managerial reluctance to control hazards (Jones 1975,196;Nader 1965,183).
- Industry seeks "voluntary compliance" because it is a means for avoiding more stringent regulation (Esposito 1970,259;Nader 1965,313;Jones 1975,99). Industry will not accept "voluntary compliance" proposals that run counter to what it judges to be its real interests (Nader 1974,117).
- When regulation is unavoidable, industry lobbies to have authority given to a sympathetic agency (Nader 1965, 322).

With regard to industry preferences, it is worth noting one interesting disagreement among authors of the studies under examination. One author claims that industry prefers known state regulators to unknown federal officials (Nader 1965,256). Another maintains that business favors the uniformity of federal regulation over unilateral state actions (Jones 1975,67). Perhaps both are true under different circumstances. Or perhaps, as public concern over technological hazards increases, industry need for a **predictable** regulatory environment will generate increasing preference for federal authority.

Aside from these active strategies, business musters support from initiatives (or, more accurately, **noninitiatives**) on the part of the regulators. Individuals in regulatory agencies may avoid stringent regulation of industry because (a) it may be their future employer (Hightower 1972,159;Verrett and Carper 1974,96), (b) professional socialization and industry contacts cause them to be them to be sympathetic to industry's needs or (c) they doubt the effectiveness of such action.

> It is not that the government decision-makers are corrupt, but that their sense of public duty is constantly eroded by industry contacts and the consideration of short-term effects on industry instead of long-term effects on consumers. (Verrett and Carper 1974,96)

> Some members of [the agency's] field staff choose not to cite violations of [standards], though they are legally enforceable, because they do not regard them as good standards. (Ashford 1976,261)

This accommodation to industry interests may be masked by an agency posture of active enforcement (Jones 1975,275;Mankiewicz and Swerdlow 1978,58;Nader 1974,116).

Even where officials feel a strong responsibility to regulate, they may be unable, due to lack of expertise or funding, to evaluate industry's technical judgments. In fact, they may become dependent on industry to conduct the necessary research and even to propose standards. Although this arrangement seems highly conducive to exploitation by business, it may not need to be so: one author asserts that when the harmful effects of pesticides became clear in England, "no segment of society [including industry] shirked its responsibilities" (Graham 1970,84).

Most of the preceding propositions have focused on the capture of government agencies by industry. But two studies imply that regulators may be trapped occasionally by other groups. In particular, advisory committees can influence policy making and sometimes come to control an agency (Boffey 1975,3;Fritschler 1975,49).

Such overweening influence reverses the usual state of affairs outlined in the case studies. Several studies view the advisory apparatus of the government as a device for giving political decisions the appearance of being technical and thus objective and apolitical (Primack and von Hippel 1974,34). Regulatory agencies frequently select advisory groups for publicity value rather than technical expertise (Turner 1976,55;Primack and von Hippel 1974,68). Moreover, agencies may support advisory committees in return for bland reports (Boffey 1975,57-58) and may supervise them closely to insure favorable reports (Primack and von Hippel 1974,56;Boffey 1975,61-62). Resolving such problems by obtaining "independent" advisors may not be easy:

> There is a school of thought in Washington that holds that it is best to fill such panels with scientists who are not personally involved in the relevant research, on the grounds that they have no axes to grind. The obvious drawback is that they also generally know very little about the subject at hand (Dotto and Schiff 1978,44).

Bureaucrats are not the only ones to capture advisors; industry is also likely to try this, say several authors. In many technical fields, industry may exert undue influence simply because it is the primary employer of experts. Business financing and/or oversight of research by "independent" scientists may also serve to insure favorable results. Industry-sponsored research **may** occasionally realize a high degree of autonomy and objectivity, but that requires very careful arrangements (Dotto and Schiff 1978,236).

On the whole, the studies cite industry and government as the major culprits in unduly influencing expert advisors, but several studies point out that no institution can claim to employ unbiased experts.

> Skepticism will often greet tests performed by manufacturers, or by others funded by them, or by government agencies whose missions promote or involve chemical-manufacturing technology. On the other side of the debate, investigators for consumer and environmental interest groups and regulatory agencies are accused of biased tests and selected data, even in the absence of profit motives. Over a period of time many "independent" laboratories have become identified with a constrained point of view. Thus the public has lost confidence in most such evaluative tests because the charges of inadequacy have so often flown back and forth—and been shown to be correct. (Carpenter 1974,32-33)

We may thus add to our portrait: hazard managers have little interest in controlling hazards under their respective jurisdictions, and they try to thwart actions by other groups. Moreover, hazard managers are more likely to act, if at all, in favor of special interests rather than in the public interest.

The Studies in Perspective

With few exceptions, the studies paint a gloomy picture of the abilities of the various hazard managers to carry out their responsibilities—whether we refer to the internal rationales of their organizations, to outside political pressures, or merely to their own approaches to analyzing a potential hazard of responding to a disaster. Before we accept this view as a true image of what is and what will be, we need to examine several important issues: the bias, validity, and effect of the 41 studies under consideration.

Bias. Virtually all of the 41 studies are litanies of failure in managing technological hazards. Does this imply that there have been no success stories? Other chapters in the present volume suggest at least partial success in the management of airborne mercury (chapter 9) and oral contraceptives (chapter 17). Such mixed accomplishment scarcely answers the call for best-case analyses (Kasperson and Morrison 1982,315) or studies of successes as well as failures in the application of decision-making methods (Covello and Menkes 1982,298), however, and triumphs are hard to come by—at least in the literature. Success stories may well be scarce because of how people decide how to define and write about hazard management (Verrett and Carper 1974,96).

First, the studies generally look at highly publicized hazards, which tend to command attention as the result of management failure rather than of its success. People do not seem to write books about successes in hazard management. This is partly because no one is angered by victory. It is also because such studies appear less interesting and less indicative of "the way things work" than do the alarums and excursions of disaster, or they seem self-serving and not worthy of publication or wide dissemination.

Second, in those fields where hazard control is consistently successful, it may have been internalized so much as to be recognizable no longer as "management." In structural engineering, for example, load stress management has been reduced to equations; hazard control has become one of the routine, taken-for-granted tools of the trade and is not written about as management.

Third, it may be possible to detect an inherent bias in the sample of studies. Indeed, the present author searched fruitlessly for studies written by businessmen.

Validity. Is technological hazard management as confused, contradictory, and uneven as the propositions suggest? Are hazard managers as ignorant, venal, and paranoid as they are portrayed?

A 1980 poll conducted by the Louis Harris organization (Marsh and McLennan 1980) features some strikingly optimistic views of hazard management performance, as offered by top corporate executives, investors, congressional personnel (legislators and aides), federal regulators, and the public. Except for regulators, all groups polled indicated that on the whole, business does a good job of protecting the public from dangerous products and substances (p. 29).

Again, with the exception of regulators, a plurality of groups attributes consumer product injuries to inappropriate use by consumers, rather than to poor product design or inadequate instructions (p. 30). A majority of congressmen, regulators, and the public believes that government standards have significantly increased occupational safety (p. 32). Despite the variations in assessments by and between groups, the poll underscores a sense of general satisfaction with the performance of hazard managers.

Yet a majority of the public and of congressional personnel, as well as a plurality of all groups save corporate executives, expect technology associated risks to be greater 20 years hence than they are today (p. 11). A tabulation of these groups' ratings of the performance of various hazard managers shows that only the medical/scientific community receives a rating of excellent or very good from a majority in all groups. Congressional personnel, regulators, and the public all registered pluralities of positive ratings for the performance of environmental groups. The public also rated consumer groups positively by a slim margin. All other managers—federal government, business, state/local government, the insurance community—received a plurality of negative (fair to poor) performance ratings from **every** group polled (p. 45). Although these opinions scarcely provide support for the overwhelming pessimism in the literature surveyed, they do reflect a widespread skepticism about hazard managers.

One may readily dismiss some of the propositions, such as those on regulatory capture, as standard assumptions from the literature on bureaucracy. Others are not so easily characterized. To refute the stereotype of narrow-minded scientists, one needs only point to the increasingly prominent role that critical science (sometimes somewhat righteously termed "public interest" science) has played in the management of technological hazards. Likewise, evidence suggests that discretionary power may result in major initiatives by some government agencies as well as relative inaction on the part of others. Moreover, some legislators have chosen to flout both stereotypes and constituencies by taking highly unpopular positions. These objections allow us only to question the overall validity of the propositions, however; without more evidence, we cannot reject them. At the very least, the consensus in some central areas is profoundly disturbing, drawing a portrait of hazard management that is not at all comforting.

Effects. The last question to be raised is, if anything, even more speculative than its predecessors, but important nonetheless. What are the effects of this parade of gloomy case studies on the hazard management process itself? Macbeth was terrified by the series of ghosts shown to him by the witches; these case studies might have an analogous effect (some seem intended to do so).

> The emergence of one issue, rather than preparing the way for the discharge of social tension—and the resolution of underlying conflict—may actually generate new foci of discontent. It may call attention to discomforts and deprivations that previously went unnoticed or were not regarded as fit subjects for public discussion. To put it

another way, political issues can create political consciousness. (Crenson 1971,171)

A proliferation of such studies might well provoke political actions intended to "improve" hazard management. Yet reading the propositions fosters considerable skepticism about obtaining short-term improvements, and cynics rarely make wise judges of potential reforms.

The burden of technological hazards as described in chapters 6 and 7 is considerable. If it is to be significantly reduced, it will require effort from all hazard managers. If the stereotypes that emerge from the studies are accurate, they offer little hope that such efforts will be forthcoming. If they are false, then the bulk of the literature on hazard management is misleading in cumulative impact, if not in detail.

> ...I backward cast my e'e
> On prospects drear!
> An' forward, tho' I canna see,
> I guess an' fear!
> Robert Burns

REFERENCES

Ashford, Nicholas A. 1976. Crisis in the workplace: Occupational disease and injury. Cambridge, Mass.: MIT Press.

Bazelon, David L. 1977. Coping with technology through the legal process. Cornell Law Review 62 no. 5 (June):817-832.

Blodgett, John E. 1974. Pesticides: Regulation of an evolving technology. In Consumer health and product hazards: Cosmetics and drugs, pesticides, food additives, vol. 2 of The legislation of product safety, ed. Samuel S. Epstein and Richard S. Grundy, 197-288. Cambridge, Mass.: MIT Press.

Boffey, Philip. 1975. The brain bank of America: An inquiry into the politics of science. New York: McGraw-Hill.

Brodeur, Paul. 1973. Expendable Americans. New York: Viking Press.

Brodeur, Paul. 1977. The zapping of America: Microwaves, their deadly risk, and the cover-up. New York: Norton.

Bupp, Irwin C., and Jean-Claude Derian. 1978. Light water: How the nuclear dream dissolved. New York: Basic Books.

Carpenter, Richard A. 1974. Legislative approaches to balancing risks and benefits in the regulation of chemicals. In Consumer health and product hazards: Chemicals, electronic products, radiation, vol. 1 of The legislation of product safety, ed. Samuel S. Epstein and Richard D. Grundy, 1-44. Cambridge, Mass.: MIT Press.

Coates, Joseph F. 1982. Why government must make a mess of technological risk management. In Risk in the technological society, ed. Christoph Hohenemser and Jeanne X. Kasperson, 21-34. AAAS Selected Symposium, 65. Boulder, Colo.: Westview Press.

Covello, Vincent T., and Joshua Menkes. 1982. Issues in risk analysis. In Risk in the technological society, ed. Christoph

Hohenemser and Jeanne X. Kasperson, 287-301. AAAS Selected Symposium, 65. Boulder, Colo.: Westview Press.

Crain, Robert L., Elihu Katz, and Donald B. Rosenthal. 1969. The politics of community conflict: The fluoridation decision. Indianapolis: Bobbs-Merrill.

Crenson, Matthew A. 1971. The un-politics of air pollution: A study of non-decision-making in the cities. Baltimore: Johns Hopkins Press.

D'Itri, Patricia A., and Frank M. D'Itri. 1977. Mercury contamination: A human tragedy. New York: Wiley.

Dotto, Lydia, and Harold Schiff. 1978. The ozone war. New York: Doubleday.

Environmental Protection Agency. 1978. Office of Planning and Management. Environmental emergency response: Volume IV, case studies. Washington: Environmental Protection Agency.

Epstein, Samuel S., and Richard D. Grundy, eds. 1974. The legislation of product safety. 2 vols. Cambridge, Mass.: MIT Press.

Esposito, John. 1970. Vanishing air: The Ralph Nader Study Group report on air pollution. New York: Grossman.

Fritschler, A. Lee. 1975. Smoking and politics: Policymaking and the federal bureaucracy. 2d ed. Englewood Cliffs, N.J.: Prentice-Hall.

Fuller, John G. 1977. The poison that fell from the sky. New York: Random House.

Goodfield, June. 1977. Playing god: Genetic engineering and the manipulation of life. New York: Random House.

Graham, Frank, Jr. 1970. Since silent spring. Boston: Houghton Mifflin.

Hightower, Jim. 1972. Hard tomatoes, hard times: The failure of the land grant college complex. Report of the Task Force on the Land Grant College Complex. Washington: Agribusiness Accountability Project.

Johnson, Branden B. 1979. A propositional inventory of technological hazard management. CENTED Background Paper no. 1. Worcester, Mass.: Center for Technology, Environment, and Development (CENTED), Clark University.

Jones, Charles O. 1975. Clean air: The policies and politics of pollution control. Pittsburgh: University of Pittsburgh Press.

Kasperson, Roger E., and Murdo Morrison. 1982. A proposal for international risk management research. In Risk in the technological society, ed. Christoph Hohenemser and Jeanne X. Kasperson, 303-322. AAAS Selected Symposium, 65. Boulder, Colo.: Westview Press.

Kavaler, Lucy. 1975. Noise: The new menace. New York: John Day.

Lawless, Edward W. 1977. Technology and social shock. New Brunswick, N.J.: Rutgers University Press.

Lowi, Theodore J., et al. 1976. Poliscide. New York: Macmillan.

Lowrance, William W. 1976. Of acceptable risk: Science and the determination of safety. Los Altos, Calif.: William Kaufmann.

Mankiewicz, Frank, and Joel Swerdlow. 1978. Remote control: Television and the manipulation of American life. New York: Times Books.

Marsh and McLennan, Inc. 1980. Risk in a complex society: a Marsh and McLennan public opinion survey. Conducted by Louis Harris and Associates. New York: Marsh and McLennan.

Nader, Ralph. 1965. Unsafe at any speed: The designed-in dangers of the American automobile. New York: Grossman.

Nader, Ralph. 1974. The regulation of the safety of cosmetics. In Consumer health and product hazards: Cosmetics and drugs, pesticides, food additives, vol. 2 of The legislation of product safety, ed. Samuel S. Epstein and Richard D. Grundy, 73-141. Cambridge, Mass.: MIT Press.

National Research Council. 1977. Committee on Environmental Decision Making. Decision making in the Environmental Protection Agency. Analytical Studies for the U.S. Environmental Protection Agency, vol. 2. Washington: National Academy of Sciences.

Nelkin, Dorothy. 1974. Jetport: The Boston airport controversy. New Brunswick, N.J.: Transaction Books.

Packard, Vance. 1977. The people shapers. Boston: Little Brown.

Primack, Joel, and Frank von Hippel. 1974. Advice and dissent: Scientists in the political arena. New York: Basic Books.

Quarles, John. 1976. Cleaning up America: An insider's view of the Environmental Protection Agency. Boston: Houghton Mifflin.

Rettig, Richard A. 1977. Cancer crusade: The story of the National Cancer Act of 1971. Princeton, New Jersey: Princeton University Press.

Rowe, William D. 1977. An anatomy of risk. New York: Wiley.

Schiff, Ashley L. 1962. Fire and water: Scientific heresy in the Forest Service. Cambridge, Mass.: Harvard University Press.

Steinhart, Carol, and James Steinhart. 1972. Blowout: A case study of the Santa Barbara oil spill. Belmont, Calif.: Wadsworth.

Stevenson, Gordon McKay, Jr. 1972. The politics of airport noise. Belmont, Mass.: Duxbury Press.

Turner, James S. 1976. The chemical feast. New York: Penguin Books.

United States Congress. 1978. House Committee on Government Operations. 1978. Dam Safety. House Report 95-880. Washington: Government Printing Office.

van der Linde, Peter. 1978. Time bomb. New York: Doubleday.

Verrett, Jacqueline and Jean Carper. 1974. Eating may be hazardous to your health. New York: Simon and Schuster.

14
Regulating Automobile Safety

Thomas Bick, Christoph Hohenemser,
and Robert W. Kates

We have described the risk factors and key elements of physical
theory underlying motor vehicle injuries and extracted from these a
prescription for individual risk reduction and a series of simply
stated highway and vehicle design criteria intended to reduce the
impact of motor-vehicle accidents. We concluded chapter 8 wondering
why, if we have the scientific knowledge to eliminate 80 percent of
present auto fatalities, we not do so. This consideration sets the
focus of the present chapter.

Our answer is framed in terms of the Clark University causal
model of hazard control presented in chapter 2. Taken together,
Table 1 and Figure 1 illustrate the application of the model to an
auto accident. Also diagrammed is a similar model drawn from the
work of William Haddon (1972), President of the Insurance Institute
for Highway Safety and first director of the National Highway
Traffic Safety Administration (NHTSA). Both models envision stages
of hazard development over time, starting with early and fundamental
causes, continuing to intermediate stages, and leading finally to
experienced consequences.

The two models define the structure of hazard causality and the
range of control actions available to managers; as such, their
purpose is primarily to encourage logical thinking. The models in
themselves do not enable us to understand why extensive scientific
knowledge has not generated greater success in reducing fatalities
and injuries. For this, it is necessary to look at the history of
auto hazard management.

The First Fifty Years of Regulation

Roadbuilding became a serious pursuit in the United States in
the late 1880s. Its purpose was primarily to provide feeder and
connector routes for the rapidly expanding rail system. In the
beginning, the building and maintenance of roads was a local responsibility. Because of the dominance of railroads, highways played
only a secondary role in the nation's transport system. In 1916 a
federal Bureau of Public Roads was established for the purpose of
providing farmers with access to markets. During World War I, U.S.
roads were still so poor that nearly all long-distance transport was
by rail. One shipment of army trucks from Detroit to New York, sent
by road rather than rail, took more than a week to arrive. Roadway

TABLE 1
Models of highway and vehicle safety management

CONTROL ACTIONS		EXAMPLES
Defined in chapter 2	Defined by Haddon	
1. Modify human wants	Pre-crash phase	Reduce car travel, reorder home/work location
2. Modify technology choice	Pre-crash phase	Substitute public transportation, other modes
3. Block initiating events	Pre-crash phase	Improve highway visibility, warning signs, driver training
4. Block outcomes	Crash phase	Median barriers, emergency brakes
5. Block conse-quences	Crash phase	Seatbelts, shatter-proof glass, removal of roadside barriers
6. Block higher order consequences	Post-crash phase	Fire-proof fuel tanks, prompt emergency medical aid

safety management during this period centered largely on dealing with the disturbing influence of a few automobiles on horse and bicycle traffic.

The beginning of the mass production of automobiles in 1920 rapidly transformed the roadway problem. Existing local roads soon became inadequate to carry the increasing traffic, and everywhere new roads were being built by the states. Along with this change came the first highway safety programs. By the mid-1920s, many states had laws dealing with safe driving, driver education, driver licensing, traffic courts, the identification and correction of hazardous highway areas, pedestrian safety, periodic vehicle inspection, and safe highway construction and maintenance.

In terms of the six hazard-control stages diagrammed in Figure 1, the principal focus of these early safety management efforts was stage 3—modifying or blocking initiating events leading to accidents. For whatever reason, early regulators of highway safety largely ignored accident outcomes (stage 4) and did little about modifying and mitigating consequences (stages 5 and 6). In

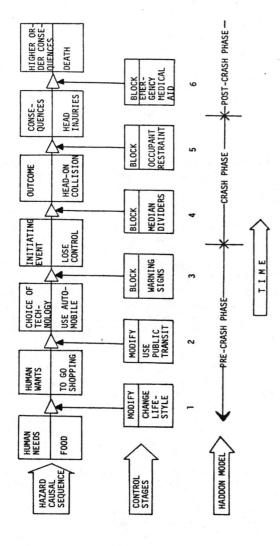

Figure 1. An illustration of the causal chain of hazard evolution. The top line indicates seven stages of hazard development, from the earliest (left) to the final stage (right). These stages are expressed generically in the top of each box and in terms of a sample motor vehicle accident in the bottom. The stages are linked by causal pathways denoted by triangles. The middle line indicates six control stages, linked to pathways between hazard stages by vertical arrows. Each is described generically as well as by specific control actions. Thus, control stage 2 should read: "You can modify technology choice by substituting public transit for automobile use and thus block the further evolution of the motor vehicle accident sequence arising out of automobile use." The third line indicates how Haddon and the Department of Transportation have structured the causal sequence of hazard. Three stages—pre-crash, crash, and post-crash—are envisioned, with various control actions and standards classified in terms of these. The fourth line shows the arrow of time. This applies to the ordering of a specific hazard sequence; it does not necessarily indicate the time scale of managerial action. Thus, from a managerial point of view, certain hazard consequences may occur first and then lead to control actions affecting initiating events.

terms of the three-phase model developed by Haddon, also shown in Figure 1, effort primarily focussed on the precrash phase, with little thought given to the crash and postcrash phases. This emphasis was to dominate the thinking of highway safety experts, with some exceptions, until the mid-1960s.

Federal Involvement

During this period, highway fatalities kept pace with the burgeoning number of vehicles. From 1910 to 1925, as the number of motor vehicles rose from one-half million to 20 million, the number of annual highway fatalities rose from 3,000 to 22,000 (Motor Vehicle Manufacturers Association 1976). In 1924 public concern about the problem prompted a response, albeit a modest one, by the federal government. Secretary of Commerce Herbert Hoover called a National Conference on Street and Highway Safety, in which specialists gathered together to discuss traffic safety management and to suggest ways of implementng this knowledge. The next ten years witnessed three similar federal conferences, none of which led to any specific action.

In the 1930s, the federal government became increasingly involved in construction and maintenance of highways, in part as a means of combatting the unemployment of the Depression. In 1935, at a time when the federal government accounted for 17 percent of total highway construction expenditures (U.S. Chamber of Commerce 1936), Congress passed its first highway safety legislation. The Motor Carrier Act of 1935 authorized the Interstate Commerce Commission to establish and enforce safety standards for motor carriers. A year later Congress directed the Federal Bureau of Public Roads, housed in the Department of Agriculture, to study the cause of highway collisions and to suggest appropriate countermeasures. The Bureau concluded that the primary causes of collisions were excessive speed, careless pedestrians, negligent drivers, poor visibility, and temporary hazards in the roadway. The report betrayed an overwhelming emphasis on collision prevention, with a special section devoted to identifying the collision-prone driver. It said nothing about improving the design of vehicles or roadways to minimize impact once a crash occurs. The report recommended that all states enact laws providing for the licensing of drivers, periodic vehicle inspection, speed limits, and the use of nationally uniform traffic control signals, accident reporting, and rules of the road (U.S. Bureau of Public Roads 1938). These recommendations, which included provision for a major federal role in highway safety management, were largely ignored until after World War II.

In 1942, the nation experienced its first major decline in motor vehicle fatalities--from nearly 40,000 per year to about half that number (see Figure 2, chapter 8). This decline was not the result of highway safety management but the consequence of drastically reduced driving and a federally mandated maximum speed of 35 miles an hour designed to save fuel and tires, as well as of the absence of thousands of young men serving in the armed forces.

By 1946, when President Truman convened the first President's Highway Safety Conference, the annual traffic fatality rate had shot back up almost to its prewar level. Participants urged the states to implement an "Action Program" designed to create a uniform, nationwide safety effort. The key aspects of this effort were basically similar to earlier proposals in their concentration on crash prevention rather than overall loss reduction; not one of the recommendations included proposals for improving occupant survivability during crashes. As before, Congress stopped short of mandating state compliance with the program.

The Interstate System

Under President Eisenhower in the mid-1950s, the nation embarked on a major new program of federal highway construction, lobbied for vigorously by highway construction interests and financed directly by federal gasoline taxes. The stated goal of the program was replacement of much of the federal highway system built in the 1930s, a system which, by this time, had become inadequate for 65 million motor vehicles and, moreover, did not permit the intercity speeds that were by then desired. The 42,500-mile system, originally called the Interstate and Defense Highway System, is now finally nearing completion after an expenditure of $132.4 billion—a far cry from the original estimate of $28 billion (Feaver 1982, A16; Motor Vehicle Manufacturers Association 1982,76). The Interstate Highway did result in fewer fatalities per vehicle mile (chapter 8), but an accompanying rise in miles traveled, increases in numbers of vehicles, and the advent of a large number of young "postwar-baby-boom" drivers partially offset this safety gain. Indeed, the annual number of deaths actually increased.

In 1954, Eisenhower called yet another White House conference on highway safety. This conference resulted in the creation of the President's Committee for Traffic Safety, the first permanent federal agency whose principal objective was the promotion of highway safety. One of the committee's first projects was a campaign for the adoption of the 1946 Action Program by every state. Though this campaign generated a marked increase in state highway safety expenditures, the response fell far short of the comprehensive nationwide program envisioned by the committee.

For the first time, in 1959 a report to Congress by the Secretary of Commerce called for comprehensive federal legislation mandating highway and vehicle design standards and for the creation of a federal agency to coordinate all federal traffic safety programs and research activities. This distinct departure from the narrow focus of previous regulatory activity foreshadowed the intensive programs of roughly a decade later.

In 1965, when motor vehicles numbered 90 million and traffic fatalities stood at 49,000 per year, a series of congressional hearings, at which the auto industry and state highway officials were roundly criticized for failing to do more to combat the highway death toll, produced renewed calls for a more active federal role. Later in the same year, Ralph Nader published Unsafe at Any Speed, a ringing indictment of the auto industry's "deliberate refusal" to make safer cars available to the public (Nader 1965). The congressional hearings and Nader's book received broad publicity, as did the conclusion of a special presidential study board that the inadequate performance of state and local highway officials was "a major reason for the current [highway safety] crisis" (Interdepartmental Highway Safety Board 1965).

This "crisis" developed more than forty years after Henry Ford first began mass producing motor vehicles and more than thirty years after the federal government became heavily involved as highway builder. In retrospect, it is clear that the preceding years were a time when highly centralized safety management, consistently ignoring the issues of automobile crashworthiness and highway safety, lagged behind scientific knowledge. Although the traffic mortality rate per unit of population remained roughly constant as vehicle numbers increased fivefold, this was no longer good enough. Congress, prodded by vocal and persuasive critics such as Nader, was finally ready to take action toward creation of a comprehensive federal role in safety management.

Federal Standard Setting Since 1966

The new era began with President Johnson's introduction of landmark legislation in early 1966. Congress subsequently divided this legislation into two bills—the Highway Safety Act and the National Traffic and Motor Vehicle Safety Act—both of which passed overwhelmingly.

The Highway Safety Act provided for an agency that would make federal funds available to the states for planning and evaluating highway safety measures. It directed the Secretary of Commerce (and later the Secretary of Transportation) to issue highway safety standards and required states to implement highway safety programs in accordance with these standards. Under the act, noncomplying states could lose planning funds and up to 10 percent of their federal highway construction subsidies. The act thus directly linked the funding of highway development and safety programs.

The National Traffic and Motor Vehicle Safety Act of 1966 also provided for a federal agency that, together with the agency established under the Highway Safety Act, was to administer a program concerned with the regulation of motor vehicles and equipment. This agency, which later became the National Highway Traffic Safety Administration (NHTSA), was charged with issuing performance standards covering all classes of new motor vehicles. The act also required automakers to notify purchasers about any safety defects. This authority was extended by a 1974 amendment that gave the Department of Transportation the authority to compel manufacturers to recall vehicles determined to be defective. In 1973, Congress passed amendments to the Highway Safety Act which for the first time earmarked federal subsidies for specific categories of highway improvements, such as roadside obstacles, pavement marking, unsafe bridges, and high-hazard locations.

Taken together, these laws set the framework for a significant departure from the predominant emphasis, which had persisted for more than forty years, on accident prevention. In particular, the second 1966 law envisioned numerous managerial strategies that could be classified as preventing outcomes and consequences in the terms of our hazard model, or as falling within the crash and postcrash phases in Haddon's schema. In addition, the two 1966 laws marked the beginning of an enormous financial commitment to reduce the death and destruction on the nation's highways. From fiscal year 1967 to fiscal year 1977, the federal government spent $6.6 billion to achieve this goal (Insurance Institute for Highway Safety 1978).

The cost to state and local governments has been even greater. In fiscal 1977 alone, states and localities spent more than $4 billion to comply with federal highway safety legislation; in the same year, new car buyers spent an additional $260 per car, or a total of about $2.5 billion, for federally mandated safety features. For the first time in history, expenditures on safety became an appreciable fraction of highway and vehicle expenditures.

How effective has been the 12-year effort during 1967–1977? One measure is the time trend of highway deaths shown in Table 2. On the basis of this trend, NHTSA claimed in 1978 that a total of 150,900 lives were saved during 1967–1977 because of the reduction of the mortality rate per 100 million vehicle miles from 5.25 to 3.25 (NHTSA 1978b). Furthermore, NHTSA strongly suggested that the observed effects bore a direct relationship to the Highway Safety Program:

> There is a way in which the value of traffic safety effort can be roughly measured. If the fatality rate had remained constant since 1967, the direct economic cost would probably now exceed **$76 billion each year**, and traffic deaths would approximate **77,000 annually**. Table 2 demonstrates what might have happened without the positive effects of the motor vehicle and highway safety standards and institution of the 55 mph speed limit. (NHTSA 1978a, 2)

In our view, there is some reason to be skeptical that safety gains during 1967–1977, which as such are undeniable, have in fact resulted from the federal highway safety program. Changes in the fatality rate might well have occurred quite independently. Thus, a decline during 1967–1977 would be likely because of the rapid entry of women into the pool of drivers and the growing proportion of urban driving, whereas an increase would be predicted based on the changing age distribution of drivers and the growing number of young drivers.

In the period 1977–1984, when little new federal regulation was instituted, fatality rates first rose but more recently dropped further to estimated rates for 1984 of 43,500 total deaths and 2.55 deaths per 100 million vehicle miles (National Safety Council 1983; Cerrelli 1984). Whereas no one is quite sure of the principal reasons behind these trends, it is obvious that an intermingling of many factors is involved, not the least of which is general decline in the U.S. economy during 1980–1982. As to the future, NHTSA argues that a number of factors will contribute to a rise in fatalities by 1990. Based on downsizing of cars, changing vehicle mix, increases in driving and drivers, erosion of the 55-mph speed limit, deterioriation in helmet laws, and increased motorcycle, bicycle, and pedestrian fatalities, 70,000 deaths and a rate of 3.55 deaths per 100 million vehicle miles are projected (NHTSA 1982).

Perhaps the most fundamental reason why gains since 1967 have not been consistently greater is societal failure to implement those strategies that by all counts are the most effective. The outcome of decisions made by the Department of Transportation, the motor vehicle industry, and the public, this situation points up the often unbridgeable gap between the science and the politics of highway safety management. To illustrate, we discuss the implementation of the two major acts passed by Congress in 1966.

TABLE 2
Lives saved due to fatality rate reduction[a]

YEAR	FATALI- TIES (BASED ON 1967 RATE)	DEATH RATE[b]	ACTUAL DEATHS	LIVES SAVED DUE TO 55-MPH SPEED LIMIT	LIFE SAVED DUE TO SAFETY PROGRAMS	TOTAL LIVES SAVED
1967	50,724	5.25	50,724		---	---
1968	53,535	5.17	52,725		810	810
1969	55,973	5.02	53,543		2,430	2,430
1970	58,487	4.72	52,627		5,860	5,860
1971	62,142	4.44	52,542		9,600	9,600
1972	66,419	4.32	54,589		11,830	11,830
1973	69,128	4.11	54,052		15,080	15,080
1974	67,376	3.52	45,196	5,070	17,110	22,180
1975	69,865	3.35	44,525	5,100	20,240	25,340
1976	73,979	3.23	45,509	4,500	23,970	28,470
1977	76,940	3.25	47,671	1,600	27,700	29,300
				16,270	134,630	150,900

[a]Adapted from NHTSA (1978a).
[b]Death rate is in units of deaths per 100 million vehicle miles.

The Highway Safety Act of 1966

Two major problems have impeded implementation of the Highway Safety Act of 1966. First, despite the mandate to broaden greatly the scope of risk management, the standards actually set tend to fall primarily within the traditional areas and to break only a modest amount of new ground.[2] Secondly, given the availability of many possible control actions, hazard managers in the Department of Transportation have been excruciatingly slow in defining priorities in terms of cost-effectiveness. Concomitantly, a number of highly cost-effective actions have been delayed and some cost-ineffective actions have been implemented on a grand scale. Thus the 1966 Act has not lived up to its potential.

Standards Set

By 1978, NHTSA had set eighteen standards, all during the period 1967-1972. Most of these standards are administered by NHTSA; three and part of a fourth are administered by the Federal Highway Administration (FHA).[3] We have classified each standard by hazard control stage (recall Fig. 1) and plotted the number of standards applicable to each stage (Fig. 2). This indicates that, the broadened perspective of the law notwithstanding, efforts have been concentrated in stage 3 of our model--that is, blocking of initiating events. Only 19 percent of the standards address control stages 4-6, or, equivalently, the crash and postcrash phases in the Haddon model. None of the standards address stages 1 and 2,

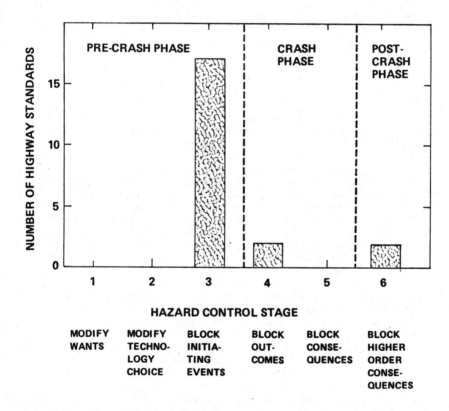

Figure 2. Distribution of highway safety managerial effort among the six stages of hazard control defined in our model (Fig. 1). Managerial effort in this case is measured by 18 highway safety standards issued by the Department of Transportation. Also shown are the boundaries of the Haddon model (dotted lines). The graph shows that only 19 percent of the highway safety standards apply to stages 4-6, none to stages 1 and 2, and 81 percent to stage 3. (The sum of the bar lengths adds up to 21 instead of 18 because we judged some standards to apply to more than one control stage and counted them under each applicable stage.)

modification of wants and choice of technology. In 1978, partly because of the states' reluctance to implement the standards, the Secretary of Transportation proposed scrapping them and substituting six principles that were, if anything, even more traditional in their focus (NHTSA 1978a, 1978b), but Congress rejected this proposal.

Costs and Benefits

Authors of the Highway Safety Act intended its implementation to occur in a cost-effective manner; that is, those measures that would forestall the greatest number of deaths and injuries per safety dollar were to be put into effect first. To estimate the cost-effectiveness of a measure, it is necessary to know the comparative costs of different actions designed to achieve an established goal. Ideally, in the case of motor vehicle safety, this goal would include fatalities, injuries, and property damage prevented. In practice, however, as in the **Highway Needs Report** (U.S. Dept. of Transportation 1976a) discussed below, the established goal has been limited to fatalities prevented, since data on injuries and property damage has in the past been too imprecise to permit realistic estimates.

Although the cost of many control actions has been roughly estimated, calculating the benefits of such actions has proved a more difficult task. To estimate the benefits of a particular control action it is necessary to know something about the causal structure of hazards (chapter 2). One might think that, in the case of motor vehicle hazards, the rather extensive information on risk, combined with underlying physical theory (as summarized in chapter 8) would be sufficient for this purpose. Yet, in 1966, it was widely believed that the number of motor vehicle deaths prevented by specific control actions could not be predicted. Thus, President Johnson in his 1966 Transportation Message said, "Our knowledge of cause is grossly inadequate. Expert opinion is frequently contradictory and confusing" (Johnson 1966).

When standards were set, therefore, they were unaccompanied by benefit estimates. Department of Transportation (DOT) planners, though insisting upon implementation of the standards, left priority ranking largely to the states. As time passed, many within DOT became convinced that the cost-effectiveness of highway safety standards could never be known with any degree of accuracy, that most control actions were so interrelated as to make separate evaluation of a given measure impossible. This attitude frustrated state highway officials who felt their states were being forced to implement, at considerable expense, federal standards of unknown effectiveness.

As a result, unhappy states began to rebel openly and to take their complaints directly to Congress. In 1973 Congress held full hearings on needed changes in the Highway Safety Act. State officials testified that the federal government was not doing all it could to evaluate the potential costs and benefits of alternative safety strategies. Congressional overseers agreed. A House report summarizing the hearings noted that there was little evidence that the money spent on the highway safety program had yet to add much to highway safety. In particular, the report noted:

One of the greatest weaknesses of the present program is the paucity of specific, up-to-date comprehensive data to support action programs. The trouble is, without adequate data, untold millions in safety monies may be spent unwisely. (U.S. Congress 1973)

To fill this gap in information, Congress in 1973 directed the Department of Transportation to complete a thorough study of the nation's highway safety needs and to prepare estimates of the costs of meeting such needs. In effect, Congress asked DOT to prepare cost-effectiveness studies and priority rankings of control actions.

The Highway Needs Report

In 1976, ten years after passage of the Highway Safety Act, and three years after the congressional mandate to provide a cost-benefit analysis for highway standards, the DOT published The National Highway Safety Needs Report (U.S. Dept. of Transportation 1976a), the first comprehensive effort to rank safety measures in terms of cost-effectiveness. The report was based on an extensive literature search and consultation with a blue-ribbon panel of 103 highway experts. From an initial list of 200 possible highway safety measures, DOT culled and analyzed in terms of cost 37 control actions of "potentially high payoff." As shown in Table 3, the 37 actions cover a wide range. It was estimated that mandatory seatbelt use, the most effective item, would save 89,000 lives over a ten-year period at a cost of $500 per life; in contrast, roadway alignment, the most ineffective action, was seen as saving only 590 lives in ten years at a cost of $7.68 million per life.

This report is useful in estimating the effect of implementing control measures in order of their cost-effectiveness—presumably the most efficient way to make decisions about how to spend taxpayers' money. Such an estimate, shown in Figure 3, indicates that beyond an annual expenditure of $1.5 to $3.0 billion (where the curve becomes essentially flat), very little is gained by further expenditure.

In addition, it is interesting to note that the 37 control actions considered by the 1976 **Highway Needs Report** appear to offer a rather well-balanced attack on the problem. As Figure 4 shows, classification of control actions by stages of our hazard model indicates that 40 percent of them address control stages 4-6. This should be compared to 19 percent in the case of standards actually issued by DOT (see Fig. 2). In this sense, then, the report is more nearly responsive to the initial directives of the enabling legislation than DOT's actions have been.

How has the report been received? Some DOT officials continue to question its value because of the methodology used in arriving at cost-effectiveness estimates. In particular, some officials argue that the report is no better than the intuitive "common sense" approach whereby highway safety experts within DOT determine priorities on the basis of their personal experience. Moreover, the authors of the report reinforce such criticism when they note that "our current information is neither sufficiently accurate nor

322

TABLE 3
Highway safety control actions ranked in order of cost-effectiveness

CONTROL ACTION	NUMBER OF FATALITIES FORESTALLED	COST (MILLIONS OF DOLLARS)	COST PER FATALITY FORESTALLED (1000s OF DOLLARS)
1. Mandatory Safety Belt Usage	89,000	45	0.5
2. Highway Construction and Maintenance Practices	459	9	20
3. Upgrade Bicycle and Pedestrian Safety Curriculum Offerings	649	13	20
4. Nationwide 55 mph Speed Limit	31,900	676	21
5. Driver Improvement Schools	2,470	53	21
6. Regulatory and Warning Signs	3,670	125	34
7. Guardrail	3,160	108	34
8. Pedestrian Safety Information and Education	490	18	36
9. Skid Resistance	3,740	158	42
10. Bridge Rails and Parpets	1,520	69	46
11. Wrong-Way Entry Avoidance Techniques	779	38	49
12. Driver Improvement Schools for Young Offenders	692	36	52
13. Motorcycle Rider Safety Helmets	1,150	61	53
14. Motorcycle Lights-On Practice	65	5	80
15. Impact Absorbing Roadside Safety Devices	6,780	735	108
16. Breakaway Sign and Lighting Supports	3,250	379	116
17. Selective Traffic Enforcement	7,560	1,010	133
18. Combined Alcohol Safety Action Countermeasures	13,000	2,130	164
19. Citizen Assistance of Crash Victims	3,750	784	209
20. Median Barriers	529	121	228

21.	Pedestrian and Bicycle Visibility Enhancement	1,440	332	230
22.	Tire and Braking System Safety Critical Inspection—Selective	4,591	1,150	251
23.	Warning Letters to Problem Drivers	192	50	263
24.	Clear Roadside Recovery Area	533	151	284
25.	Upgrade Education and Training for Beginning Drivers	3,050	1,170	385
26.	Intersection Sight Distance	468	196	420
27.	Combined Emergency Medical Countermeasures	8,000	4,300	538
28.	Upgrade Traffic Signals and Systems	3,400	2,080	610
29.	Roadway Lighting	759	710	936
30.	Traffic Channelization	645	1,080	1,680
31.	Periodic Motor Vehicle Inspection—Current Practice	1,840	3,890	2,120
32.	Pavement Markings and Delineators	237	639	2,700
33.	Selective Access Control for Safety	1,300	3,780	2,910
34.	Bridge Widening	1,330	4,600	3,460
35.	Railroad-Highway Grade Crossing Protection (Automatic gates excluded)	276	974	3,530
36.	Paved or Stabilized Shoulders	928	5,380	5,800
37.	Roadway Alignment and Gradient	590	4,530	7,680

NOTE: Control actions include only those authorized by the Highway Safety Act. They do not include vehicle performance standards, which are authorized under the Motor Vehicle Safety Act.

Source: U.S. Department of Transportation (1976a,p.VI-8). The number of fatalities forestalled is based on a 10-year period. All figures are subject to substantial uncertainties, as described in Note 4 and in the text.

324

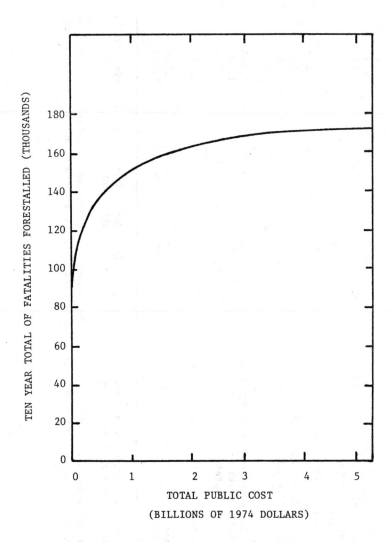

Figure 3. The figure illustrates a principal result of the 1976
National Highway Safety Needs Report (NHTSA 1976a). It shows the
relationship that would exist between public expenditure in highway
safety and the number of fatalities forestalled if control actions
were implemented in order of cost-effectiveness (Table 3). The plot
is useful because it indicates beyond what level of expenditure no
significant gains would be made in fatalities forestalled. This
level appears to be in the neighborhood of $1.5 to $3 billion.
Note, however, that this is true only if the assumptions of the plot
are satisifed; that is, if the measures are put into effect in increasing order of cost per fatality forestalled, starting with the
lowest.

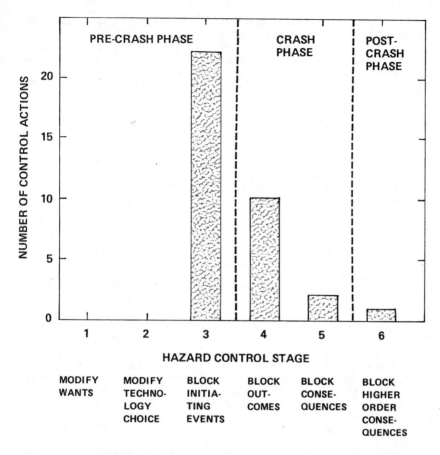

Figure 4. Distribution of 37 highway safety control actions analyzed in the 1976 National Highway Safety Needs Report (NHTSA 1976a)
among the six stages of hazard control. The figure indicates a more
even distribution than the actual standards issued (Fig. 3). For
example, 40 percent of the actions fall into stages 4-6, as compared
to 19 percent for the standards issued. (The sum of the bars adds
to 40 and not 37 because some actions, judged to apply to more than
one control stage, were counted under each applicable stage.)

conclusive to support a truly definitive analysis" (U.S. Dept. of
Transportation 1976a,I-4).

In our view, these are valid objections if they are focussed on
the effectiveness of particular actions. Yet the wide range in
effectiveness indicated by the report is a result unlikely to be
altered qualitatively by better data.[4] Of particular interest is
the fact that a number of control actions appear to have costs substantially below the average of $200,000 to $300,000 per life normally obtained in economic valuations of life (Rice, Feldman, and
White 1976;Faigin 1976).

For the DOT official confronting the present situation, the 1976 **Highway Needs Report** should prove a valuable consensus document and an important step toward transcending the squabbles between various individual experts. In particular, it illustrates how a cost-effectiveness framework can potentially facilitate the process of resource allocation in highway safety management. But how does such facilitation work in practice?

Seatbelts: A Failure

If the **Needs Report** makes one thing clear, it is that mandatory seatbelt use is by far the most cost-effective control action when compliance is high. A seatbelt law and/or an educational campaign effective enough to promote 80 percent seatbelt use by front-seat passengers could save 89,000 lives in a ten-year period at the trivial cost of $500 per life:

> Since the effectiveness of belt systems is so great, and almost all cars now have these belt systems installed, it could be said that the country's greatest highway safety need, above all others, is for a countermeasure program which will achieve higher belt wearing rates by car occupants. (U.S. Dept. of Transportation 1976b,A-257)

In this context it is worth noting that most research places seatbelt usage in the U.S. at about 11 percent (Claybrook, Gillan, and Strainchamps 1982,29-31).

What obstacles have stymied the implementation of much higher belt-usage rates? Or, put another way, why have so few states enacted mandatory seatbelt-use laws similar to those now in effect in 33 nations around the world (Grimm 1984), despite many attempts in state legislatures? Only in the area of child restraints have such laws been forthcoming (Mann 1982; Sinclair 1982). Perhaps now that all 50 states and the District of Columbia have enacted some form of child-restraint legislation (Insurance Institute for Highway Safety 1985), the long-term prospects for increased adult use are brighter.

A detailed analysis of this entire seatbelt issue, interesting as it may be, is beyond the scope of this review. Instead, we provide in Table 4 a synopsis of major issues. This summary suggests that the single major obstacle to seatbelt-use laws in the United States is popular rejection of such laws. On the basis of experiments on risk perception and acceptance by the general public, it is possible to understand why and how the United States public arrives at its opposition (Slovic, Fischhoff, and Lichtenstein 1978). It is more difficult, however, to generalize about official conceptualizations of the problem when countries as diverse as Australia, France, Ivory Coast, Japan, and Sweden have mandated use of belts. What is clear is that DOT managers have vacillated over time between behavioral approaches (such as education and coercion) and technological approaches (such as passive restraints).

Other countermeasures that are high on the **Needs Report** cost-effectiveness list but remain unimplemented or underimplemented may be analyzed in a manner similar to the seatbelt case. Enforcement of the 55-mph speed limit (fourth in cost-effectiveness) and

TABLE 4
The debate over compulsory seatbelt use laws

ARGUMENTS AGAINST COMPULSORY USE	REBUTTAL
Even if such laws were enacted nationwide and resulted in 80 percent use, they would not significantly reduce motor vehicle accident mortality.	There is substantial agreement that 80 percent use would reduce motor vehicle accident mortality by 20-25 percent.
Mandatory belt use laws would not significantly increase current levels of use, which are about 14 percent.	Extent of compliance depends on degree of enforcement. Some nations, including France, have achieved 75-80 percent use through enforcement of seatbelt laws.
The prospective installation of passive restraints makes seatbelt laws a moot issue.	It will take 10 to 15 years to equip all cars with passive restraints. Thus, belt-use laws passed today could still save thousands of lives.
A broad educational campaign would significantly increase voluntary use, thus obviating the need for compulsory use.	So far educational campaigns, including highly intensive media campaigns in specific communities, have failed to achieve significant improvements.
The expense and difficulty of enforcing seatbelt laws makes them impractical.	Enforcement costs are included in cost-effectiveness estimates given in the 1976 National Highway Needs Report and, even so, seatbelt laws come out as a control action that is 40 times cheaper than any other.
Mandatory seatbelt laws cannot be considered politically feasible.	This is the position presently taken by the Department of Transportation, and it is supported by public opinion polls. Thus Gallup found 76 percent opposed compulsory seatbelt laws. Yet given the success of such laws in some other countries, it is not clear that public acceptance is unobtainable.

state laws requiring the use of motorcycle helmets (thirteenth in cost-effectiveness) both fall into this category. In both cases, the major obstacles appear to be sociopolitical. For 55-mph enforcement the major block to implementation appears to be the resistance of state highway officials to federal intrusion, coupled with the open defiance of many long-distance truckers. In the case of enforcing the wearing of motorcycle helmets, opposition comes from highly effective special interest groups composed mainly of motorcycle enthusiasts.

Periodic Motor Vehicle Inspections: Too Successful

The **Needs Report** is also valuable for singling out actions of low cost-effectiveness that nevertheless consume relatively large shares of the highway safety dollar. One example is the requirement that states conduct periodic motor vehicle inspections (PMVIs), which ranks thirty-first of 37 possible safety measures, at $2.1 million per life saved.

Prior to 1976 the concept of PMVI had been one of the most persistently advocated vehicle safety measures. In the 1966 House Report that accompanied the Highway Safety Act, PMVI was named as one of the seven highway safety areas requiring a national standard (U.S. Dept. of Transportation 1977a). Thus, PMVI was one of the first federal safety standards and, since 1970, approximately $120 million per year has been spent on its implementation by the states. If the standard were fully enforced by every state in 1980, the total cost would probably be close to $300 million per year (NHTSA 1978a).

Many states, citing inconclusive evidence of the benefits to be derived, have resisted implementation of the PMVI standard. The decade from 1966–1976 was marked by repeated efforts by DOT to encourage state compliance—efforts that included listing PMVI as one of DOT's "must-items in 1972" and one of its five "bottom line emphasis elements" in 1974. In 1975 DOT announced its intention of beginning sanction hearings against thirteen states that had failed to implement all the requirements of the PMVI standard. Congress, in reaction to DOT's threat to use its sanction power, later cancelled these hearings.

Assuming the **Needs Report's** cost-effectiveness estimate to be reasonably reliable, the past emphasis on this measure represents a significant misallocation of federal and state safety resources. To its credit, DOT has, since publication of the **Needs Report**, deemphasized PMVI. It has, for example, recommended making implementation of the PMVI standard optional (U.S. Dept. of Transportation 1977a), suggesting the standard serve only as a guideline to those states wishing to maintain inspection programs voluntarily. DOT has also resisted the recommendation of the General Accounting Office that it undertake further priority research into the effectiveness of PMVI. By 1979 PMVI was compulsory in only 26 of 50 states, and though DOT was continuing to push for PMVI, the urgency had clearly gone out of a once avid campaign.

Judging from Table 3, PMVI is by no means the only case of gross misallocation of safety funds. Enormous sums have been spent annually in recent years on measures that the Needs Report rates as thirty-sixth and thirty-seventh in cost-effectiveness. These are paved or stabilized shoulders, at $5.8 million per life saved, and roadway alignment, at $7.7 million per life saved. Why, considering these low levels of cost-effectiveness, are states continuing to spend limited safety funds on these countermeasures? One answer is that the funds in question are not really that limited, nor are they specifically designated for safety purposes.

According to some DOT officials, continued spending results from the availability of large amounts of highway construction money in both federal and state highway trust funds and from the powerful influence exerted at the state level by a consortium of highway construction interests, including labor unions, highway contractors, and suppliers of materials and equipment. A particularly clear example of this phenomenon was an attempt in 1976 by Governor Michael Dukakis of Massachusetts to return to Washington funds earmarked for superhighway shoulder-widening (Lewis 1976). Arguing that such measures are a waste of the taxpayers' money, Dukakis met with a storm of protest, from the construction industry and its supporters, and that finally forced him to do the politically expedient thing, that is, rescind his earlier order to return the funds.

Dealing rationally with low-priority highway safety items, then, is every bit as difficult as dealing with high-priority items. The 1976 Highway Safety Needs Report in both cases serves as a valuable framework for formulating hazard decisions. In some cases involving blatant political interference with hazard-management policy, it can also serve as an uncomfortable reminder that all that is good for a particular interest or group may not be sound safety policy. Yet, as a purely logical and analytical framework, the report does literally nothing to resolve these problems. The solution ultimately involves balancing the varied and complex forces set in motion by public perception and acceptance, official dogma, and the private benefits derived from the public hazard-management dollar.

The Motor Vehicle Safety Act of 1966

The administration of the 1966 Highway Safety Act, as outlined in the previous section, has generated considerable controversy. Yet, by comparison, the other 1966 law, the National Traffic and Motor Vehicle Safety Act, has been even more controversial. The act for the first time gives the federal government authority to issue vehicle performance standards mandated for all new cars sold in the United States. These standards are to be designed not only to prevent vehicles from crashing but to protect vehicle occupants and other crash victims once a crash occurs. Hence, like the Highway Safety Act, the Vehicle Safety Act represents a historic shift from a narrow emphasis on crash prevention toward the broader perspective of crash prevention plus crash survivability.

Standards Set

By 1978, DOT had issued 50 vehicle standards which, as a group, show a remarkable concentration in the areas of the hazard causal sequence previously ignored in motor vehicle accident management. Thus, as seen in Figure 5, 80 percent of the standards affect control stages 4-6 in our model, and 24 of the 50 standards are located within the crash or postcrash phases of the Haddon model.

330

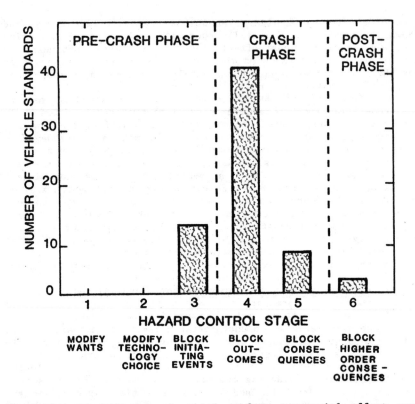

Figure 5. Distribution of vehicle safety managerial effort among the six stages of hazard control defined in our model (Fig. 1). Managerial effort in this case is measured by 50 vehicle standards issued by the Department of Transportation. The distribution of effort is seen to be heavily weighted toward stages 4-6, with 80 percent of the standards issued falling in this region.

Typical of the new vehicle standards are regulations about energy-absorbing steering assemblies, padded dashboards, occupant restraint systems, and fuel-system safeguards--all designed to block consequences or prevent their exacerbation into higher-order consequences.

As the highway standards discussed earlier, vehicle standards can have widely varying costs and benefits. In contrast to most highway standards, however, the cost of vehicle standards is largely than the public sector of the economy. As such, the cost of vehicle standards is less "hidden" than that of highway safety standards. Some individual car owners who think of themselves as safe drivers feel the equipment may not be necessary. The automobile industry fears the added costs will affect the demand for new cars.

In addition, serious doubts can easily arise about the cost-effectiveness of a particular standard. These doubts are particularly compelling for those who are opposed on principle to any mandated private expenditure on public safety. Prior to a Reagan executive order **requiring** cost-benefit analysis (1981), DOT decision makers were not required by their enabling legislation to make economic factors the controlling consideration in their decisions. Rather, they were required to "consider" economic impacts and other "costs" of vehicle standards, but they could issue new standards, whatever their cost, as long as they were "practicable," an undefined term in the act that gave broad discretion to the agency. This is scarcely tantamount to requiring "a showing of a quantified excess of benefits over costs" (Claybrook, Gillan, and Strainchamps 1982,4). Yet the Reagan order does just that, and the result has been revocation of or inaction on existing or proposed safety standards.

Passive Restraints

The complexity of DOT's vehicle standards program is perhaps best illustrated by the most important, and probably the most cost-effective, example of the fifty vehicle standards issued to date—that dealing with passive restraints. Passive restraints, with minor differences, have the same purpose and function as manually operated seatbelts: they assure that the stopping distance of front-seat passengers in frontal crashes is at least as long as that of the crashing vehicle (chapter 8). The difference is that passive restraints do not require voluntary action by vehicle occupants. Instead, the restraint works without any intervention on the part of the occupant. Two popular forms of passive restraints are airbags, which inflate during frontal crashes on signal from an impact sensor, and passive belts, which automatically move into place around the occupants of the front seat.

Because of their functional similarity to seatbelts, passive restraints, if fully deployed, would save about the same number of lives—12,000-13,000 per year. Initial cost is estimated to be $27 and $120 (in 1976 dollars) for passive belts and airbags, respectively.[5] (These cost figures assume high-volume mass production; on occasion, industry quoted much higher figures for optional equipment produced in low volume.) Assuming a ten-year lifetime for new vehicles equipped with airbags, this means their respective cost-effectiveness is about $30,000 and $130,000 per life saved.[6] This makes them comparable to some of the most effective control actions listed in the 1976 Highway Safety Needs Report (see Table 3). Despite a higher initial cost than seatbelts, passive restraints enjoy a high level of public acceptance (NHTSA 1978c). For example, a June 1977 Gallup poll showed the public favoring the installation of air bags in all new cars by 46 to 37 percent, while a 1979 GM study showed that 70 percent of drivers would select the airbag as their choice of restraint system (Insurance Institute for Highway Safety 1977;NHTSA 1979).

What then is the reason that passive restraints, though fully developed technologically for at least ten years, have been installed in so few vehicles? The answer is a case study in frustrated hazard management which has probably cost about as many American lives as the Viet Nam War. The opponents of passive restraints in this drama run from the principal auto makers, to Ronald Reagan, to a variety of conservative groups advocating a "government-keep-hands-off" policy. The proponents are some members of DOT, Ralph Nader, Joan Claybrook, and allied public interest groups, and a number of private groups such as the Insurance Institute for Highway Safety.

The story begins in 1942 when Hugh DeHaven, working at Cornell Medical College, began publishing papers showing, first, that properly packaged, the human body is extremely resistant to transient decelerative forces and second, that the provision of such proper "crashpackaging," would greatly reduce injuries and deaths among occupants of motor vehicles and aircraft (DeHaven 1952). This insight was brought to the new highway safety program in 1966 by its first director, William Haddon. The restraint of vehicle passengers during crashes soon became one key part of DOT's packaging strategy. Vehicle Standard 208, adopted in 1966, required the installation of lap and shoulder belts in all passenger cars. Since so few people voluntarily used these restraints, however, mere installation did not solve the problem.

Haddon's forceful advocacy soon convinced DOT officials that passive restraints were preferable to active ones for the simple reason that they do not depend on the cooperation of distracted or unwilling drivers. Accordingly, as early as 1969 DOT issued an advance notice of a proposed amendment to Vehicle Standard 208 that would require impact-inflating airbags in all new passenger cars by 1972. The notice specifically called for installation of "inflatable occupant restraint systems," but it was soon changed to "passive restraints" in order to permit automakers to choose between airbags and passive belts, or any other system that would perform equally well.

From the beginning, though the technology was available off the shelf, particularly at General Motors (U.S. Congress 1976), automakers vigorously opposed mandatory passive restraints for a variety of reasons, including cost, lack of reliability, and infringement on the right of the public to choose. They took their case to the White House, to Congress, and via a concerted media campaign, to the American people. Succumbing to this unprecedented lobbying effort, DOT in 1970-1971 thrice delayed the effective date of the rule. In addition, in 1971 NHTSA decided to give automakers a choice between passive restraints and ignition-belt interlock systems (which prevent cars from starting unless seatbelts are fastened). This, in effect, mandated the installation of interlocks in all 1974 cars, because interlocks were far cheaper than airbags.

Both the ignition interlock option and the delay on passive restraints may well have originated with the Nixon White House. The President's interest in the interlock system began soon after Henry Ford II visited the White House to complain about the cost of government regulation and to extol the virtues of the interlock system as an alternative. Soon thereafter, following a White House meeting with presidential aides Peter Flanigan and John Ehrlichman, then Secretary of Transportation John Volpe informed DOT officials that the President wanted DOT to defer the restraint standard. DOT subsequently proposed the interlock option as an allowable substitute for the airbag.

In 1972 Ralph Nader and the Center for Auto Safety filed suit in Federal District Court to block the interlock option on the basis that it resulted from rulemaking based on secret, **ex parte** communications, not included in the proper DOT rulemaking docket, between the White House and the Secretary of Transportation. The suit was unsuccessful, and all 1974 cars rolled off the assembly lines equipped with the interlocks.

As might have been expected from public rejection of "buckle-up" campaigns, the interlock system proved unpopular. This prompted skeptics, Nader included, to suggest that the interlock system was part of a devious scheme hatched in Detroit and supported by the White House to mobilize the public against any type of mandatory restraint for automobile occupants. Indeed, the scheme was a short-term success. Soon after the 1974 models began filling dealers' showrooms, Congress responded to the resulting public outcry not only by repealing the interlock amendment but by reserving for Congress the power to veto within sixty days any DOT rule dealing with passive restraints.

The effective date of the passive restraint standard was indefinitely suspended. DOT, interpreting the public opposition to interlocks as a reaction against any mandatory seatbelt-use laws, scuttled remaining efforts in this direction, including a plan to award federal incentive grants to states that enacted such laws.

But the issue was not dead. Safety advocates continued to push for a passive restraint rule. Early in 1976 Secretary of Transportation Coleman assumed direct control over the controversial issue. He commissioned a special study and held a series of hearings to solicit the opinions of experts on both sides of the issue. In December 1976 he reported his findings (U.S. Dept. of Transportation 1977b), concluding that passive restraints are cost-effective and workable. Nevertheless, he decided not to require their installation in new cars but opted instead for a demonstration program to begin in 1978.

In 1977 Brock Adams took over from Coleman as Secretary of Transportation and immediately called for new hearings on the issue. announced in June 1977 his decision to overrule his predecessor and to mandate automatic protection for front-seat occupants for all new cars beginning in 1981. Congressional resolutions to veto the new ruling failed, and Adams's order became "final" on 1 September 1977.

Both sides in the fight immediately responded with lawsuits. The Pacific Legal Foundation, a conservative group opposed to mandatory safety regulation, sued to overturn the order as an abuse of DOT rulemaking discretion, but their suit was denied by a unanimous appellate court decision. On the other side, Ralph Nader and his group, Public Citizen, challenged the order for permitting unreasonable and unnecessary extension of the effective date. Meanwhile, congressional opponents led by Representative Bud Shuster (R-Pa), successfully attached an amendment to the 1979 DOT appropriations bill prohibiting the use of any funds for implementing or enforcing any regulation requiring the use of passive restraint systems other than belts.[7] Despite these efforts, as of the end of 1980 Brock Adams's order stood, and automakers were preparing to introduce passive restraint systems in all 1981 models. Victory for passive restraints appeared inevitable.

But victory was not to be. In one of its first acts in office the newly elected Reagan administration rescinded the Adams order (October 1981), only to have the court reinstate it (June 1982), agree to review its own reinstatement (November 1982), and finally submit to yet another postponement. By February 1983, the U.S. Court of Appeals had granted manufacturers at least until late 1984 to comply with the standard. On 11 July 1984, a decision by Secretary of Transportation Elizabeth Dole effectively prolonged the delay and reopened the gate for eventual rescindment. Dole ruled that all automobiles sold after 1 September 1989 must have passive restraints (air bags or automatic seatbelts) unless states representing more than two-thirds of the population of the United States enacted mandatory seatbelt laws by 1 April 1989. Indeed, it is difficult to conjure up a more complete frustration of hazard management.

Yet the case of passive restraints, although the most widely disputed, is not exceptional. The pattern repeats itself in many other cases.

The Air Brake Standard

DOT Vehicle Standard 121, the Air Brake Standard, is premised on the simple physical principle that the more uniform the braking distances of all highway vehicles, the lower their collision rate will be, as well as on the need to prevent skidding. The basic requirement of the standard is that a vehicle equipped with air brakes be capable of stopping in a limited distance without leaving its traffic lane or locking its wheels. As such, the standard is aimed primarily at large trucks and buses. (In 1972, multiunit trucks and commercial buses, which constitute less than one percent of all vehicles, were responsible for 6.5 percent of all fatal accidents (NHTSA 1978a).

To meet the antilock provision of the standard, manufacturers have developed electronic control braking systems that cost about $1,200 per vehicle (NHTSA 1978a). At that cost, a rough estimate of the standard's cost-effectiveness is $500,000 per life saved.[8] This places it well above the cost of passive restraints yet still below many of the control actions envisioned in the 1976 Highway Safety Needs Report (see Table 3).

As might be expected from the high initial cost of the required equipment, opposition to the proposed standard has been intense. The standard first issued in 1971 was to take effect in September 1974. As that date drew near, industry's lobbying efforts to weaken the standard intensified. Despite pressure from industry, DOT went ahead with the standard, though it postponed the effective date until January 1975.

The controversy continued after the provision became effective. Trucking firms complained that the delicate braking mechanism frequently malfunctioned and, additionally, that it was extremely difficult to maintain. To respond to these concerns, and to stave off the political opposition underlying them, DOT relaxed the standard somewhat by lowering brake performance levels. The strategy seemed to work. Following congressional hearings, the House rejected industry's pleas for further relaxation of the antilock requirement (U.S. Congress 1976).

But the debate persisted. As of January 1, 1978, the standard was to become applicable to buses as well as trucks. In response, bus manufacturers and operators joined the trucking industry in an intensive lobbying effort on Capitol Hill. On the other side, equipment manufacturers, drivers (including the Teamsters Union), and public-interest activists lined up in support of the standard.

Reports that the antilock systems occasionally malfunctioned pitted manufacturers of the devices against the trucking lines. At DOT, the debate generated an intra-agency squabble.[9] As in the case of passive restraints, the issue finally became so hot that the Secretary decided to assume control of it. Thus, early in 1978, Brock Adams issued orders that retained the standard for truck tractors but rescinded it for truck trailers and refused any further delay for the application of the standard to buses, with the exception of school buses. Adams's position on the standard represented a major compromise. It pleased the trucking industry by modifying the standard to exempt trailers and it pleased safety advocates by including buses for the first time. In fact, early in 1978 it appeared that the steam had been taken out of the controversy.

But then came an unexpected court decision that surprised participants on both sides. A U.S. Appeals Court invalidated the standard (Paccar 1978). The court apparently accepted the claim of the plaintiff representing the trucking industry. Reams of technical documents lent support to the claim that vehicles equipped with antilock devices could actually be more dangerous than those without. The Supreme Court's subsequent refusal to hear the case on appeal upheld the appellate court decision.

A History of Failure

The stories of the passive restraint and air brake standards are thus tales of failures in hazard management. Indeed, both accounts qualify easily as "tales of woe" (chapter 13) that attest to significant public opposition and to the ability of big industry to delay safety-oriented rulemaking. Neither the passive restraint standard, first proposed in 1969, nor the air-brake standard, proposed in 1971, is in effect today. By using the courts, the Congress, and the White House, the auto and trucking industries and their allies among the general public have successfully resisted what most hazard managers regard as cost-effective standards, capable of substantially reducing fatalities, injuries, and property loss.

The opposition of industry in these cases is perhaps easier to understand than that of the public. Profit-oriented industry is likely to view suspiciously any "private cost" mandated by government, and in the absence of a corresponding "private sense of responsibility," to oppose such measures on narrow economic grounds. Only strong public disapproval of such industry policy can prevent this response.

But why does the public exhibit such a widespread lack of concern with the country's enormous highway loss problem? Why do most ordinary citizens fail to buckle up, and why do many oppose passive restraints in new cars? The answer cannot lie in the process of government rulemaking and intervention as such, since the public readily accepts—indeed, demands—government regulation of air and water quality and a host of other areas. Rather, the answer must involve deeper questions of risk perception and acceptance as they relate to the automobile (chapter 5).

Problems and Prospects

Why are close to 50,000 deaths and over 2,000,000 injuries caused by motor vehicles tolerated each year in the United States? Why are we not doing better? Our analysis suggests four major conclusions that partly answer these questions:

- Basic knowledge of motor vehicle hazards, compared with our knowledge of other technological hazards, is extensive and advanced. This basic knowledge, reviewed briefly in chapter 8, suggests that considerable improvement is possible. Following our general prescription for hazard reduction could reduce risks as much as five times.

- Historically, there have been major improvements in decreasing the fatality rate per vehicle mile. The years from 1926 to 1961 witnessed a steady decline in the number of deaths per vehicle mile. Yet, because of an even more rapid increase in the number of vehicles and miles driven, the total number of deaths per year increased slightly over the same period. Beginning in the early 1960s, for the first time in decades, the death rate per vehicle mile began to increase (see chapter 8) probably as a result of the increasing number of young "baby-boom" drivers and record production of high-powered cars and of convertibles. The death rate per 100,000 population also rose substantially during the sixties. The political crisis that developed and which led to major federal intervention in 1966 was thus as much a crisis of increased driving and of demographics as a crisis of inadequate safety management.

- The crisis of 1966, together with accompanying major changes in bureaucratic organization, resulted in a significant shift in the scope of contemplated control actions. Compared to pre-1966 safety management, much more emphasis was placed on the later stages of hazard control. At these stages, which we designate as blocking of outcomes, consequences, and high-order consequences, and which Haddon terms crash and postcrash phases, the need is for preventive action after the crash itself is in progress. Theoretically, this shift should have brought remarkable safety improvements. Yet actual improvements in safety since 1966 have been far less than those theoretically achievable. This may be attributable in part to the demographic changes previously mentioned. When one relates traffic deaths to the number of vehicles being driven (so as to control for these demographic changes), it appears that some real safety improvements have been made during the past decade (see Fig. 6).

- The failures since 1966 have been political failures. Large sums spent on measures such as motor vehicle inspection and highway widening serve the needs of bureaucracies and service industries, but they offer little in the way of reduction of the death and injury toll. On the other hand, measures such as passive restraints and truck brake standards, which are known to be highly effective, have drowned in a deluge of delaying tactics, political manipulation, and endless litigation. A complex mix of forces has orchestrated this defeat. Failure can be attributed not only to the weakness of the federal bureaucracy in the face of industry pressure, the direct interference of the President, and an indifferent public; it is also rooted in the persistence of the preconceptions that dominated hazard management philosophy in the pre-1966 era.

Thus, whereas the science of motor vehicle risk reduction is highly advanced, the politics of safety management must bear much of the responsibility for our not doing better. In addition, the change of direction in 1966 was conceptually so fundamental, indeed

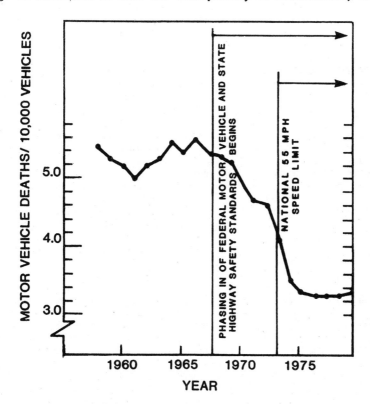

Figure 6. Beginning in the early 1960s, as the postwar baby boom came to driving age, motor vehicle fatalities increased markedly. When the death rate is given per 10,000 vehicles (so as to control for the growing driving-age population), it can be seen that the rate began to decline in the late 1960s following the implementation of the new highway and motor vehicle safety laws. Source: Haddon (1980).

revolutionary, that its full implications are just beginning to make an impact on the immense inertia of the political process. The 1966 changes, at least in an optimistic view of the problem, may simply need more time to be absorbed. Thus our suggestions for the future prescribe making use of the framework provided by the spirit, if not the letter, of existing legislation.

✓ **Broadening Control Actions.** Though the spectrum of possible control actions has been hugely broadened since 1966, it is highly desirable to continue this process. The insights that could be derived from risk studies and further application of physical analysis have so far been only partly converted to practical proposals for action. For example, the effects of large size differences in vehicles, the importance of **crushability** of vehicles, and the effect of increasingly large speed differentials among vehicles though all factors recognized by researchers--find insufficient expression in hazard-management practice. In addition, attention to the early control stages, such as modifying wants and choice of technology, is strikingly lacking in current management practice. A search could be made to find acceptable actions (other than recessions) that might have the effect of reducing the demand for motor vehicle travel.

✓ **Cost-Effectiveness.** Though beginnings have been made in establishing cost-effectiveness priorities, this has by no means become a routine approach. Without some consensus in this area, resource allocation will continue to be plagued by disabling controversy, leading ultimately to frustration of the management process. Cost-effectiveness is a useful tool for rational resource allocation even though it must often be based on rough projections from inadequate data.

✓ **Public Financing.** A principal problem in implementing certain control measures could be overcome if some private costs were transferred to the public sector. This might require, for example, that tax-supported trust funds be used to subsidize, either directly or indirectly, some of the highly effective vehicle-related actions we have discussed. This action could help overcome resistance to those mandated controls that now require private expenditure and could achieve a more even distribution of costs.

✓ **Research.** It is clear that motor vehicle safety management is particularly resistant to government regulatory action, even at a time when other areas such as clean air and water quality are not. Important for understanding this fact are not only the details of the regulatory process but also the method by which the public arrives at its own risk assessment. For the public's assessment of risk is eventually fed back through Congress into the regulatory process, in some cases resulting in the blocking of what would be highly effective measures. The basic problem is that, at present, members of the public are willing to accept **for themselves** a loss pattern in death and injury, which when taken in the aggregate, carries an extremely high price in medical costs, lost productivity, and property damage. Given that our society is already heavily burdened by the medical and welfare costs entailed by an increasingly older and unemployed population, the cost of highway losses at present levels may be more than society can or should bear, quite independently of the degree of personal suffering involved. Dealing with this issue will require not only a better understanding of how the public perceives risks, but, once achieved, this understanding must be followed by a concerted effort to educate the public, at least to the degree that some mandatory actions in support of the general welfare are accepted.

Despite these suggestions, which are based on an optimistic extension of the present legislative context, our final prognosis is not a hopeful one. If the economy continues to expand despite rising oil costs and chronic inflation, driving will also continue to increase, and the two-car household will become standard. Thus, much improvement in hazard management will, as in the past, be vitiated by the risks created by increased driving and may, because of the introduction of smaller, less crashworthy cars, even exceed present loss projections. There are other scenarios, of course, that could lead to a less dour future: a more socially responsive motor vehicle industry might develop a car that would achieve major safety goals at an acceptable cost; or a major catastrophe, possibly involving trucks, could lead to renewed efforts to improve truck safety; or a major shift in our perception of driving risks could cause us to regard them as involuntary rather than voluntary. But these are not probable outcomes. The motor vehicle, more than any other modern technology, is so rooted in our economy, our lifestyle, and our psyches that even the anguish of 50,000 yearly victims has had little effect on long-established trends.

NOTES

1. This chapter is a revised version of "Target: Highway Risks (Part II): The Government Regulators," Environment 21 no. 2 (March 1979):6–15,29–39, a publication of the Helen Dwight Reid Educational Foundation. Reproduced by permission.

2. It is required by 23 U.S. Code Section 402(a) that standards address, though not be limited to, the following areas: improvement of driver performance (including driver education), pedestrian performance, accident investigation and reporting, vehicle registration, operation, and inspection, highway design and maintenance, identification and correction of high-hazard locations, traffic control, and emergency services.

3. It is noteworthy that the Federal Highway Administration itself has established many standards for highway safety matters under its jurisdiction and has been increasingly active in recent years in pushing for their implementation on federally aided highways.

4. This statement must be moderated by recognition that part of the range in cost-effectiveness arises from the fact that costs are defined in an incremental or marginal way. To illustrate, consider the two measures **mandatory seatbelt use** and **median barriers**, rated at $506 and $228,000 per life, respectively. The first includes only the cost of enforcement and presumes the existence of hardware. The second includes total construction costs required for modification of existing roads. If initial

costs were included for seatbelts, the cost of mandatory use would jump fifty times, to about $25,000 per life. Which is the "correct" cost of mandatory seatbelt use? The answer really depends on the questioner's point of view. For the DOT hazard manager confronting the present situation, the incremental costs for seatbelts seem most relevant; for someone wishing to compare seatbelt deployment to another, equivalent measure, use of total costs is more logical. Thus, when using the figures in Table 3, the reader is advised to take some care in recognizing how they were calculated.

5. The usual number given for the lives saved by passive restraints is 9,000 (See, for example, NHTSA 1978a). This does not include the 3000 lives saved by the present use of nonpassive belts. Hence, the installation of passive restraints in all cars would result in a total saving of 12,000 lives.

6. To obtain these estimates, take the total cost of equipping 140 million vehicles and divide by the total number of lives saved over a period of ten years.

7. The Senate rejected a limitation similar to the Shuster amendment, but a House/Senate conference committee restored it. At the same time, the conference committee added another amendment that effectively nullified the impact of the Shuster spending limitation. The latter amendment, introduced by Representative Silvio Conte (R-Mass.) specified that the DOT was not prohibited from using funds for any research and development activity relating to occupant restraint systems; this in effect narrowed the Shuster amendment to a funding limitation on enforcement activities only. Since no enforcement was even planned in fiscal 1979, the Shuster amendment is probably meaningless.

8. An estimate for the cost-effectiveness of the air brake standard may be obtained as follows. There were 1.25 million multi-unit trucks and commercial buses in 1976, all presumably with air brakes. These were involved in 3,000 highway fatalities (NHTSA 1978a). According to a NHTSA estimate, obtained through a study at the University of Michigan, the standard could prevent up to 19 percent of these fatalities, or 570 in 1976. Assuming that vehicles equipped with the new braking system have a ten-year life, and that the maintenance costs of the equipment are equal to the initial costs, 5,700 lives would be saved at a cost of $2,400 per vehicle, which is equivalent to $520,000 per life.

9. Manufacturers claimed that the antilocks were not being properly maintained by the shippers, a contention supported by organizations representing the drivers. The shippers, in turn, blamed the malfunction on inadequate design. To solve this difficulty required action by two agencies within DOT: NHTSA, which issues standards and can enforce installation but cannot deal with maintenance problems; and the Bureau of Motor Carrier Safety (BMCS), which among other things is charged with maintenance enforcement. In the case of antilocks, NHTSA urged BMCS to issue maintenance regulations and to establish a program of

enforcement. BMCS refused, claiming that since antilocks were still experimental, such regulation would be premature. NHTSA, in order to get results, finally overrode BMCS and took its case to Adams.

REFERENCES

Cerrelli, Ezio C. 1984. Preliminary report: 1984 traffic fatalities (January–June). DOT HS–806–618. Washington: National Highway Traffic Administration, October.

Claybrook, Joan, Jaqueline Gillan, and Anne Strainchamps. 1982. Reagan on the road: The crash of the U.S. auto safety program. Washington: Public Citizen.

DeHaven, Hugh. [1952]. Accident survival: Airplane and passenger-automobiles. Paper presented at the annual meeting of the Society of Automotive Engineers, January 1952. New York: Crash Injury Research Project.

Faigin, Barbara M. 1976. 1975 societal cost of motor vehicle accidents. DOT HS 802-119. Washington: Department of Transportation, National Highway Traffic Safety Administration.

Feaver, Douglas B. 1982. Roadblocks, part 1: Feeling America's mobility crumble under our wheels. Washington Post (21 November), Al, Al6–Al7.

Gallup Poll. 1977. 45% of public favors automobile air bags. New York Times 24 July,1.

Grimm, Ann C. 1984. Restraint use laws by country, as of August, 1984. UMTRI Research Review 15 no. 1 (July–August):1–7.

Haddon, William Jr. 1972. A logical framework for categorizing highway safety phenomena and activity. Journal of Trauma 12:193–207.

Insurance Institute for Highway Safety. 1977. Background manual on the passive restraint issue. Washington: The Institute, August.

Insurance Institute for Highway Safety. 1978. Vehicle standards get small share of safety budget. Highway Loss Reduction Status Report 13 no. 7 (31 May):5–6.

Insurance Institute for Highway Safety. 1985. Child restraints in all 50. Highway Loss Reduction Status Report 20 no. 3 (16 March):7.

Interdepartmental Highway Safety Board. 1965. Special report on federal policy and program for highway safety. Washington: The Board.

Johnson, Lyndon B. 1966. Transportation for America: The president's message to the Congress, March 2, 1966. Weekly Compilation of Presidential Documents 2 (7 March):304–312.

Lewis, Anthony. 1976. A make-work nation? New York Times, 7 June, p. 29.

Mann, Judy. 1982. Child safety. Washington Post 22 December, B1.

Motor Vehicle Manufacturers Association. 1976. MVMA motor vehicle facts and figures '76. Detroit: MVMA.

Motor Vehicle Manufacturers Association. 1982. MVMA motor vehicle facts and figures '82. Detroit: MVMA.

NHTSA (National Highway Traffic Safety Administration). 1978a. Highway safety: A report under the Highway Safety Act of 1966 as amended, January 1, 1977-December 31, 1977. Washington: NHTSA.

NHTSA (National Highway Traffic Administration). 1978b. Motor vehicle safety 1977: A report on the activities under the National Traffic and Motor Vehicle Safety Act of 1966 and the Motor Vehicle Information and Cost Savings Act of 1972, January 1, 1977-December 31, 1977. Washington: Dept. of Transportation.

NHTSA (National Highway Traffic Safety Administration). 1978c. Public attitudes toward passive restraint systems: Summary report. DOT HS-803 567. Washington: Peter D. Hart Research Associates for NHTSA, August.

NHTSA (National Highway Traffic Safety Administration). 1979. Motor vehicle safety 1979. Washington: Dept. of Transportation, 1980.

NHTSA (National Highway Traffic Safety Administration). 1982. Traffic safety trends and forecasts. DOT-HS 805-998. Washington: Dept. of Transportation, October.

Nader, Ralph. 1965. Unsafe at any speed: The designed-in dangers of the American automobile. New York: Grossman.

National Safety Council. 1983. Special to the New York Times: '82 ratio of mileage to deaths in traffic termed record low. New York Times, 16 March 1983,A16.

Paccar, Inc. 1978. Paccar, Inc. et al. vs. NHTSA and the Department of Transportation, 573 F 2d 632 (Ninth Circuit).

Reagan, Ronald. 1981. Regulatory reform. Executive Order 12291. 17 February.

Rice, Dorothy J., Jacob J. Feldman, and Kerr L. White. 1976. The current burden of illness in the United States. Occasional Paper. Washington: Institute of Medicine, National Academy of Sciences.

Sinclair, Molly. 1982. State child restraint laws urged. Washington Post, 8 December, B1 and B6.

Slovic, Paul, Baruch Fischhoff, and Sarah Lichtenstein. 1978. Accident probabilities and seatbelt usage: A psychological perspective. Accident Analysis and Prevention 10:281-285.

U.S. Bureau of Public Roads. 1938. Motor vehicle traffic conditions in the United States. Washington: Author.

U.S. Chamber of Commerce. 1936. National conference on highway financing. Washington: Author.

U.S. Congress. 1966. Highway Safety Act of 1966: Report from the Committee on Public Works to accompany H.R. 13290, July 15, 1966. House Report 1700. 89th Congress, 2d. session.

U.S. Congress. 1973. House. Federal Aid Highway Act of 1973: Report together with additional, supplemental and minority views of the Committee on Public Works to accompany S.502. House Report 93-118. 93rd Congress, 1st session. 10 April.

U.S. Congress. 1976. House Committee on Interstate and Foreign Commerce. Subcommittee on Oversight and Investigations. Federal regulation and regulatory reform: Report. . . 94th Congress, 2d sess. Subcommittee Print. Washington: Government Printing Office. October.

U.S. Department of Transportation. 1976a. The national highway safety needs report. Washington: Dept. of Transportation.

U.S. Department of Transportation. 1976b. The national highway safety needs study. Washington: Dept. of Transportation.

U.S. Department of Transportation. 1977a. An evaluation of the highway safety program. Report to Congress from the Secretary of Transportation. Washington.

U.S. Department of Transportation. 1977b. The Secretary's decision concerning motor vehicle protection. Federal Register 42 (27 January):5071.

15
Controlling PCBs

Abe Goldman

Of the hazardous chemicals that have been the focus of management efforts, few have received more scrutiny or been subject to a more fully elaborated set of regulations than polychlorinated biphenyls--PCBs. Despite the effort, however, it is not clear that the PCB hazard has fully been brought under control, or that it will be in the near future. The reasons why the problems continue to be so refractory have to do not only with the chemical and physical nature of PCBs but also with the social and economic context of their use.

This chapter consists of four sections. The first portrays the PCB hazard in terms of uses, toxicity, and total dispersion. The second introduces four causal chains (see chapter 2) involving a range of releases, exposures, and consequences. The third reviews the management of the PCB hazard and concludes that although authorities of various regulatory agencies often overlap, neither management effort nor management success has been evenly distributed. It is found that the history of the hazard control effort has been marked by repeated disappointments in what were hoped to be definitive controls, and as a result the steps taken have been increasingly costly and have increasingly interfered with the use of PCBs as such. The final section evaluates the hazard management experience and discusses themes that are relevant to the generic problems of hazard management (see also chapter 3).

Nature of the PCB Hazard

Uses of PCBs

PCBs, a family of industrial chemicals that were first developed in the 1930s, have since been used in a wide variety of industrial and consumer products and processes. Most are viscous oils, varying chemically in the number and position of chlorine atoms attached to a biphenyl molecule. They were manufactured in the United States solely by Monsanto and were marketed as mixtures (known in the United States as Aroclors) containing a specified average chlorine content. The various Aroclors exhibit a range of physical and chemical properties that have made them highly useful for a large number of applications, including electrical equipment, industrial machinery, and a variety of consumer goods.

PCBs are extremely stable and highly resistant to breakdown by heat and most chemical and physical agents. As discussed below, these same characteristics also make them so potent an environmental hazard. PCBs also exhibit low electrical conductivity, low flammability, high heat capacity, low solubility in water, low vapor pressure, and a high dielectric constant.

Initially, PCBs were used in a number of "closed system" applications in which they were not directly susceptible to environmental release or spills. Most important of such applications were PCB use in capacitors[1] and transformers[2], which together accounted for 77 percent of total U.S. use in the period 1930-1975. Virtually all liquid capacitors manufactured in the United States, Europe, and Japan, and most transformers located in areas where fire is an important hazard, have contained PCBs (Versar, Inc. 1976). Other closed-system applications have included hydraulic fluids and heat transfer systems.

Spurred by Monsanto's advertising, use of PCBs in "open system" applications increased sharply in the 1960s. These include such products as plasticizers for paints, pigments, inks, and adhesives, carbonless copy paper, and casting waxes for high-grade machine tools (Peakall and Lincer 1970). U.S. production of PCBs peaked in 1970 at over 85 million pounds, with domestic sales totalling over 73 million pounds (Versar, Inc. 1976).

Cumulative usage of PCBs in the United States for 1930-1975 is summarized by major categories in Table 1 (NIOSH 1977). Annual U.S. domestic sales from 1957-1971 are indicated by type of end use in Figure 1, which shows the rapid reduction of sales in 1971, when Monsanto voluntarily terminated PCB sales for open-system applications (see below).

TABLE 1
Cumulative PCB usage, 1930-1975, by major categories

	AMOUNT (MILLION LBS.)	PERCENT OF TOTAL
Total U.S. production (1930-1975)	1,400	
Total U.S. usage	1,253	100.0
Capacitors	630	50.3
Transformers	335	26.7
Hydraulic systems and lubricants	80	6.4
Carbonless copy paper	45	3.6
Heat transfer systems	20	1.6
Petroleum additives	1	0.1
Other plasticizer uses (paints, pigments, adhesives, etc.)	115	9.2
Miscellaneous industrial	27	2.2

Source: Versar, Inc. (1976,7).

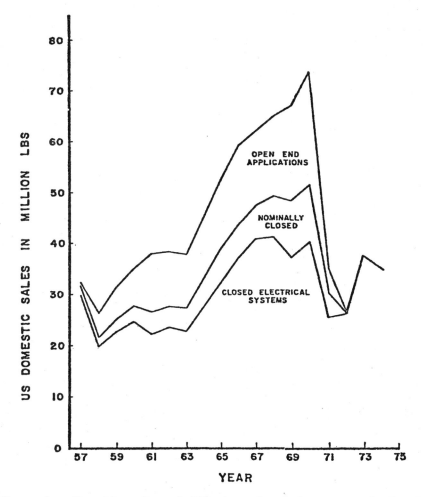

Figure 1. Domestic sales of PCBs by end use for open and closed applications. Source: Versar, Inc. (1976,202).

PCBs as Hazards

That PCBs are hazardous has been known since shortly after their development in 1930. Not until the discovery of their environmental distribution and effects, however, were control actions undertaken in the late 1960s and early 1970s. PCBs are chemically related to DDT and other chlorinated hydrocarbons, and they appear to be even more persistent in the environment than is DDT (Versar, Inc. 1976;Murphy 1977). They have been shown in high doses to cause a variety of acute ailments—particularly skin and related disorders—as well as long-term organ damage, especially to the liver. There is also some evidence, though not conclusive, of carcinogenicity and detrimental effects on fetal and infant development. Like

DDT and other chlorinated hydrocarbons, PCBs can accumulate in animal tissue and are not generally biodegradable.

Although some human health effects are difficult to establish, a number of studies have implicated PCBs as causes of specific ailments ranging from liver damage to chloracne (NIOSH 1977,123-124). The evidence of the harmfulness of PCBs comes from a fairly long history of occupational exposure, supplemented by animal experiments, and a major incident in Japan in 1968 where PCBs leaked into rice oil. Over a thousand people who consumed the oil were affected with symptoms of what came to be known as "Yusho" (rice oil) disease (Higuchi 1976). The main effects included various skin and related ailments, which in some cases were quite severe--chloracne, hyperpigmentation, swelling of the eyelids, and eye discharges. Victims also often suffered liver damage, digestive disorders, numbness of the extremities, menstrual abnormalities, fatigue, headaches, fever, and other symptoms. There has been some evidence among Yusho patients of increased cancer, particularly of the liver and stomach. Children were affected by Yusho disease either through direct ingestion of the contaminated oil, or through placental transport or breast feeding. Some children later suffered learning impairments and behavioral disorders. Frequently, the Yusho symptoms have persisted for years after the original ingestion of PCBs.

Most PCBs--whether ingested, inhaled, or absorbed through the skin--remain in the body for years, stored in fatty tissue. They can be gradually released from this storage--particularly during periods of rapid weight loss--and this may result in the appearance of acute symptoms. One of the main means by which PCBs (and other chlorinated hydrocarbons such as pesticide residues) can be discharged from the body is through breast milk. Recent studies have shown that small amounts of PCBs are virtually universally present in the U.S. population and that concentrations in human breast milk average over one part per million (ppm), frequently exceeding the FDA action levels for cow's milk (Rogan, Bagniewska, and Damastra 1980;EPA 1978;FDA 1979a). By the end of the nursing period, the PCB level in the infant's body will typically exceed that of the mother (Mosher and Moyer 1980). This is of particular concern because infants are especially vulnerable to PCBs.

The effects of long-term low-dose exposures are not well known, but a report by the National Institute of Occupational Safety and Health (NIOSH) concludes that no threshold has been observed below which damage to liver tissue is entirely absent (NIOSH 1977). Animal studies have shown that PCBs can cause cancer in rats and mice, but human data at present are still inconclusive (Allen and Norback 1977;FDA 1979a;NIOSH 1977).

Various animal species are also especially sensitive to PCBs. Even low dietary concentrations, for example, can have severe reproductive effects on mink and rhesus monkeys (NIOSH 1977). PCBs are also toxic to various species of birds, fish, and other aquatic life (National Conference... 1976), and they play a synergistic role in the accumulation of cadmium and other toxic metals and organic compounds in mink and other species (Olsson et al. 1979).

Of course, other chemicals can produce equally or even more severe consequences. What makes the PCB hazard of particular concern is the confluence of a social fact--the extensive usage of PCBs--with a series of features resulting from their chemical and physical characteristics, three of which are particularly ominous: the capacity of PCBs to be transported over long distances in the air or water, their resistance to breakdown by physical or biological systems, and the extent to which they can accumulate and concentrate in biological tissue.

Dispersion of PCBs

The dispersion of PCBs is now essentially global—they have been found even in Antarctic ice and in the tissues of animal species remote from human settlement. The highest environmental concentrations, though, occur in the industrialized countries where PCBs have been manufactured and processed for decades, and in particular in areas where sizable releases have occurred.

The status of 1,403 million pounds of PCBs produced in the United States since 1930 is summarized in Table 2 (Versar, Inc. 1976). Only 55 million pounds, or 4 percent, are estimated to have been destroyed so far. One hundred fifty million pounds, about 11 percent, are free in the environment, released to the soil, air, water, or sediments. It is this relatively small amount that comprises almost all of the traces of PCB so far detected in biota and the ambient environment. Another 290 million pounds (20 percent) are already in landfills and dumps. Except in cases where contamination of a particular site is especially serious (e.g., when the area's water supply is threatened), it will probably not be feasible to recover these already discarded PCBs. Moreover, the EPA itself has projected that as PCBs escape from dumpsites, environmental levels will increase from 150 million pounds to 540 million pounds (Nader, Brownstein, and Richard 1981,201).

Causal Structure of the PCBs Hazard

Partly because of the wide range of uses, the human hazard of PCBs involves a number of release and exposure pathways. Using the causal structure model described in chapter 2, one can group these pathways into four major sequences involving:

TABLE 2
Disposition of PCBs produced in the United States, 1930–1975

DISPOSITION	AMOUNT (MILLION LBS.)	PERCENT OF TOTAL
Currently in service	758	54
In landfills and dumps	290	21
Exported	150	11
Free in environment	150	11
Destroyed	55	4
TOTAL U.S. PRODUCTION	1,403	

Source: Versar, Inc. (1976,7).

- direct exposure, usually through occupational contact;
- accidental contamination of food or feed through leaks from PCB-containing equipment;
- environmental release during manufacture of PCBs;
- disposal of used PCBs and PCB-containing products.

Figure 2 illustrates the four sequences, each of which is described in further detail below.

Occupational Exposure

The occupational pathway has the longest history and is the one where indications of the toxicity first appeared in 1937 (NIOSH 1977). Occupational exposure usually occurs through inhalation in the workplace, where PCB concentrations may be a thousand to a million times normal background levels (Versar, Inc. 1976;NIOSH 1977). Direct skin contact also occurs, but ingestion is rare. Chemical and electrical workers and others who work on the production of PCBs or the manufacture of PCB-containing equipment bear the greatest risks. According to one estimate, 12,000 such workers in 1977 were subjected to occupational exposure (NIOSH 1977,35;Lloyd et al. 1976). Other exposed occupational groups include people who service transformers, pesticide applicators (prior to 1968), railway employees, miners, workers with plastic and paint products, and rural highway workers involved in road-oiling.

Although most occupational exposure to PCBs has involved people who work directly with PCBs and PCB products, a recent accident has revealed the potential for large-scale, possibly catastrophic exposure of office workers and others. On 5 February 1981, an electrical fire in the basement of a modern government office building in Binghamton, New York burst the building's transformer, which contained about 180 gallons of PCB coolant (Montgomery 1981;Hilts 1981;Magiera 1981). The heat volatilized the PCBs, which together with smoke and ash were spread throughout the building by the ventilating system. Fortunately, the accident occurred in the middle of the night when the building was empty. Even so, the Binghamton office building was almost fully contaminated, not only with PCBs but also with PCDFs (polychlorinated dibenzofurans) and dioxin compounds that were produced by the fire, and it has remained closed since the accident. Decontamination, which is still incomplete, has already cost several million dollars.

Direct Food Contamination

Direct contamination of food or animal feed is the pathway with the greatest catastrophic potential--as demonstrated by the Japanese experience. This potential results from the ability of the modern food system to distribute food products rapidly and widely and the difficulty of detecting contamination early in the process. The Yusho symptoms were caused by PCB concentrations in oil that averaged about 2000 ppm (Higuchi 1976), yet it took over a year to determine the source of contamination and to establish that PCBs were the cause of the widely observed symptoms. Accidents similar to the Yusho case have occurred in the United States, though never with consequences of comparable severity. A list of major incidents

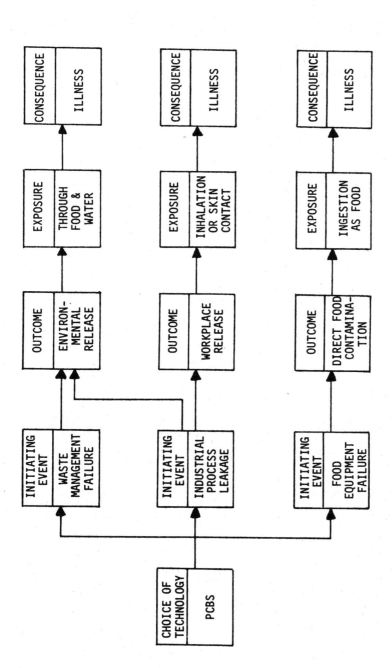

Figure 2. Causal sequences involved in the PCB hazard. Note that there are in fact distinct sequences involving: (1) environmental release through waste-management failure; (2) workplace release through industrial-process leakage; and (3) direct food contamination from food-equipment failure.

appears in Table 3 (FDA 1979b;Highland 1976). All these occurrences have in common the vulnerability of the food system to inadvertent contamination from unexpected and generally uncontrolled sources.

Environmental Releases

Environmental discharges of PCBs from industrial production have excited the greatest public and regulatory concern of any of the four main pathways. The principal concern here is with the aquatic environment. Because of the persistence of PCBs, most kinds of releases can lead eventually to water. In fact, it is now believed that the oceans and other large bodies of water are the main reservoirs of released PCBs (National Research Council 1979;EPA 1979a). PCBs deposited on land eventually enter ground water, even though they may remain bound to soil particles for a long time. PCB vapors in air are absorbed onto dust particles, transported through the atmosphere, and deposited through rain into oceans and lakes. Some recent studies of the Great Lakes indicate that atmospheric fallout is the major source of PCB entry, exceeding direct discharges by various industrial polluters (Murphy 1979;Eisenreich and Hollod 1979).

Once in the aquatic environment, PCBs readily enter the food chain because they are more soluble in fats than in water. At the higher trophic levels they bioaccumulate to concentrations that are a million times greater than ambient water (EPA 1979a). Human exposure to environmentally released PCBs follows through direct consumption of fish and seafood. Populations that consume large amounts of fish are likely to have especially high dietary intake of PCBs. For example, a Japanese study finds that 80 percent of dietary PCBs come from fish or seafood, 6 percent from meat, 6 percent from eggs, and the remainder from other foods (Higuchi 1976).

It is useful to distinguish two types of environmental releases—intensive and pervasive. The former involve point discharges, which can result in heavy local contamination. The latter, though often individually small, are cumulatively significant, and the bodies of water affected, usually large lakes or oceans, may be quite distant from the original points of release.

Intensive environmental releases have come mainly from the manufacture of PCBs and PCB components and machinery and from related accidents. For 1971-1975, the EPA identified twenty spills ranging from leaks of storage drums or transformers to major truck accidents (Kopp 1976). Table 4 lists some of the more notorious intensive pollution sites. The main source of pervasive releases has been from open-ended manufacturing processes, such as PCBs in plasticizers and carbonless copy paper. For comparison, estimated quantities are also shown in Table 4.

The Disposal of PCBs and PCB Products

This last of the four main hazard pathways overlaps somewhat with environmental releases, but different management approaches apply. Included are the continuing disposal of used PCBs from transformers and other equipment, contaminated materials and equipment from PCB production and processing, and the many discrete quantities that are in use or that reside in obsolete appliances and

TABLE 3
Major incidents of PCB food contamination in the United States,
1969-1979

DATE	CONTAMINATED PRODUCT(S)	MEANS OF CONTAMINATION AND RESULTS
1969	Milk	Herbicide containing PCBs sprayed on pastures used by grazing dairy cattle.
1979	Milk	PCB-containing sealants used in dairy farm silos in about six states; silage contaminated with PCBs.
1970-71	Chickens	Food wrappers containing PCBs fed to chickens along with stale bakery products. High PCB levels detected by FDA surveillance; over 140,000 chickens destroyed.
1971	Meat meal	PCB leakage from pasteurization equipment used in processing meat by-products into meal for animal feed. Contamination discovered before distribution of meal.
1971	Fishmeal; poultry and eggs	Heating system leak in pasteurization plant in Wilmington, North Carolina caused PCB contamination of fishmeal. After FDA investigation 123,000 pounds of egg products and 88,000 chickens in several states destroyed.
1971	Turkeys	High PCB levels found in Swift & Co. turkeys; approximately 50,000 turkeys kept from market.
1977	Fishmeal	PCBs in stored transformers in Puerto Rico warehouse released during fire. Fishmeal in warehouse contaminated with PCBs and distributed as animal feed. Over 400,000 chickens and hundreds of thousands of eggs destroyed when contamination discovered.
1979	Beef cattle	Feedlot cattle in Kansas contaminated with PCBs in insecticide oil. Oil had been purchased from salvage dealer in 1971 and contained spent transformer fluid. About one-third of cattle died; remainder of herd destroyed.
1979	Animal feed, grease; poultry, eggs, hogs, bakery goods, etc.	Leak from stored transformer contaminated meat by-products used for animal feed and grease. Distributed feed and grease resulted in contamintion of poultry, hogs, eggs, and bakery products. FDA investigation led to destruction of about 800,000 chickens, 3.8 million eggs, 4,000 hogs, 74,000 bakery goods, and 800,000 pounds of animal feeds and ingredients.

Source: FDA (1979b).

TABLE 4
Major PCB releases in the United States

RELEASED FROM	RELEASED INTO	QUANTITY (LBS.)
INTENSIVE RELEASES		
General Electric (G.E.) capacitor plants Fort Edwards and Hudson Falls, NY	Hudson River	1,500,000
G. E. transformer plant Pittsfield, MA	Housatonic River	42,000
G. E. transformer plant, Rome, GA	Coosa River	
Aerovox & Cornell-Dubilier Corp. New Bedford, MA	New Bedford Harbor	200,000
PERVASIVE RELEASES		
PCB production, 1974	general environment	1,100 per year
Plasticized materials containing PCBs	general environment	2,000,000 per year

Sources: BNA (1980b), Hammond (1972), Kleinert (1972), Versar, Inc. (1976).

machinery. This causal stream represents the most long-term segment of the PCB hazard. The consequences, both human and environmental, are difficult to pinpoint and involve long delays. The hazard is usually pervasive rather than intensive, with little likelihood of acute effects on one site or population. The volume of material involved is immense, however, and thus represents a substantial global-scale threat. The 758 million pounds of PCBs in service represent 54 percent of the total cumulative production in the U.S. since 1930 (Table 2).

History of PCB Management

The history of the management of PCBs falls into three periods, each marked by a growing recognition of some aspect of the hazard and culminating in a seemingly successful definitive control action. In both of the first two periods, the control action proved ultimately to be only partially successful and the process had to be repeated. We are currently in the third period.

The first period extends from 1930 to 1971. Management of acute occupational hazards, the first hazard sequence to be recognized, generally took the form of steps to prevent extreme exposure. Failure to deal adequately with the longer-term occupational hazard, however, has left a residue of chronic damage among those who worked with PCBs. Nor were the other pathways recognized or controlled until the last few years of this period. What finally moved PCBs onto the hazard management agenda were two events in the late 1960s. The first was the surprising discovery of their ubiquity in the environment. A Swedish scientist, Sören Jensen, who had been searching for environmental DDT residues and consistently finding traces of other contaminants, eventually determined that these were PCBs, and in 1966 he published an article on their environmental dispersion (Jensen 1966). Other research confirmed his findings, and by 1969 and 1970 the popular media began to publish articles about the dangers of PCBs in the ecosystem. Among other discoveries was the finding that the reproductive failure of commercial mink in the Midwest was caused by PCBs in the Lake Michigan salmon they were being fed (Auerlich, Ringer, and Iwamoto 1973).

The other event was the outbreak of Yusho disease in Japan in 1968. By 1973, 1200 patients had been identified. In order to forestall a similar disaster, the U.S. Food and Drug Administration (FDA) and the U.S. Department of Agriculture (USDA) initiated early in 1970 a food surveillance program and established action levels for PCBs in milk, poultry, fish, and other foods. Large-scale food seizures followed the discovery of high PCB levels in chickens, eggs, turkeys, and other foods (see below). In October 1970, the USDA ordered the elimination of PCBs from pesticide mixtures. A number of government agencies and executive panels began to study the PCB problem, and international conferences were held as European nations also became concerned.

In 1971, after some informal government pressure, Monsanto announced that it was terminating sales of PCBs for open-system uses (in plasticizers, sealants, paints, papers, etc.) and for any uses directly related to food or animal feed. The company also reduced production of the higher chlorinated mixtures, substituting lower chlorinated PCBs that are more easily decomposed (Burger 1976; Papageorge 1975). Because the growth in open-system uses during the 1960s (Figure 1) had brought a considerable increase in dispersion, the Monsanto action was a significant step. Between 1970 and 1971, Monsanto's PCB sales dropped 60 percent (Versar, Inc. 1976). At the time, this appeared to be a major control action—an example of government and industry cooperation in successful management of an environmental hazard (Burger 1976). It was hoped that it would greatly reduce, if not eliminate, the importance of PCBs as a hazard. Certainly, there was reluctance to consider a similar restriction on the use of PCBs in capacitors, transformers, and other nominally closed systems because of the lack of appropriate substitutes and the expense of new product development (Papageorge 1975; Regenstein 1982). The removal of PCBs from equipment that was already in use would have been even more expensive.

In the five years between 1971 and 1976, the second period of PCB management, the reassurance created by the Monsanto ban was quickly dissipated by further alarming findings (National Conference . . . 1976; NIOSH 1977). These eventually led to the passage of the

Toxic Substances Control Act (TSCA) in 1976--which for the first time gave EPA authority to regulate use of toxic substances such as PCBs, and the issuance of EPA regulations on PCBs in 1979. Among the more important developments of 1971-1976 were:

- PCBs were regularly found in human tissue in the United States (NIOSH 1977);
- PCBs were found in mother's milk in the U.S., Canada, Germany, Japan, and Sweden (New York Times 1976);
- Evidence that PCBs might be carcinogenic appeared in animal studies by 1973 (Allen and Norback 1977);
- Further major accidents occurred with spills of substantial amounts (Table 3;Kopp 1976);
- General Electric plants were found to be continuing the discharge of PCBs into waterways (Boyle and Highland 1979;EDF and Boyle 1980);
- Extremely high levels (up to 575 ppm) were found in Hudson River fish. Subsequently the Hudson was closed to most commercial fishing, and some sections were closed to all fishing (Hellman 1976;EDF and Boyle 1980; Severo 1980).

This continuing history of PCB dispersion and PCB health effects culminated in a national conference on PCBs in November 1975 (National Conference... 1976). Conference participants reviewed levels of human exposure and health effects, PCB uses, environmental levels, ecological effects, and substitutes, and the economic aspects of PCB use and control. One of the main findings was that whereas PCB levels in most foods were lower than before, environmental levels were as high or higher, despite the decline in most open uses.

As a result of these events and findings, PCBs were explicitly included in the TSCA legislation passed in October 1976. Section 6(e) of TSCA mandated that the EPA issue by 1 July 1977: disposal and marking rules for PCBs; by January 1978: prohibition of all PCB manufacture, processing, and distribution in commerce for all uses that were not totally enclosed; by January 1979: a ban on all manufacture; and by July 1979: a ban on all PCB processing and distribution in commerce. EPA met only the last of these deadlines (GAO 1981,6-7), and was 18 and six months late, respectively, for the first two. Moreover, exemptions were permissible in the absence of "an unreasonable risk of injury to health or environment" and if efforts had been made to find a safe substitute (TSCA 1976). Continued use of PCBs in enclosed systems--for example, in capacitors and transformers--was also permitted.

In setting up its new Office of Toxic Substances (OTS) the EPA regarded the PCB regulations as a pilot program for future regulation on chemical substances--though with the distinction that Congress had already rendered an implicit cost-benefit judgment on the use of PCBs (Jellinek 1978). Unfortunately, issuance of the final regulation was delayed by over a year, partly because of problems in organizing the OTS and partly because of the many comments from business and environmental groups.

In sum, the period 1971-1979 saw major progress in recognition and understanding of the PCB hazard and an end to manufacture and to most new uses. The PCB hazard had not been eliminated, however, despite some sanguine beliefs to the contrary. What remained was the problem of what to do about the PCBs still in service and those already in the environment, particularly at sites of intensive pollution. This third and continuing phase of PCB control has had to contend with these problems and will require judgments about risks and appropriate control actions. Also lurking in the background is the broader issue of PCB substitutes and their environmental impacts, the significance of which have only recently become apparent (Svanberg and Lindén 1979).

Hazard Management for Four Causal Chains

To gain insight into society's priorities in dealing with chemical hazards, I review next the salient aspects of the management effort for each of the hazard chains identified for PCBs. Details of the management effort differ markedly among the four hazard chains.

The Occupational Hazard

Beginning with the first reports of chloracne and other symptoms in the early 1930s, a dominant response to the occupational hazard has been to reduce exposure through improved ventilation, and later, engineering of the work environment (NIOSH 1977). Such actions generally proved successful in reducing acute symptoms. At a later stage, exposure levels were measured and guided by quantitative standards. These standards were recommended by several states, but they were in no case mandatory. For example, by the 1940s a number of states recommended a maximum of 1 mg/m^3 in workplace air. In 1937 actual average workplace concentrations varied from 0.5-1.5 mg/m^3. As late as 1959 a study of airborne PCBs yielded industrial air concentrations in the range of 0.2-10.5 mg/m^3 (NIOSH 1977).

During the 1960s, despite increased awareness of the PCB hazard, little change took place in available standards. For example, in 1962 the American Conference of Governmental Industrial Hygienists (ACGIH 1962,26,27) recommended that the acceptable levels be set at 0.5 mg/m^3 for PCBs of high chlorine content and 1.0 mg/m^3 for those with lower chlorine content, largely since the former were more persistent in the body. These standards, adopted as federal standards in 1968, have remained as the only formal standards to this date.

With the many alarming events of the 1970s, NIOSH undertook a comprehensive study of PCBs, and recommended in 1977 a reduction of the occupational standard by a factor of 1000 to 1 $\mu g/m^3$. The NIOSH recommendations were based on data that had by that time come to light, viz.: (a) there was no demonstrated threshold of exposure for liver damage; (b) there was circumstantial evidence (animal data) that PCBs are carcinogens; and (c) it had been found that PCBs may cause reproductive effects in humans (NIOSH 1977). NIOSH also recommended other measures, including medical surveillance of workers, advising mothers of possible dangers of transmitting PCBs to infants, and a number of engineering controls.

Today, in 1983, OSHA has not yet acted on the 1977 NIOSH recommendations. At the same time, because of Monsanto's action to reduce production in 1971, and the more recent (1979) total ban on PCB production, occupational exposure has declined drastically. Thus, in effect, the most important action to reduce occupational exposure was initiated not through a standard or through industrial engineering, but because environmental concerns external to the manufacturing process led to the much more radical step of first curtailing and then stopping production.

By way of summary, the causal-chain diagram in Figure 3 illustrates the three major ways of controlling occupational exposure for PCBs. Note that the sequence of control interventions followed since the mid-1930s runs in a direction opposite to the causal sequence.

Direct Food Contamination

In contrast to the desultory rate of regulatory action on occupational hazards, the response to direct food contamination has been relatively active and wide-ranging. The various instances of destruction of contaminated foodstuffs, often in large quantities, constitute the most dramatic actions. As described above, the contamination occurred when food or feed came into contact with PCBs or PCB-containing materials, generally during processing or storage.

In contrast to the occupational hazard, the population at risk from food contamination is diverse and widely dispersed. All socioeconomic classes are potentially at risk, and individual action is not blunted by the possibility of placing one's job in jeopardy. And perhaps most important, as the Yusho incident demonstrates, the possibility exists for a disastrous event with widespread acute consequences. Together these characteristics no doubt account for the greater regulatory and public attention that has been accorded the food contamination hazard.

Of the range of potential control actions, the most common to date has been exposure prevention, which has required regular monitoring of foodstuffs by various agencies. Such monitoring began in 1970, partly in response to the Yusho incident and to other incidents in this country. The FDA initially set PCB tolerance levels for various foods and food packaging in 1973. Seizures and destruction of contaminated foods have also occurred, particularly in the early 1970s. It is likely, though, that cases of food contamination must have occurred and gone unreported, undetected, or both, even earlier.

The food monitoring-seizure mode of control has a continuing importance, but in view of the complexity and interconnectedness of the U.S. food system, it is inevitably costly, inefficient, and highly susceptible to failure. A 1979 contamination incident dramatically illustrates the inadequacies (FDA 1979a;GAO 1979,14-15). In July 1979, a routine USDA test in Utah found a hen with a PCB fat content of 15 ppm, five times the 3-ppm FDA standard. After various delays in processing the tests and an arduous process of tracing the contamination, a food processing plant in Billings, Montana, was identified as the source. Sometime in June 1979 about 200 gallons of transformer fluid containing PCBs had leaked into the plant's drainage system and eventually contaminated approximately a million pounds of meat meal. Subsequent FDA tests showed that a June batch of meat meal from the plant contained over 1000 ppm PCBs. By the

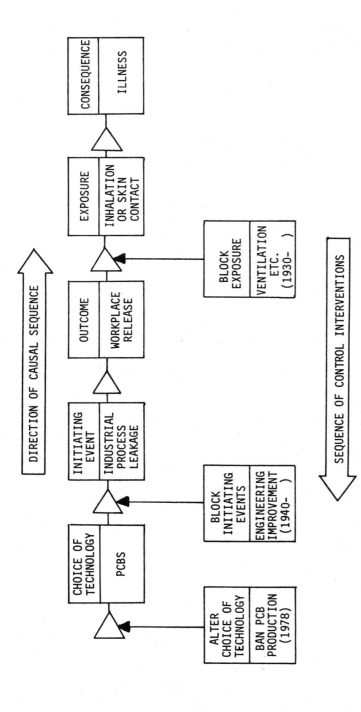

Figure 3. Causal sequence for occupational exposure, with management interventions. Note that sequence of interventions has moved increasingly "upstream" in time.

time the plant had been identified, contaminated meat meal, animal feed, and grease had been distributed throughout at least 19 states and two foreign countries. By the end of October 1979, the overall federal investigation had resulted in the destruction of about 800,000 chickens, 3.8 million eggs, 4000 hogs, 74,000 bakery items, 800,000 pounds of animal feeds and feed ingredients, and 1.2 million pounds of grease—worth $2.7 million (FDA 1980;GAO 1981,5).

Since the incident, the FDA, EPA, and USDA have joined in proposing comprehensive regulations to prohibit the use of virtually all PCB-containing equipment—including transformers, large capacitors, and heat exchange systems—in plants that manufacture, process, or store food, animal feed, fertilizers, agricultural chemicals, or food packaging materials (USDA, FDA, and EPA 1980). This involves removal of such equipment from a large number of food and agriculture facilities throughout the country.

The other significant recent regulatory action in this area was the revision of the FDA food tolerance levels in 1979. The tolerances for milk and dairy products, poultry, eggs, and fish and shellfish were lowered, but for all except the last the usual levels detected are considerably below the new tolerances. Much controversy, however, has centered about the 2-ppm level set for fish, which is often exceeded.

The ways of controlling direct food contamination are summarized in Figure 4. As in the case of occupational hazard, this shows a trend to move control action "upstream" in the causal sequence as downstream interventions—for example, exposure blocks—prove ineffective.

Environmental PCBs

Control of environmental releases of PCBs involves at least three approaches—limiting release at the source; blocking environmental pathways that lead to human exposure; and controlling human intake of contaminated biota.

In many cases, the most effective means of control is exercised at the source. Experience with factory effluents, which have been easily reduced once governmental standards or other constraints have been applied, demonstrate this successful control. General Electric plants were able to reduce daily discharges into the Hudson River from 30 pounds to one pound per day soon after heavy state pressure was applied in 1976, and they achieved further reduction later (Hellman 1976). Such reductions are brought about by process design, effluent treatment, and discharge control measures in which recovered PCBs become part of the managed waste stream. Other cases of potential source control involve releases from existing equipment, cases of illegal dumping, and the continuing release from millions of small components. These cases are far more difficult than factory effluents, and none are currently effectively managed. It is, to a large extent, because of this failure that PCB levels in the Great Lakes are expected to decline only slowly in the near future (Murphy 1979;Eisenreich and Hollod 1979).

Once PCBs have been deposited in the environment, it is difficult but possible to block partially pathways to eventual human uptake. Dredging of sedimentary deposits, which has been tried on the upper Hudson, is one such method. Of the 1.5 million pounds of

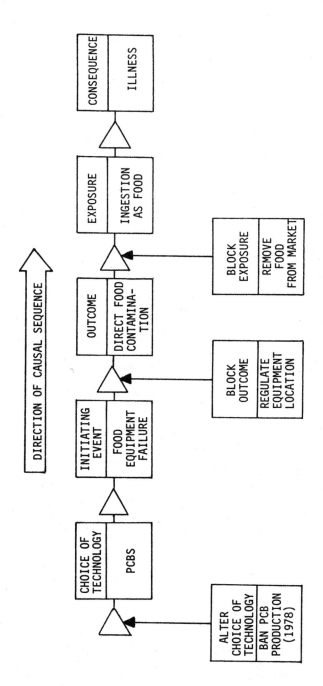

Figure 4. Causal sequence for direct food contamination, with management interventions.

PCBs believed to be in the Hudson, about 0.3 million pounds have been dredged (Severo 1980; Boyle and Highland 1979). Current estimates are that removal of most of the accessible remaining deposits would cost at least $30-40 million, with some estimates running to five times that amount. The estimated cost of dredging New Bedford harbor alone is $130 million (Ackerman 1983,1;Jahnke 1980). Neither state nor federal governments are willing to provide such funds. Aside from cost, the dredging process is problematical since it can release PCBs and other toxic chemicals that might otherwise remain bound in bottom sediments.

A final approach to control of environmental PCBs is to block human exposure to contaminated biota. The principal source of such exposure is fish and seafood consumption, though other important sources iclude meat, milk, eggs, and other animal products. According to 1973 FDA regulations, the tolerance level in fish and shellfish is 5 ppm. Estimates that nearly all trout and salmon in Lake Michigan that are longer than twelve inches exceed this limit led to the termination of commercial fishing and continuing battles over the enforcement of the standard (Boyle and Highland 1979). To make matters worse, the FDA proposed in 1977 that the tolerance level in fish be reduced to 2 ppm, with the expected result that up to 25 percent of all inland commercial fishing would be affected. The new standard was finally issued in 1979, but a flood of objections have stayed its adoption (FDA 1979a). Whatever eventually evolves in regard to this standard, it is clear that controlling fishing will not be a simple problem. For one thing, the FDA standard applies only to commercial fishing and excludes game fishing which is currently regulated by the states. In addition, it may not be sensible to apply restrictions only to a given contaminated body of water. For example, migrating striped bass, which spawn in the Hudson and may become heavily contaminated, can then range all over the East Coast, from Delaware to Maine.

Figure 5 includes the three methods of controlling environmental PCBs. As in the other two cases (Figures 3 and 4), upstream control options appear to be the more effective, whereas downstream options are both costly and incomplete. This conclusion is supported by a National Academy of Sciences (NAS) report (National Research Council 1979). This report finds that at the 2-ppm limit, it costs about $125,000 to eliminate one kilogram of PCBs from fish by imposing fish marketing restrictions. In contrast, an equivalent effect can be obtained for $120,000 by dredging, and for $400-$14,000 by source-control techniques.

Disposal of PCBs and PCB Products

The proper disposal of PCBs and PCB products presents one of the most difficult management problems. EPA regulations of 1978 and 1979, which were mandated by TSCA, established detailed guidelines for the disposal of PCB mixtures containing 50 ppm or more of PCBs (EPA 1979b,31516). The guidelines specified two acceptable methods of disposal: burial in an approved chemical landfill or incineration at extremely high temperature. Incineration, insofar as it destroys PCBs rather than merely storing them, is the preferred method. Because no commercial incinerators were available when the regulations appeared, however, over one million gallons of PCBs sat

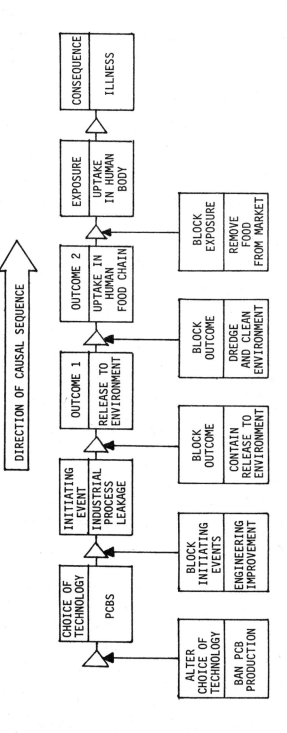

Figure 5. Causal sequence for environmental release due to industrial process failure with management interventions. The sequence for waste management failure is identical except for initiating events (see Fig. 2).

awaiting incineration (Electric Power Research Institute 1980).
Meanwhile, a series of technical, economic, and political problems
has hampered implementation.

The private sector has betrayed an understandable reluctance to
establish expensive and unpopular facilities. The public, wary both
of emissions and potential accidents in transport, handling, and
storage, has resisted the siting of disposal facilities. Technical
snags have only prolonged the delay. By 1982 the EPA had approved
only nine landfills, four noncommercial incinerators, and two commercial incinerators (both within the same EPA region). Chemical
destruction processes, which are relatively inexpensive and sometimes mobile, are just now emerging as promising alternatives (GAO
1981,27-29). In short, the EPA has failed to solve the PCB disposal
problem.

That failure has sparked a rash of indiscriminate and illegal
dumpings, leading to serious contamination of soil and ground water.
In effect, this has converted the disposal problem into one more
input to the causal sequence, already illustrated in Figure 5, that
applies to environmental releases. The difference here is that some
cases of illegal disposal are of a magnitude that dwarfs even some
of the uncontrolled effluents of industrial processes. A notorious
incident of illegal disposal occurred in July and August of 1978
when 35,000 gallons were dumped along the sides of 200 miles of
rural highway in North Carolina (Brown 1979,241-247). It is estimated that 40,000 cubic yards of soil were severely contaminated.
The most effective control measure would have cost $12 million and
would have required transporting the soil to a federal waste disposal site in Alabama. The state of North Carolina decided to choose
between two cheaper alternatives—either treating the PCBs in situ
with activated charcoal, or shipping the contaminated soil to a
nearby landfill in North Carolina.

Summary and Conclusion

The PCB Hazard Story

As we have seen, the main regulatory actions aimed at PCBs have
included manufacturing, use, and distribution bans under TSCA; disposal regulations, also under TSCA; FDA food-tolerance limits; federal workplace standards; and the ban on PCBs used in food-related
equipment. The results of these actions, for now and for the future, may be summarized as follows:

1. There has been substantial reduction in occupational
   exposure, though some occupational exposure will continue since PCBs in existing components will remain.
   The differential between occupational and public exposure will decline, but it will continue to hold for
   some workers.
2. The chance of further localized pollution caused by
   production wastes has been sharply reduced but not
   eliminated. Accidental releases and deliberate illegal
   dumping are two principal reasons for this.
3. PCBs levels in food, including fish, have apparently
   declined (FDA 1979a), as have human exposures because

of fishing bans and monitoring actions. Elimination of PCBs from the diet is probably impossible for the foreseeable future; further, the success achieved to date has come at the expense of substantial fishery resources.

4. If the proposed regulations banning food-related uses of PCBS (USDA, FDA, and EPA 1980) go into effect, the probability of accidental, large-scale food contamination will decrease further.

5. Some reduction in low-level pervasive releases may be occurring, principally because of termination of open-ended uses. Yet, this may be a temporary reduction that will be reversed as the current inventory of PCBs in existing equipment reaches landfills and dumps.

6. A large portion of the PCBs still in use may be discarded properly, in accordance with new waste-disposal regulations. In this context it must be remembered that most of the PCBs produced between 1930 and 1977 still reside in present equipment inventories (Table 2).

7. The contamination of the state office building in Binghamton with PCBs and other by-products (see above) has revealed an unanticipated new hazard pathway with the potential for disastrous consequences; and such surprises may occur again.

8. Finally, the PCB burden already in human adult tissues represents a continuing hazard to breast-fed members of the next generation. Indeed, of the three groups that are most affected by PCBs--electrical and other exposed workers, fish eaters, and infants who are breast fed--the last may bear the most substantial risk in the future.

Thus, although some aspects of the PCB hazard have been controlled, significant risks still remain. In arriving at the current point, the delays in managing the PCBs hazard have been of crucial importance. The specific causes of delay exemplify some common obstacles to successful hazard management. During the first period (1930-1971), delays stemmed largely from lack of knowledge; in the second period (1971-1978), institutional inadequacies predominated; and in the current period it appears that unresolved conflicts about the distribution of costs and risks play the key role.

The tactics of the current period range from lawsuits, to regulatory appeals, to outright defiance. Yet the basic goals are similar: to allocate costs in a manner favorable to the particular party involved without direct violation of the regulatory framework. Each of these conflicts will probably require separate resolution. Yet unless the issues of cost and risk distribution receive attention at an earlier stage and unless generic procedures for dealing with those issues evolve, control of hazardous substances will become an impossible endeavor.

Lessons for Hazard Management

A veritable "tale of woe" (chapter 13) in many respects, the half-century history of PCB use has at least been rich in insights on the needs and problems of controlling hazardous substances. It has helped reveal both a major category of chemical properties that can generate high risk and a set of common managerial oversights and obstacles that require societal attention. It also helped dramatize the need for comprehensive toxic substance legislation and was a major spur for the adoption of TSCA.

The most obvious lesson of the experience has been that substances that are environmentally persistent and that have a tendency for bioaccumulation can pose a substantial hazard and require close regulation, even if their acute toxicity is low. A much more difficult lesson is that it is necessary to anticipate the extent and conditions of end use for potentially hazardous chemicals before they become widely distributed. For PCBs it was the massive extent of their use and the rapid proliferation of open-ended applications that made the hazard so serious and difficult to manage. TSCA contains at least partial recognition of this problem when it permits the EPA to consider potential end uses before issuing permits for production. In the same spirit, the EPA can also review new uses or significant production increases of existing chemicals.

A related question of growing concern is whether some of the substitutes that have been developed for most PCB uses are actually benign or are almost as hazardous as the original PCBs. Chlorinated paraffins (CPs) have been among the main substitutes for the prior open uses of PCBs as plasticizers and lubricating oils (Svanberg and Lindén 1979). World production of CPs has increased significantly since the early 1970s and now totals 230,000 metric tons (about 506 million pounds) per year. Although not quite as environmentally persistent as PCBs, CPs can be transported in the environment in similar ways and they are apparently bioaccumulated. Unfortunately, methods for their detection are not well developed, and there is little data about their chronic effects, though they do not seem acutely toxic. The EPA has, however, selected them along with nine other categories of substances for intensive review (BNA 1980a). The most widely used substitute for PCBs in capacitors has been dioctyl phthalate. Pthalates have simlarly been selected by the EPA for close review because of evidence of health and environmental effects (BNA 1980a).

This illustrates one of the major difficulties with attempts to control hazardous substances on a chemical-by-chemical basis, particularly when the very properties that make a substance useful can also render it hazardous. The experience of PCBs shows how difficult it can be to regulate just one chemical even when its risks are well-established. It is thus even more distressing when the result of regulation has been to promote the use of a variety of substitutes that may be almost as hazardous as the original substance but whose properties and consequences are far less-known.

Beyond its pilot role for the EPA, the PCB case has important lessons for the chemical industry. The resources of this industry for testing chemicals far exceed those of the regulatory bodies, and this gives industry an intrinsic advantage in instituting early controls, if desired. Further, the experience of PCBs suggests that giving up a major part of a market or even foregoing production of a chemical may not be crippling to a large diversified producer such as Monsanto. There appears, therefore, to be no basis for the common view that production restrictions will debilitate industry; and hence no argument against substantial self-restraint by producers.

In contrast, use restrictions have had a strong impact on industries that incorporate PCBs into fixed capital such as transformers. The implication is that it is extremely important for producer and consumer industries to cooperate in controlling chemical hazards early in the process of development before growth of technological applications creates intractable "downstream" problems such as those discussed for PCBs. Once a hazardous chemical is firmly embedded in fixed capital and once releases have occurred that threaten the livelihood of other groups, particularly groups who do not benefit directly from the technology in question, then conflict, delay, and frustration of hazard control are virtually inevitable. Indeed, to the extent that such conflicts become endemic, they are likely to stymie the entire effort to control technological hazards.

NOTES

1. The PCB content of capacitors, which are used to regulate current in most electrical machinery and appliances, may range from less than an ounce to over 80 pounds, depending on size and function. Millions of capacitors are manufactured each year; in 1974 alone, 100 million capacitors were manufactured, containing some 21 million pounds of PCBs (Versar, Inc. 1976).

2. Transformers are used to change voltage during electrical transmission. An estimated 5 percent of the transformers in service in the United States in 1976 contained PCBs, with an average of 230 gallons per transformer (Versar, Inc. 1976).

REFERENCES

ACGIH (American Conference of Governmental Industrial Hygienists). 1962. Committee on Threshold Limit Values. Documentation of the threshold limit values. 1st ed. Cincinnati: ACGIH.

Ackerman, Jerry. 1983. New Bedford still awaits PCB aid. Boston Globe, 13 March, pp. 1 and 16.

Allen, J.R., and D.H. Norback. 1977. Carcinogenic potential of the polychlorinated biphenyls. In Origins of human cancer, ed. H.H. Hiatt et al., Book A:173-186. Cold Spring Harbor Conferences on Cell Proliferation, vol. 4. Cold Spring Harbor, New York: Cold Spring Harbor Laboratory.

Auerlich, R.J., R.K. Ringer, and S. Iwamoto. 1973. Reproductive failure and mortality in mink fed on Great Lakes fish. Journal of Reproduction and Fertility 19 (Supplement):365-376.

BNA (Bureau of National Affairs). 1980a. Chemical Regulation Reporter 4 (7 March):1883-1893.

BNA (Bureau of National Affairs). 1980b. Study of Housatonic River confirms PCB concentrations in impoundments. Chemical Regulation Reporter 4 no. 16 (18 July):391-392.

Boyle, Robert H., and Joseph H. Highland. 1979. The persistence of PCBs. Environment 21 no. 5 (June):6-13ff.

Brown, Michael. 1979. Laying Waste. New York: Pantheon.

Burger, Edward J. 1976. A case study: Polychlorinated biphenyls. In Protecting the nation's health: The problems of regulation, 187-201. Lexington, Mass.: Lexington Books.

EDF (Environmental Defense Fund), and Robert H. Boyle. 1980. Malignant neglect. New York: Knopf.

EPA (Environmental Protection Agency). 1978. PCB chemicals in human breast milk reported by EPA. Environmental News (14 September).

EPA (Environmental Protection Agency). 1979a. Water-related environmental fate of 129 priority pollutants, vol. 1. Washington: Office of Water Planning and Standards, EPA.

EPA (Environmental Protection Agency). 1979b. Polychlorinated biphenyls (PCBs): Manufacturing, processing, distribution in commerce, and use prohibitions. Federal Register 44 (31 May): 31514-31542.

Eisenreich, S.J., and G.J. Hollod. 1979. Atmospheric inputs of polychlorinated biphenyls to the Great Lakes: Impact on the Lake Superior system. Paper presented at the American Chemical Society Conference, September.

Electric Power Research Institute. 1980. The delicate disposal of PCBs. EPRI Journal 5 no. 2 (March):20-21.

FDA (Food and Drug Administration). 1979a. Polychlorinated biphenyls: Reduction of tolerances. Federal Register 44 (29 June):38330-38340.

FDA (Food and Drug Administration). 1979b. PCB contamination of food in the western United States. Unpublished report, November.

GAO (General Accounting Office). 1981. EPA slow in controlling PCBs. CED-82-21. Washington: GAO, December 30.

Hammond, Allen L. 1972. Chemical pollution: Polychlorinated biphenyls. Science 175:155-156.

Hellman, Peter. 1976. For the Hudson: Bad news and good. New York Times Magazine, 24 October, 16-20ff.

Highland, Joseph H. 1976. PCBs in foods: A look at federal government responsibilities. In National Conference...(1976), 443-450.

Higuchi, Kentaro, ed. 1976. PCB poisoning and pollution. New York: Academic Press.

Hilts, Philip J. 1981. The deadly business of cleaning up. Washington Post, 26 February, p. A16.

Jahnke, Art. 1980. PCBs in New Bedford harbor. Real Paper (Boston), 27 November.

Jellinek, Steven J. 1978. TSCA: The mandate is clear. In Toxic substances control: Proceedings of the second annual Toxic Substances Control Conference, December 8-9, 1977,65-71. Washington: Government Institutes.

Jensen, Sören. 1966. A new chemical hazard. New Scientist 32: 612.

Jensen, Sören. 1972. The PCB story. Ambio 1 no. 4 (August):123-131.

Kleinert, Stanton J. 1976. Sources of polychlorinated biphenyls in Wisconsin. In National Conference...(1976), 124-126.

Kopp, Thomas E. 1976. PCB disposal, reclaiming and treatment. In National Conference...(1976), 108-123.

Lloyd, J.W., R.M. Moore, B.S. Woolf, and H.P. Stein. 1976. Polychlorinated biphenyls. Journal of Occupational Medicine 18: 109-113.

Magiera, Frank E. 1981. PCB, a dangerous toxin, is common here. Evening Gazette (Worcester, Mass.), 17 March,23S.

Montgomery, Paul L. 1981. PCB cleanup shuts upstate tower for months. New York Times, 21 February, p. 25.

Mosher, Marcella, and Greg Moyer. 1980. PCBs and breast milk. Nutrition Action 7 no. 11 (November):10-13.

Murphy, Thomas. 1979. Atmospheric inputs of polychlorinated biphenyls to the Great Lakes. Paper presented at the American Chemical Society Conference, September.

NIOSH (National Institute for Occupational Safety and Health). 1977. Criteria for a recommended standard: Occupational exposure to polychlorinated biphenyls. NIOSH 77-225. Washington: Government Printing Office.

Nader, Ralph, Ronald Brownstein, and John Richard. 1981. Who's poisoning America? San Francisco: Sierra Club Books.

National Conference on Polychlorinated Biphenyls. 1976. Proceedings of the National..., November 19-21, 1975, Chicago, Illinois. EPA-560/6-75-004. Washington: Environmental Protection Agency.

National Research Council. 1979. Committee on the Assessment of Polychlorinated Biphenyls in the Environment. Polychlorinated biphenyls. Washington: National Academy of Sciences.

Olsson, B., et al. 1979. Cadmium and mercury contaminations in mink after exposure to PCBs. Ambio 8 no. 1:25.

Papageorge, William. 1975. A case study: Polychlorinated biphenyls. Paper written for the Study on Decision Making for Regulating Chemicals in the Environment. Washington: National Research Council, National Academy of Sciences.

Peakall, David B., and Jeffrey F. Lincer. 1970. Polychlorinated biphenyls: Another long-life widespread chemical in our environment. Bioscience 20:958-964.

Regenstein, Lewis. 1982. PCB's: Present in every living creature. In America the poisoned, 293-306. Washington: Acropolis Books.

Rogan, Walter, Anna Bagniewska, and Terri Damastra. 1980. Pollutants in breast milk. New England Journal of Medicine 302 (26 June):1450-1453.

Severo, Richard. 1980. Hudson: Portrait of a river under attack. New York Times, 9 September, pp. C1 and C3.

Svanberg, Olof, and Eva Lindén. 1979. Chlorinated paraffins: An environmental hazard. Ambio 8:206-209.

TSCA (Toxic Substances Control Act). 1976. Public Law 94-469. Section 6(e): [PCB Regulations]. 11 October.

USDA (U.S. Dept. of Agriculture), FDA (Food and Drug Administration), and EPA (Environmental Protection Agency). 1980. Restrictions on use of polychlorinated biphenyls. Federal Register 45 (9 May):30980-30993.

Versar, Inc. 1976. PCBs in the United States: Industrial use and environmental distribution. EPA 560/6-76-005. PB 252-012/3WP. Springfield, Virginia: National Technical Information Service.

16
The Consumer Product Safety Commission[1]

Roger E. Kasperson and Thomas Bick

When the Consumer Product Safety Commission (CPSC) opened its doors in 1973, the American consumer movement hailed the agency as a powerful new instrument for creating a healthier, safer household environment. Established by the Consumer Product Safety Act (1972), the CPSC was to be open and accessible to the public, insulated from outside political interference, and vested with authority over a broad range or products, most of which had been previously unregulated.

Today the CPSC's early supporters look back on the agency's first five years in bitter disappointment. Representative John E. Moss, a leading advocate of the Commission in 1972, recently called the agency "one of his biggest disappointments" and characterized its performance as a "miserable record" (Thomas 1978). Spokesmen for industry and consumer groups alike joined the chorus of criticism, some even going so far as to call for the dismantling of the Commission.

A Twofold Experiment

The CPSC experiment is twofold: it constitutes both an effort to control a broad range of technological hazards and an ambitious attempt at regulatory reform.

The CPSC is empowered to intervene in a broad sector of the private economy to combat the hazards associated with a vast range of products. In this respect the act is part of a general trend in federal hazard management, one reflected in other recently enacted federal health and safety laws. As with current regulatory efforts in the areas of air and water pollution, occupational health, and toxic substances, the CPSC's mission is enormously extensive and complex: the hazards it seeks to combat are highly varied and often poorly understood, whereas the potential impacts of its actions on the nation's economic health are great. It has been estimated that the CPSC's regulatory domain embraces more than 10,000 different types of products, more than 2,500,000 firms (almost half of all U.S. businesses), an annual toll of 30,000 fatalities, 16 to 21 million injuries, and $5.5 billion in product-related injury costs (U.S. Congress 1976a,197).[2]

As daunting as this breadth of mission is, it is the degree of regulatory reform embodied in the legislation creating the agency that makes the CPSC experiment unique. The congressional sponsors of the act sought to establish a model of regulatory reform—an agency that would be powerful yet open, its broad authority constrained by the active participation of industry and consumer groups in its regulatory processes. It was to be the most accessible bureaucracy in Washington, yet at the same time insulated, as no agency before it had been, from politically motivated pressure and manipulation. To this end the Commission was conceived as an arm of Congress rather than an extension of the Executive Branch, so that its day-to-day operations would be independent of the political vagaries of the White House.

How It Works

The Consumer Product Safety Act established four major goals for the agency:

- protecting the public against unreasonable risks of injury associated with consumer products;
- assisting consumers to evaluate the comparative safety of consumer products;
- developing uniform safety standards for consumer products and minimizing conflicting state and local regulations;
- promoting research and investigation into the causes and prevention of product-related deaths, illnesses, and injuries.

To accomplish these objectives, Congress provided the CPSC with an unusually broad and potent set of regulatory tools.[3] The agency can, for example, issue mandatory standards, require industry to finance testing procedures, ban or require the recall of products, direct that consumers be notified of hazards, specify labeling requirements, seize and destroy hazardous goods, require manufacturers to provide performance data, and inform the public of the comparative safety of products within different categories. CPSC may enforce its rules and orders by seeking stiff civil or criminal penalties against those who fail to comply with its regulations.

For the most part, only the requirement that CPSC's rules be "reasonable" (a test that makes court challenges to the rules on nonprocedural grounds extremely difficult) limits this wide-ranging authority.[4] Although the agency is required to consider the economic impacts of its actions, there is no legal requirement that such impacts be the decisive factor in the choice of whether, or how, to act.

Organization

The CPSC is headed by four commissioners and a chairman. All major decisions require a majority vote of the commissioners. Thus the commission is a "collegial" body. It is also a "matrix" organization in that it has both program and functional units. The Office of Program Management comprises eight "program areas": fire and thermal burns, electric shock, acute chemical and environmental hazards, chronic chemical and environmental hazards, tools and housewares, structural hazards, toys, and sports (with the last four collectively designated as "mechanical hazards"). The functional units are organized into five directorates: hazard identification and analysis, engineering and science, compliance and enforcement, field coordination, and administration. Each directorate comprises "teams" that correspond to the eight program areas. The organizational structure is a matrix in the sense that each team member answers to two supervisors--the program manager of the team, who coordinates hazard strategy formulation, and the head of the functional unit, who is responsible for hiring, firing, and quality control.

Managing the Hazards

Hazard management by the Commission involves a number of steps (Figure 1). First the CPSC identifies the most frequent and severe product-related injuries and determines their causes. This is done primarily through the National Electronic Injury Surveillance System (NEISS), a network of telecommunications terminals located in the emergency rooms of 119 statistically representative hospitals across the country. In-depth investigations of injuries (4,000 to 5,000 investigations yearly) and screening of death certificates, consumer complaints, controlled laboratory tests, and data furnished by other agencies supplement this information. In addition, any interested person or group discovering a product hazard may petition the Commission to initiate a hazard control action.

Next, the Commissioners (as of 1977) arrange product hazards into priority groupings. Once priorities have been established, strategies for managing the high-priority hazards must be formulated. The hazard is assigned to one of the eight program areas. The program manager draws upon technical expertise from the various CPSC directorates to prepare a "briefing package" summarizing the information and proposing one or more management strategies, listing the pros and cons of each.

The Commission reviews the briefing package, chooses a particular strategy, or returns the package for further work. Not uncommonly, a briefing package appears before the Commissioners several times before they make a final strategy decision. If the decision is to promulgate a safety rule (a standard or ban), they initiate rulemaking procedures. If however, the chosen strategy is public education, the staff prepares an information and education plan.

After a control strategy is implemented, the Commission evaluates its effectiveness. The Directorate of Hazard Identification and Analysis monitors the agency's injury reporting sources to gauge the injury-reducing impact of the managerial strategy. Meanwhile, the Directorate of Compliance and Enforcement determines the extent of industry compliance.

Public Participation

The CPSC has two important public participation features--the "offeror" and the citizen petition processes.

The offeror process is unique to the CPSC. Section 7 of the Act requires the Commission, when it decides to issue a product safety standard, to announce this intention publicly and to invite

374

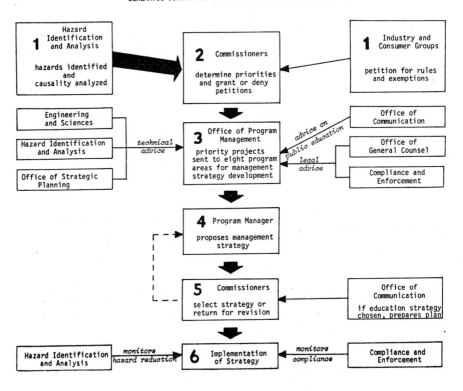

Figure 1. Schematic diagram of the CPSC hazard-management process. The flow of decisions proceeds from stages 1-6, as indicated on the diagram. To initiate action the Commissioners receive input from hazard identification units within the agency or from industry and consumer groups on the outside. After a decision to take action, the issue passes on to the Office of Program Management, which deals with the problem via the matrix organization described in the text. This eventually leads to a proposal for action that goes back to the Commissioners, who may decide to implement the proposal or return it to the staff for revision. In the last stages of the process, the Office of Communication and the Office of Compliance and Enforcement may enter the process as needed, either to prepare educational materials or to monitor and enforce compliance with a completed ruling.

any interested person to submit an existing standard for adoption by the CPSC or to offer to develop a standard. The CPSC must accept one or several of these offers if the offeror is technically competent, is capable of developing an appropriate standard within a specified period, and agrees to comply with the procedures (including public participation) specified by the Commission (section 7(d) of the Act). The Commission then reviews the proposed standard prepared by the offeror and, after analyzing its estimated economic and environmental impacts and its hazard-reducing potential, adopts, amends, or rejects it. If there are no qualified offerors or if the only acceptable offeror represents the industry to be regulated, the Commission may develop its own proposed standard. Once a standard is adopted, anyone may request judicial review of the standard within sixty days.

In addition, the Act provides for early public participation in its decisionmaking via the citizen petition process. Such a petition must set forth the facts supporting the need for a rule and a description of the rule sought. The Commission may deny the petition unless it seeks to regulate a consumer product that "presents an unreasonable risk of injury" and the denial "unreasonably exposes the petitioner or other consumer to a risk of injury" (section 10(e) of the Act). The petitioner, however, may challenge such a denial in court.

CPSC Performance

Despite its broad regulatory powers and the early enthusiasm of its sponsors and staff, the Commission has developed a reputation as a poorly run, unaggressive, and largely ineffectual bureaucracy. A 1976 House oversight committee report spoke for many of the agency's critics when it concluded that, after three years, the Commission had not utilized its broad regulatory powers, had been slow to develop safety standards, and had "yet to demonstrate its capacity to plan, to prescribe administrative rules and guidelines, and to set clear priorities" (U.S. Congress 1976a, 195). Hearings held in the House and Senate in 1977 and 1978 further highlighted the failures and deficiencies of the agency. In 1978 the President's Office of Management and Budget (OMB) indicated its intention of seeking to have the Commission abolished. To what extent does the record support the negative assessments of so many of the CPSC's observers?

It is important to note at the outset that the Commission is not without its positive achievements. Most observers would agree that the agency's open-door policy has made it a model of a publicly accessible bureaucracy. Most of its meetings have been open to the public and, unlike the situation in most other federal agencies, Commission policy has made disclosure of its records the rule rather than the exception. Even many of the Commission's most vocal critics agree that its system of injury surveillance is among the best hazard identification systems in the country. Most also agree that the actions the Commission has taken in requiring industry to notify it of substantial product hazards have also been successful. In its first two and a half years, the Commission processed 350 notifications from manufacturers, resulting in the correction (largely through informal negotiation) of four million defective consumer products.[5] Product standards and bans issued as of 1 January 1978, though criticized by many as too few and too weak, could ultimately prevent an estimated 553,600 annual consumer injuries (Table 1) and thousands of deaths.

Finally, the continuing threat of federal regulation has undoubtedly led industry to impose product-safety standards of its own, although the actual extent of such actions is, of course, not known. The consumer product industry is also well aware that a growing number of private attorneys are using the Commission's injury data to help win product liability suits. This is an unpublicized but important contribution of the hazard identification system and may well prove one of the Commission's most significant impacts on product safety.

As significant as these achievements are, they are more than matched by the agency's many deficiencies. The Commission has in

TABLE 1
Estimated safety contribution of 8 CPSC regulations[a]

PRODUCT	ESTIMATED ANNUAL IN- JURIES (IN THOUSANDS)	TYPE OF STANDARD ENVI- SIONED	ESTIMATED INJURY REDUCTION (IN %)	ESTIMATED PRODUCT LIFE[b] (IN YRS.)	ULTIMATE NO. OF IN- JURIES PRE- VENTED (IN THOUSANDS)
Architectural glass	464.9	MSR[c]	70	30	325.4
Bicycles	1,211.5	MSR	17	7	206.0
Matchbooks[d]	24.1	MSR	20–30	4	4.8–7.2
Refuse bins	28.4	Ban	48	10	13.6
Fireworks	15.6	Ban & MSR	50	1	7.8
Pacifiers	N.A.	Ban & MSR	30–70	1	N.A.
Cribs	23.5	Ban & MSR	20	10	4.7
Swimming pool slides[e]	N.A.	MSR	N.A.	N.A.	N.A.
TOTAL	1,768.0				553.6

[a]This table does not include two standards, for baby rattles and cellulose insulating materials, that became effective on August 21, 1978 and September 8, 1978, respectively; nor does it include a ban on self-pressurizing products containing vinyl chloride that became effective in June 1978.
[b]Product Life provides an estimate of the time required before maximum injury prevention possible is achieved.
[c]Mandatory Safety Rule.
[d]Estimates do not reflect the impact of a recent court decision [D. D. Bean and Sons v. CPSC, --F.2d-- (1st Cir., 31 March 1978)] invalidating some parts of the matchbook standard.
[e]Most of this standard has been invalidated by the courts (Aqua Slide 'n' Dive 1978).

the past consistently backed away from forceful regulation. Thus far it has enacted only 11 standards and bans. The first safety standard did not become effective until some three and one-half years after the Commission opened its doors for business; even worse, the standard was for swimming pool slides—hardly a product at the top of the average consumer's "most hazardous" list. To add insult to injury, a federal appeals court recently invalidated most of the standard on the basis that it was unsupported by the evidence available to the Commission (Aqua Slide 'n' Dive 1978).[4] The Commission has repeatedly opted for less controversial labeling requirements whenever its staff has proposed standards or bans. It has almost totally neglected one of its most important responsibilities—to provide consumers with comparative information on the safety of specific products. Despite explicit authorization in the act, the Commission has failed to promulgate rules requiring manufacturers to keep records, submit reports, or provide performance and technical data.

So far the agency has utilized one of its most important powers—its right to move against "imminent hazards" under Section 12—only four times. Further, the Commission failed until 1978 to develop a policy for dealing with chronic hazards (products having health effects that show up only after prolonged exposure to the product or after a long latency period following exposure). As a result of this delay the Commission's management of chronic hazards has been beset by confusion and ad-hoc responses. For example, despite overwhelming evidence of carcinogenicity and the receipt of formal citizen petitions for action, the Commission procrastinated over a year before banning TRIS in children's pajamas and more than two years before banning asbestos in patching compounds and fireplace logs. A chronic-hazards policy has now been adopted by the Commission.

In short, the record more than justifies the disappointment of the Commission's founders and supporters. Though recent signs point to a revitalized, more activist Commission, the agency must, on the whole, be considered a failure. What explains the magnitude of this failure and what lessons can be learned that may be applicable to other efforts at federal hazard control?

False Explanations

The citizen-petition process was blamed by the Commission's first chairman, William Simpson, for much of the agency's misplaced effort during its first three years. The former chairman claimed that he interpreted Section 10 as giving the CPSC no choice but to act upon any citizen petition that sought the regulation of any unreasonably hazardous product. This predicament, he alleged, led to a petition domination of hazard management and a consequent lack of the resources needed to regulate the many "high-hazard" products not the subject of petitions (Simpson 1976,12-12,27,91).

Even during Simpson's tenure, however, the petition process consumed only a small portion of the agency's total resources. Furthermore, the CPSC responded to relatively few petitions in its first three years—leaving a huge petition backlog that has only recently been reduced to manageable proportions. In addition, neither the language of the act nor its legislative history supports Simpson's interpretation of Section 10.

The collegial nature of the CPSC (whereby decisions are made by the five-member Commission rather than a single administrator) has also been blamed for many of the agency's shortcomings.[6] It is noteworthy, however, that other collegial bureaucracies, such as the Securities and Exchange Commission, are among Washington's most efficient regulators. Although some of the CPSC's regulatory delays could perhaps have been avoided under the leadership of a single administrator, most of the sluggishness and mismanagement that has plagued CPSC rulemaking occurred at the staff level, before proposals were brought to the attention of the Commissioners. In short, the ineffectiveness of the Commission has stemmed not so much from the way it made its decisions as from the decisions it made.

Common Regulatory Problems

A number of the CPSC's problems are common to other regulatory agencies. Since its inception, woefully inadequate funding has seriously hampered the effectiveness of the Commission. In fact, the level of funding has consistently been only two-thirds of that authorized by Congress (Table 2). The first year budget request of $30.9 million by the Office of Management and Budget (OMB) was fully $24 million less than what CPSC sponsors considered the absolute minimum needed. The budget reduction in the second year (FY 1975) was the largest inflicted on any federal agency and forced the CPSC to reduce drastically its second-year operating plan. OMB budget requests for the Commission increased only negligibly over the next three years, less in fact than the rate of inflation--a situation

TABLE 2
Funding of the CPSC

FISCAL YEAR	AMOUNT AUTHORIZED	CPSC REQUEST (IN MILLIONS OF	OMB REQUEST DOLLARS)	AMOUNT APPROPRIATED
1974	59.0	30.9	30.9	34.8
1975	64.0	42.8	42.8	37.0
1976	51.0	49.2	36.6	39.6
Transition quarter	14.0	12.9	9.1	10.0
1977	60.0	54.9	37.0	39.8
1978	68.0	41.1	40.2	39.1

[a]Indicates amount actually available to the Commission.
[b]$3.9 million transferred from the Food and Drug Administration 1974 appropriation.
[c]The original $38.3 million appropriation was vetoed by President Nixon.
[d]CPSC request later modified to $50.4 million.
[e]CPSC request later modified at urging of OMB to $41.1 million.

that Representative Moss characterized as "deregulation through budget slashing" (Moss 1976,1-2).

Seriously compounding the fiscal problem has been a lack of organization and efficiency. As portrayed in a recent management survey, the working atmosphere in the agency has been "confusing, frustrating, counter-productive, and inefficient" (Thomas 1978,14). Until recently, CPSC functional units considered rulemaking proposals sequentially: only after all units had in turn analyzed and commented on a proposal would the proposal appear before the Commissioners for final decision. This organizational procedure, finally modified in a 1977 reorganization, contributed to much of the delay that has plagued the CPSC's rulemaking efforts in its first five years.

The Pitfalls of Innovation

These obstacles have, to some extent, bedeviled most federal agencies. It is therefore not surprising that many observers of the Commission identify them as the underlying causes of the agency's malaise. Much less attention has been directed to the sources of those CPSC failures that stem from its experimental nature as a hazard manager and as a new kind of regulator. Two of CPSC's major shortcomings result from the attempt to regulate a broad universe of diverse hazards within the catchall framework of consumer products. These two are: the failure to allocate resources effectively and the inability to manage chronic hazards.

Setting Priorities

The misallocation of resources has been largely a result of the agency's inability to establish priorities. As manager of an enormously varied hazard domain, the newly formed agency immediately confronted the difficult question of where and how to begin. Neither Congress nor existing federal agencies provided much guidance.

Until recently, the typical federal regulator was required to act only within a limited range of authority to achieve narrowly circumscribed hazard-management objectives. Federal health and safety laws tended to focus on a particular industry (railroads, meat processors), a particular type of product (drugs, cosmetics), or a particular kind of hazard (children swallowing poison or becoming trapped in abandoned refrigerators). During the past decade, however, federal regulation has expanded into much broader arenas (chapter 19): all sources of air and water pollution, all hazards in the workplace or household, all toxic substances. Such broadscale intervention requires careful analysis of the hazards to be regulated. Here is the place to employ a taxonomy (chapter 4) to facilitate priority-setting.

A statement of policy soon after the Commission's inception suggests a considered approach to the problem of priorities:

> The Commission will deal first with those products which pose the greatest risk of injury to the public. The Commission will set (and will periodically reevaluate) its priorities, taking into consideration the number of injuries associated with a particular product, the severity of

those injuries, the consumer's likelihood of exposure to that product, and any other factors which the Commission considers important (U.S. Congress 1974,88).

Despite this declaration of intent and despite the availability of an effective hazard-identification system (see Figure 1), the Commission failed to establish priorities until well into its fourth year. As a result, it became mired in unproductive work, allocating its limited resources haphazardly among both serious and trivial hazards. Chairman Simpson ruefully acknowledged to Congress in 1976 that "last year we estimated 75 percent or more of our activities were reactive as opposed to planned" (Simpson 1976,15).

It was not until mid-1976 that a priority policy, listing criteria for ranking product hazards, finally emerged. The criteria employed and the weightings attached by the staff (Table 3) suggest the difficult judgments and the enormous information burden confronting any agency charged with managing such a broad hazard universe. Table 4 shows the annual injuries, possible injury reduction, and 1978 projected goals for each of the 29 high-priority hazards finally recognized by the Commissioners.

These data, when compared with the results of CPSC regulations enacted as of January 1, 1978 (Table 1), suggest the extent to which misplaced effort is possible in the absence of clear priorities. Whereas two of the agency's safety rules (for architectural glass and bicycles) should forestall an estimated 200 to 300 thousand injuries a year, the other rules all have much lower injury-reducing potentials. Had the Commission focused its limited resources on higher-priority hazards first, it could have provided much greater protection for the American consumer. For example, the Commission estimates that a safety standard for power lawn mowers could save up to 88,000 annual injuries; for bathtubs and showers, 88,000; for public playground equipment, 46,000; for upholstered furniture (a flammability standard), 45,000. Instead, the agency chose to develop standards for such products as matchbooks, fireworks, pacifiers,

TABLE 3
CPSC priority criteria

CRITERION	PRIORITY WEIGHT*
Frequency of injury	24
Severity of injury	25
Chronic illness and prospect of future injuries	16
Unforeseen nature of the risk	14
Vulnerability of population at risk	9
Probability of exposure to the hazard	12

*Priority weights were arrived at through a staff process at the CPSC. Larger weights indicate assignment of greater importance by the CPSC staff.

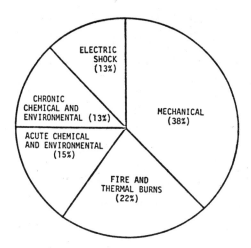

Figure 2. CPSC budget allocations among programs for fiscal year 1978. Note that acute hazards (electric shock, mechanical accidents, fire and thermal burns, and chemical and environmental risks) take up 87 percent of the budget, leaving only 13 percent for managing chronic hazards. The total budget for 1978 was $39.2 million.

and swimming pool slides, each of which will likely forestall no more than 10,000 annual injuries. In short, the absence of clear priorities has made the CPSC an erratic hazard manager.

Chronic Hazards

During the first five years of its existence, the CPSC concentrated its managerial effort on acute hazards. For FY 1978, for example, the Commission allocated only 13 percent of its budgetary resources to chronic hazards (Figure 2) despite the growing contribution of these hazards to overall mortality in the United States and the greater costs involved in their evaluation and control. Also, until very recently, the Commission has had no policy or procedure to guide its response when evidence of such hazards comes to its attention.

Further, as noted above, the Commission's major hazard identification program (the hospital emergency-room reporting system) is oriented to acute hazards. Its chronic hazard identification system consists of a computer system that draws on several nationally available listings of hazardous materials (MEDLINE, TOXLINE, CHEMLINE and the Clearinghouse on Mutagens and Carcinogens). A recent review by the National Research Council found, however, that these data bases omitted information on many chemicals present in household products and that the CPSC suffered a serious lack of high level technical expertise--epidemiologists, biostatisticians, toxicologists, and medical scientists (National Research Council 1977). It is not surprising, therefore, that by the beginning of 1978, the Commission had regulated or proposed to regulate only six substances

TABLE 4
The high-priority hazards of the CPSC

CPSC PRIOR- ITY	PRODUCT	ANNUAL INJURIES (IN THOUSANDS)	ANTICIPATED ANNUAL INJURY REDUCTION (IN THOUSANDS)	1978 PROJECTED GOALS
1	Asbestos	N.A.	N.A.	Preparation of Final Ban of Additional Products Containing Asbestos
2	Lawn Mowers	160.0	56-88	Final Standard Completed: Begin Development of Proposed Certification Rule
3	Gas Space Heaters	4.8	.7-1.2	Publish a Proposed Ban for Unvented Gas Space Heaters
4	Communication Antennae	0.4	N.A.	Completion of 75 percent of Sec. 7 Rule Feasibility Study; Complete Enforcement Plan for Labeling Standard
5	Public Playground Equipment	228.9	11.4-45.7	Publish Proposed Technical Requirements for Regulation
6	Chlorofluoro- carbons in Aerosols	N.A.	N.A.	Proposed Regulation of Uses of CFC's Other Than Non-Essential Aerosol Propellants Is Published, Economic Impact to Be Studied
7	Architectural Glazing	464.9	325.4	Promulgate Final Certification Reg.; Propose a Weathering Amendment
8	Unstable Refuse Bins	28.4	13.6	Issue and Enforce Ban on Unstable Refuse Bins
9	Lead in Paint	N.A.	N.A.	Issue and Enforce Final Banning Regulation for Paint Containing .06 Percent Lead
10	Baby Pacifiers	N.A.	N.A.	Effective Date of Banning Reg. is 2/26/78
11	Sharp Points & Edges on Toys	44.7	7.2-28.9	Publish Final Technical Requirements on Sharp Points
12	Methyl Alcohol	N.A.	N.A.	Issue Final Regulation
13	Upholstered Furniture	89.7	31.4-44.9	Remedial Options Analyzed to Determine Strategy; Take Appropriate Action
14	See #11 above			Issue Final Regulation on Metal and Glass Edges in Toys
15	Children's Sleepwear	N.A.	N.A.	Publish Sleepwear Enforcement Policy

16	Miniature Christmas Tree Lights	1.9	0.9-1.4	Issue Final Mandatory Safety Standards
17	Television Sets and Stands	51.6	20.6-25.8	Participate in Appropriate Standard
18	Aluminum Wire	N.A.	N.A.	Section 15 Activities Continue on Old Technology; A Voluntary Standard Will Be Considered for New Technology
19	Ranges and Ovens	69.6	13.9-20.8	Support and Evaluate Efforts of Voluntary Standard Groups (UL.; ANSI)
20	Skateboards	88.0	N.A.	After Remedial Options Analysis Is Completed (1977) Decision on Further Action
21	Extension Cords and Trouble Lights	8.7 (cords) N.A.(lights)	0.4-0.7 (cords) N.A (lights)	Through Offeror Process, Publish Proposed Mandatory Safety Standard
22	Bicycles	1,175.0	217.8-276.5	Amend Reg. to Include Wet Brake and Pedal Slip Criteria
23	Matches	24.0	4.8-7.2	Implement Standard May 1978
24	Ladders	186.0	26.0-65.1	Support and Monitor ANSI Work on Revised Voluntary Standard for Metal Ladders and for Draft Voluntary Standard on Wood Ladders
25	Energy Conservation	N.A.	N.A.	Hazard Analysis Completed; Initiate or Participate in Appropriate Standards Activities
26	Bathtubs and Showers	159.1	39.8-87.5	Support and Participate in Completion of Voluntary Standard
27	Smoke Detectors	4.1	N.A.	Decide on Feasibility of Sec. 7 and Remedial Strategy Research Complete; Take Appropriate Action
28	Children's Football Helmets	131.6	N.A.	Remedial Options Complete; Option Pursued as Selected by Commission
29	Small Parts in Toys	7.9	.4-2.4	Publish Proposed Regulation
	TOTAL (for Products with Injury Estimates)	2,929.3	770.3-1,035.1	

(spray adhesives, vinyl chloride, lead-based paint, asbestos in patching compounds and fireplace logs, TRIS-treated sleepwear, and chlorofluorocarbons) that pose long-term health risks, and in only one of these actions did the initiative come from the Commission itself.

The tendency has been to attribute the Commission's impotence in chronic hazard control largely to administrative mismanagement. Some of the criticism is undoubtedly well placed, but it obscures the much more fundamental problems that this class of hazards presents to a single small government agency. Since 70 to 80 percent of all cancer defies scientific explanation, the assignment of specific carcinogens to specific cancers is the exception, and the causes of heart disease are still poorly charted, regulating long-term hazards, even with plentiful resources, is bound to be an enormously difficult task.

Complicating the job of the CPSC is the fact that some consumer products are formulated from existing stocks of chemicals well along in the productive process. It is therefore questionable whether even improved management and augmented resources will permit satisfactory control of chronic hazards and whether the CPSC should be in this business at all. Although the Commission has recently formulated a chronic hazards policy,[7] there is a strong argument that such hazards either should be the responsibility of some coordinated intergovernmental program (as suggested by the current activities of the Interagency Regulatory Liaison Group and the Toxic Substances Strategy Committee) or that the legislative mandate of the CPSC should be revised to vest all chronic-hazard control in some umbrella agency (a Toxic Substances Commission?) that could deal with these hazards on a much more comprehensive basis. This would free the CPSC to do what it does best—manage the acute hazards associated with consumer products.

A Frustrated Attempt

In creating the CPSC its congressional sponsors sought to establish a model of regulatory reform. They included in the Consumer Product Safety Act innovative provisions designed both to insulate the agency from outside political pressure and to make it open, accessible, and responsive to the public. After more than five years the CPSC has fulfilled neither expectation.

The Act includes a number of "insulating provisions." The agency's five Commissioners do not serve at the pleasure of the President who appoints them; rather, once appointed to staggered seven-year terms, they continue in office unless removed for neglect of duty or malfeasance. No more than three of the five Commissioners may be from the same political party. The hiring of new personnel, other than Commissioners, is not subject to White House review or approval.[8] The CPSC submits its budget requests and legislative proposals simultaneously to the Congress and the President's Office of Management and Budget. This innovation marked the first time a federal agency was not required to submit its proposals to the OMB for review and modification prior to submission to Congress, the intent being to reduce the CPSC's dependence on White House fiscal decisions. The act also forbids policy-making CPSC employees from taking jobs with industries within CPSC jurisdiction for at least one year after leaving the agency, a provision recently favored by President Carter to slow the "revolving door" between regulators and regulated industries. The Commission is also required to include in its annual report a log of all meetings between Commission officials and persons affected by its regulatory activities.

Despite these insulating safeguards, White House interference in the affairs of the CPSC has been pervasive. President Nixon did not fill the five Commissioner positions until five months after the creation of the agency. Most of the Commissioners finally selected, including Chairman Simpson, a Republican businessman, shared Nixon's distaste for the federal regulation of private industry. As Nixon searched for commissioners, his Office of Management and Budget (which had advised Nixon to veto the Act) almost singlehandedly set up the CPSC's organizational structure, and arranged the wholesale transfer of 530 personnel from other agencies to fill most of the 768 original CPSC positions.[9] Despite the innovative budget submission requirement, the OMB also dominated the agency's funding process. Congressional appropriations committees, unaccustomed to receiving budget requests directly from an agency rather than through the OMB, paid greater deference to the latter's recommendations. The debilitating budget appropriations detailed in Table 2 were the result. Substantial third- and fourth-year funding increases, confidently anticipated by CPSC planners, were never forthcoming.

In sum, the effort to establish the CPSC as a politically independent hazard manager within the federal government was a failure, a fact that attests both to the power of the executive branch to assert its authority over administrative agencies and to the inadequacy of the provisions in the act designed to counter that authority. In addition, Congress itself is a potential source of political pressure on the agency whenever affected industries lobby to have bans or standards modified or set aside.[10]

Public Participation

Many would agree that the CPSC has become one of Washington's most accessible and open agencies but fail to link the shortcomings of the CPSC as hazard manager to the CPSC as a stimulator of public participation. It is assumed that, in a well-ordered world, these two roles must be somehow compatible, indeed reinforcing.

The CPSC's offeror process requires the Commission to accept the offer of one or more qualified persons or groups to develop a needed safety standard. The intent was that both industry and consumer groups would become directly involved in the Commission's most important rulemaking function: the issuance of product safety standards. To date, however, industry has almost totally dominated the process. Only one standard (for power mowers) has been developed by a consumer group (Consumers Union) and one other (for miniature Christmas tree lights) by a joint industry-consumer group offeror.

In retrospect, it is also apparent that the act's congressional sponsors sorely underestimated the time and money needed to draft comprehensive safety standards. Some offerors, and others participating in the process, have spent millions of dollars, and four to five years, working to draft acceptable standards. Consumers Union found its four-year involvement with the power mower standard so costly that it stated in a letter to the Commission that it would probably never again serve as an offeror (Karpatkin 1976,364).[11] Although consumer representatives must, by law, be added to all standard-development committees, they usually lack the technical expertise and funding to contend effectively with well-heeled industry groups. The Commission, for its part, has failed to provide the resources needed for effective consumer participation—though it clearly has the power to do so (U.S. Congress 1976a,221).[12]

The other measure intended to assure public involvement in CPSC rulemaking—the citizen-petition process—presents the agency with the dilemma of how to reconcile the need for effective citizen participation with the need for efficient hazard management. Recently the CPSC has moved toward a policy of summarily denying petitions not related to its high-priority hazards. Such a policy not only potentially weakens the "regulatory reform" portion of the CPSC experiment but also may undermine the greatest substantive virtue of the petition process—its ability to alert the Commission to the occasional hazard that is unidentified or underestimated by normal agency processes.

An even larger issue, however, is involved in determining the long-term effectiveness of the CPSC. Facing a regulatory domain occupied by powerful industry groups, the Commission depends for its success on the existence of an informed and active consumer constituency. But after displaying an unprecedented degree of cohesiveness in the late 1960s and early 1970s, the consumer movement is now dispersed and disorganized, its political clout noticeably diminished. Though consumer groups have the support of the chairmen of the primary CPSC oversight committees in Congress (Wendell Ford in the Senate and John Moss in the House), support for the Commission on Capitol Hill appears to be dwindling (see Epilogue). Congressional approval, over the opposition of almost every major consumer group, of John Byington as CPSC Chairman in 1976 underscored the weakness of both the agency and its consumer constituency.

The consumer movement may well fall victim to the nation's overriding concern with inflation. A recent Harris poll on consumerism, for example, reveals a public that desires increased protection from "dangerous consumer products" but is also greatly concerned about the high prices of consumer goods and hostile to increased government regulation (Louis Harris and Associates 1977,29 and 69).[13] The need for an informed and involved constituency is one that is apparently recognized by the new CPSC Chairman, Susan Bennett King, who in a recent interview stated that one of the most important things the Commission has to do is to tell its own story better (King 1978).

The Problem

For the first five and a half years of its existence, the CPSC was unaggressive and ineffective in meeting its responsibilities to inform and protect the American consumer. It consistently backed away from forceful regulation, devoted itself to trivial as well as serious hazards, and repeatedly opted for actions less aggressive than those recommended by its staff. The CPSC failed to provide consumers with comparative safety information for products so that individuals can use their own purchases to create safer home environments. Finally, the Commission has failed to develop the technical or fiscal resources to control the chronic hazards presented by consumer products, despite the fact that such hazards account for an increasing proportion of mortality in the United States and are the subject of growing public concern.

It is apparent that the Commission shares many problems with other regulatory agencies—legislative inadequacies, serious underfunding, executive interference, a conflict-of-interest problem in the expert advice it receives, and powerful industry resistance. But the Commission has exacerbated these with a number of problems distinctively its own—a degree of administrative mismanagement and disorganization spectacular even in the context of large bureaucracy, ineptitude in its rulemaking process, responsibility for a vast and varied hazard domain, failure to set clear priorities for Commission efforts, and lack of any real capacity to manage chronic hazards.

Some Suggested Solutions

To improve CPSC, we recommend the following four specific changes:

- Congress should reconsider the appropriateness of the Commission's domain of hazard responsibility. It should either remove the Commission's responsibility for chronic hazards and vest it in a new central governmental institution or, alternatively, mandate for chronic hazards uniform national policies and procedures that would apply to all federal regulatory agencies.
- Congress should provide more guidance on the criteria for setting priorities and should specifically indicate how the Commission's resources should be allocated in relation to those priorities. Congress should also specify milestones for demonstrated achievement in creating a safer environment in and about American households.
- The Commission should immediately undertake an ambitious program, adequately funded and staffed, to determine the relative safety of various products within different classes of products. It should then mount a vigorous campaign to inform the public so that this information can be used by consumers in making decisions about their purchases.
- The Commission should take effective action to counter those political influences in its regulatory process that are clearly inimical to the protection of the American public from the dangers of consumer products. Specifically, the offeror process should be revamped to reduce industry dominance, substantially increased funding should be provided to support enlarged consumer participation in the agency, and the appointment of a full time "consumer advocate" within the agency should be reconsidered.

The Larger Lessons

An examination of the difficulties that have plagued the CPSC experiment suggests that the lessons to be learned have implications which extend far beyond the agency itself. For example, it seems likely that the trend toward more ambitious federal efforts to control the hazards of technology will mean that government agencies will increasingly be confronted with the management of broad and varied hazard domains. The CPSC experiment indicates that there is a pressing need for the classification of hazards—a veritable call for the kind of taxonomy proposed in chapter 4—and for a defensible assignment of priorities. Ad-hoc, case-by-case response is the quicksand of modern hazard management.

A second lesson can be learned from the CPSC's mixed report as an experiment in regulatory reform. The experiment indicates how little is known about how to give the public an effective voice in bureaucratic decisions—and suggests that even less is known about how to design innovations that contribute to rather than detract from the substantive work of an agency.

Finally, given that some benefits to society must usually be foregone in order to reduce hazards, a regulatory agency can travel down the road of hazard control only so far without a "safety constituency." Long-term success in creating safer households, workplaces, or other environments depends upon an unambiguous public resolve both to demand protection and to pay the price, together with a congressional commitment to implement change even in the face of troubled economy.

Epilogue

Since the foregoing article was published in 1978, the Consumer Product Safety Commission has made important progress on a number of problems. In particular, the Commission has instituted changes, such as more centralized authority, that have upgraded its administrative efficiency. The Neal amendments of 1978, when the Commission was reauthorized, provided the Commission with greater discretion in the use of the offeror process, and this has improved the Commission's work on the development of standards. The same amendments encouraged the Commission to pursue actively the resolution of safety problems through voluntary standards, and the Commission has responded with a generally effective program. The squandering of Commission resources on minor safety problems has been substantially rectified through a more soundly based priorities system, one that now recognizes some 12 priority hazards rather than the 100 or more in this category prior to 1978. Similarly, the Commission has sought to improve its capability in responding to chronic hazards through coordination with other agencies in the Interagency Regulatory Liaison Group (during the Carter administration) and its participation in the National Toxicology Program.

These changes have produced results. The Commission did not issue its first mandatory rule until some three and a half years into its existence. Between 1978 and 1982, however, the Commission promulgated or put into effect some seven mandatory rules (dealing with unstable refuse bins, CB and TV antennas; sharp points, edges, and small parts in toys; unvented gas space heaters; and lawn mowers), with an estimated risk reduction of 100 deaths and 90,000 injuries annually.

During the same period, the Commission staff has also participated in some 60 voluntary standards development projects (of a total of 83 in the Commission's entire history). At the same time, the Commission has continued its participatory mandate. Between 1977 and 1981, the Commission communicated (in writing, through its hotlines, and at regional offices) with more than one million consumers, distributed more than 12 million copies of safety publications, issued a regular listing of recalled products to some 6500 media editors, and broadcast radio and TV public service announcements.

This increased management activity has been accomplished in the face of declining resources. Whereas we noted in 1978 that a lack of technical and financial resources constituted one of the most serious barriers to improving the Commission's performance, the Commission's funds, measured in real dollars, declined by over 20 percent between 1978 and 1982 as a result of a relatively constant budget and high inflation rates (Figure 3). The nation's price tag for the Commission's work, excluding the costs passed on in higher product prices, reached approximately 25 cents per annum per capita by 1982. Measured another way, in cost effectiveness of risk reduction, the Commission estimated that its 1982 projects would prevent

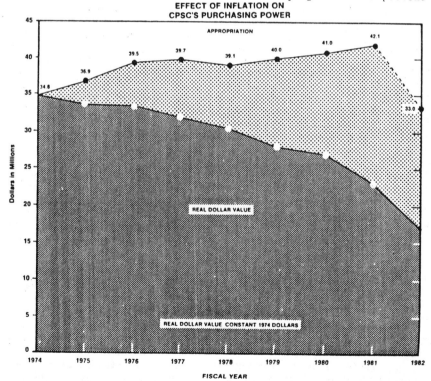

Figure 3. Effect of inflation on CPSC's purchasing power. Source: U.S. Congress (1981,254).

more than 3,000 deaths and over one-half million injuries in that year alone, at an estimated cost of $13,000 for every life saved. This figure, if it is accurate, compares very favorably with cost-effectiveness measures for other technological risks where figures in the $250,000-$1,000,000 range are not uncommon. In addition, the Commission estimates that between 1978 and 1982 its recall actions averted over one million possible injuries.

Despite these apparent improvements in the performance of the Commission, its prospects are bleaker in 1984 than they were in 1977. This situation is due particularly to the antiregulatory efforts of the Reagan administration, supported in the Congress, which have been more severe for the Consumer Product Safety Commission than for most other federal regulatory agencies. As already noted, the Commission has undergone an erosion of real financial resources. Beyond this, the Reagan administration cut the Commission budget by 30 percent in FY 1982 alone. A reduction of this magnitude in an already overextended agency will have the result, in the words of its past chairman, that "the Nation's product safety efforts....will move toward their pre-CPSC status.... characterized by reaction to specific preventable tragedy" (Statler 1981,253). The budget cut eliminated some 25 percent of the Commission's personnel, largely in the areas of enforcement and investigation, reduced oversight of product recalls, cut drastically the Commission's already inadequate Chronic Hazards program, eliminated two-thirds of the Commission's regional and district offices and all its public service announcements, and weakened the Commission's hazard identification capability by reducing from 119 to 74 the number of participating hospitals in the National Electronic Injury Surveillance System (NEISS), widely regarded as one of the nation's most effective hazard-monitoring systems.

The Commission also has continuing problems that the budget cuts exacerbate. As noted above, substantial reductions are under way in the Commission's efforts on chronic hazards. In 1978, we noted the Commission's lack of capability to deal with the formidable problems presented by this class of hazards. Since then, the Commission has made modest efforts to improve its performance, largely through coordination with other federal agencies, research groups, and industry. But the Commission has continued to allocate only a small part (under 5 percent) of its staffing resources to assessing and developing standards for these hazards. Its total budget for chronic hazards has totaled only $3-3.5 million, as compared, for example, with a budget of $180 million for EPA's Office of Pesticides and Toxic Substances. Even before the Reagan cuts, the CPSC employed only 55 professionals in this area, and it neither conducted nor sponsored any basic research. It is abundantly clear that the Commission in 1984 lacks the capability to mount an effective program even to identify and assess, much less control, this class of hazards.

Similarly, the informational and educational program of the Commission remains underdeveloped, yet it would appear to be essential for hazards that are often susceptible to individual control. The Commission itself remains unconvinced that informational and educational efforts can comprise an effective strategy for hazard reduction. As with the chronic hazards program, few resources (some 8 percent or less of its budget) have been allocated to this work.

And it is undoubtedly more difficult, both because of resources and the current political climate, for the Commission to be more aggressive in disseminating candid information on industry's failings.

The fact that the Consumer Product Safety Commission has been singled out for deeper cuts than other federal agencies speaks to a continuing basic problem. In our 1978 analysis, we noted the lack of a "safety constituency," arguing that the Commission's long-term success depended upon "an unambiguous public resolve both to demand protection and to pay the price, together with a congressional commitment to implement change even in the face of troubled economy." Whereas the public interest in protection remains strong, the absence of an effective consumer constituency and a congressional resolve is more striking in 1984 than 1977.

A classic Hemingway tale, "The Short, Happy Life of Frances Macomber," recounts the untimely death of the protagonist at the moment of fulfillment. Hemingway would appreciate the irony in an agency with such a troublesome adolescence confronting emasculation at the very time that it is emerging from its problems as an effective manager of hazards.

NOTES

1. This chapter originally appeared as "The Pitfalls of Hazard Management." Except for minor revisions appropriate to this volume, the article is reproduced with permission from Environment 20 no. 8 (October 1978):30-42. The article presents the authors' view of the Consumer Product Safety Commission as a then five-year-old experiment in regulatory reform. In an epilogue, Kasperson updates the performance of the Commission since the original 1978 analysis and reflects on its clouded future in the 1980s.

2. The Commission may regulate any product that is used in or around a residence or school for the personal activity, comfort, or enjoyment of a consumer. Specifically excluded from CPSC's jurisdiction are motor vehicles, fuels, nuclear materials, pesticides, aircraft, boats, food and drugs, medical devices, cosmetics, tobacco products, and firearms and ammunition—all of which are regulated to some degree by other federal agencies.

3. For a comprehensive breakdown of the CPSC's legal authority, see George Washington Law Review (1975).

4. Almost all successful challenges to the Commission's rules have been based not on the substance of the rules but on failures of the Commission to follow proper rule-making procedures.

5. This claim should be viewed in light of the fact that some 2.5 million firms sell 10,000 consumer products annually. Only 350 notifications may indicate the failure of some firms to report such hazards (U.S. Congress 1976a,211).

6. The so-called "Ash Report" (United States 1971), commissioned by President Nixon in 1971, for example, thoroughly criticized collegial decision making. Former CPSC Chairman Simpson has also criticized decision-by-commission. In a response to questions submitted by Represenative John E. Moss, Simpson referred to collegial decisionmaking as "an unworkable alternative for effective and productive leadership of an organization."

7. The first "phase" of the CPSC's chronic-hazards policy—the classification of potentially dangerous household chemicals—has recently been challenged in the court on procedural grounds by Dow Chemical Company.

8. This 1976 Amendment to the Act grew out of a dispute (between the CPSC and the Civil Service Commission) resulting from the CPSC's refusal to approve the hiring of CPSC employees who had not received political clearance from the White House. This amendment clarified CPSC's authority to bypass the White House when hiring new personnel.

9. Although the sponsors of the Consumer Product Safety Act envisioned the transfer of some personnel, they hardly contemplated that 70 percent of the CPSC's original manpower would come from ongoing government programs. This influx of employees, who had operated under the policies and practices of their former organizations, contributed significantly to the CPSC's start-up problems.

10. According to Rhoda H. Karpatkin, Executive Director of Consumers Union, the lawnmower industry has lobbied Congress extensively regarding power-mower standards.

11. As a result of this letter, the Commisson reimbursed Consumers Union for most of its out-of-pocket expenses. The CPSC in 1978 is in its fifth year of work on power-mower safety standards and has already spent over 11,000 person-hours, $800,000 of CPSC resources, and $166,000 provided to Consumers Union. The $1.2-billion lawnmower industry, for its part, estimates that it has spent $4 million in research and response. Meanwhile, Combined Insurance Company of America projects some 165,000 injuries and several deaths this year from power mowers (Hartford Courant 1978).

12. The Commission now has before it a rule-making petition seeking a regulation providing for funding of consumer participants.

13. Rhoda Karpatkin of Consumers Union comments that she believes there is increasing public demand for product safety and consumer protection and considers the results of the Harris poll "heartening" and an indication of the dual desire for better, safer products and an end to inflation.

REFERENCES

Aqua Slide 'n' Dive Corporation v. CPSC. 1978. 569 F.2d 831. 3 March.

Consumer Product Safety Act. 1972. PL 92-573 15 USC 2051 et seq. 27 October.

Earley, Pete. 1982. Injury-reporting system gets hit again. Washington Post, 1 November, A13.

George Washington Law Review. 1975. Special Issue: The Consumer Product Safety Commission, 43 no. 4 (May).

Hartford Courant. 1978. Power-mower injuries climb: Agency debating safeguards. 24 May.

Karpatkin, Rhoda H. 1976. Letter, dated 9 January 1976, to Sadye Dunn, Secretary, CPSC. Reprinted in U.S. Congress (1976b, 364).

King, Susan Bennett. 1978. Interview in Boston Sunday Globe, 24 September.

Louis Harris and Associates. 1977. Consumerism at the crossroads. Conducted for Sentry Insurance by Louis Harris and Associates and Marketing Science Institute. [Stevens Point, Wisconsin: Sentry Insurance].

Moss, John E. 1976. Testimony in U.S. Congress (1976b).

National Research Council. 1977. Committee on Toxicology. A review of the role of health sciences in the Consumer Product Safety Commission. Washington: National Academy of Sciences.

Simpson, William. 1976. Testimony in U.S. Congress (1976b).

Statler, Stuart M. 1981. Statement in U.S. Congress, House Committee on Commerce, Science, and Transportation, Subcommittee for Consumers. Consumer Product Safety Commission reauthorization: Hearings..., 282-294, 307-339. 97th Congress, 1st session. Serial 97-35. Washington: Government Printing Office.

Thomas, Jo. 1978. Performance of consumer agency disappoints its early supporters. New York Times, 30 January, pp. 1, 14.

United States. 1971. President's Advisory Council on Executive Organization. A new regulatory framework: Report on selected independent regulatory agencies. Washington: Government Printing Office.

U.S. Congress. 1974. House Committee on Interstate and Foreign Commerce, Subcommittee on Commerce and Finance, and Senate Committee on Government Operations. Joint hearings on Consumer Product Safety Commission oversight. 93rd Congress, 2d session. Serial 93-100. Washington: Government Printing Office.

U.S. Congress. 1976a. House Committee on Interstate and Foreign Commerce, Subcommittee on Oversight and Investigations. Federal regulation and regulatory reform: Report.... 94th Congress, 2d session. Washington: Government Printing Office, October.

U.S. Congress. 1976b. House Committee on Interstate and Foreign Commerce. Subcommittee on Oversight and Investigations. Hearings on Regulatory Reform, vol. 4. 94th Congress, 2d session. Serial 94-83.

U.S. Congress. 1981. Senate Committee on Commerce, Science, and Transportation, Subcommittee for Consumer. Consumer Product Safety Commission reauthorization: Hearings...April 1, 2, 3,

394

and 7, 1981. 97th Congress, 1st session. Serial 97-35. Washington: Government Printing Office.

17
Contraceptives:
Hazard by Choice

Mary P. Lavine

Exercising control over reproductive processes to space pregnancies or limit family size is a desire that spans time and culture. Traditionally, fertile couples have relied upon coitally related barrier methods, interrupting the sexual act, limiting sexual activity to periods of presumed infertility, or abortion as the primary methods of fertility control. Success levels and user satisfaction were often disappointing. As newer contraceptive methods offered both more effective pregnancy prevention and enhanced user satisfaction, profound changes occurred in contraceptive preferences, extent of contraceptive use, and fertility patterns. Only belatedly has acknowledgment of the risks accompanying these new technologies initiated risk management efforts.

The new contraceptive methods responsible for these important changes were oral contraceptives ("the Pill"), Intrauterine Devices (IUDs), and improved techniques of male and female sterilization. Postcoital and injectable contraceptives, have yet to make significant contributions to overall fertility control. Improved sterilization techniques and wider acceptance of abortion also helped curb fertility when contraceptive practices failed to prevent pregnancy. The Pill, the IUD, and sterilization have displaced both the fairly effective traditional methods of contraception (diaphragm and condom) and the much less effective traditional methods (douching, rhythm, withdrawal, and the like). Future reproductive expectations influence an individual user's choice of method. Couples intending more children favor oral contraception, whereas those who have completed their families increasingly choose sterilization. Preferred contraceptive method also varies with the user's age, education, race, and religious preference. Choice may also be constrained by prior medical history.

An absolute increase in the practice of contraception has accompanied adoption of the newer contraceptive methods. Over the past twenty years, the widespread acceptance of the Pill, the IUD, and improved sterilization techniques has accelerated the rate at which contraception is practiced. Among married white women of reproductive age during the decade 1965-1975, contraceptive use increased from just over 65 percent to nearly 80 percent (Westoff and Jones 1977,154) and appears to be increasing still. Among those who intended no further pregnancies (as distinct from those trying to space pregnancies), contraceptive use increased from 75 percent to over 90 percent during the same ten-year period.

Fertility has declined in response to increased contraceptive use and effectiveness. Declines in birth rates, fertility rates, and family size were already under way in the United States as early as the late nineteenth century. The greater reliability and ease of use afforded by the newer contraceptive techniques accelerated these use trends. Birth rates are now about half those of the 1950s baby boom, with unwanted fertility declining from one birth in five in 1965 to one in twelve by 1975 (Planned Parenthood 1977,3-4). These changes are well documented and their causes—only partly attributable to the methods themselves—are moderately well understood.

Recognition and management of risks associated with the use of the Pill and the IUD lagged until their prescription and use had become a well-established practice. Thus the experience with these products, along with the more recent "morning-after" and injectable contraceptives, followed a history of technology management that parallels that of many controversial, nonmedical technologies. Startling revelations and frightening speculations about undisclosed risk shattered initial expectations and early praise. Hailed as a miracle drug during its early days, the Pill has become shrouded with doubt over its relationship to the risks of stroke, heart attack, blood clotting, liver disease, and cancer. The IUD, promoted as the safe, nondrug alternative to oral contraceptives, has had its safety record compromised by the continuing litigation that surrounds the Dalkon Shield and the incomplete answers to frequent problems of pelvic inflammatory disease (PID). The use of Diethylstilbestrol (DES) as a postcoital contraceptive ("morning-after pill") has become inexorably linked with the experiences of the "DES daughters" and with the U.S. Food and Drug Administration (FDA) ban on the use of DES in animal feed. The decision by FDA not to authorize sale of the injectable progestogen-based Depo Provera for contraceptive purposes has prompted as many questions as the debate over its cancer risks.

Despite this gloomy perspective, the hazard management experience with contraceptives offers insights into innovation in at least three major aspects of hazard management: identifying hazards, considering consequences, and informing about risks. This chapter will consider the experience with each of these, but first it is necessary to review technological innovation in contraceptive practice and present a brief history of regulatory management in this area.

A Review of Four Innovations in Contraceptive Technology, 1960–1980

Oral Contraceptives

Developed during the mid-1950s by Dr. Gregory Pincus of the Worcester Foundation for Experimental Biology at the urging of Margaret Sanger (Davis 1978), oral contraceptives (OCs) appeared in the United States market in late 1960. "The Pill" most commonly consists of estrogen and progestogen in combination. When taken in a daily dosage (averaging less than 50 micrograms of estrogen in present formulations), OCs prevent pregnancy primarily by preventing ovulation, but also perhaps by producing changes in cervical mucus

and the endometrium, actions that impede sperm action and implantation, respectively (Burke, Crosby, and Lao 1977;FDA 1978a;Kols et
al. 1982). Used correctly, the Pill is a highly effective method of
contraception, with a failure rate of less than one pregnancy per
100 women-years of use. (The progestogen-only minipill is somewhat
less effective, with a failure rate of about three pregnancies per
100 women-years of use.) In 1975, the Pill ranked first among U.S.
couples practicing contraception, although it has since been
relegated to second place by sterilization.

The quick acceptance of oral contraceptives during the 1960s
(Figure 1) has resulted in a total exposure, during the first twenty
years, of perhaps 20 to 25 million women in the United States. Current use in the United States is probably 8 to 9 million women (Kols
et al. 1982,A-192; based on 1980 retail sales of 67 million prescriptions, plus some estimated 2 million women who receive the
Pill through nonprescription sources, primarily clinics and hospitals). This represents a decline from the 1974-1975 high of approximately 10 million users.

Studies of American and British women generally acknowledge
that the use of oral contraceptives increases the risk of blood
clots in the veins, ischemic heart disease (including heart attack),
stroke, and hypertension. The most serious of these risks are heart
attack and stroke (Kols et al. 1982,A-199). Some uncertainty exists
about potential birth defects attributable to use up to three months
prior to pregnancy or inadvertent use during pregnancy. Also, the

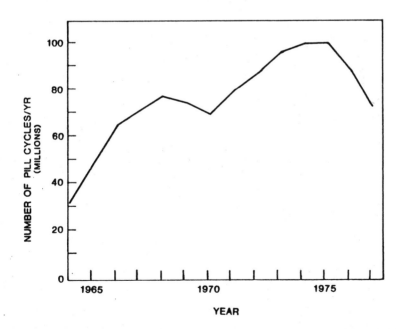

Figure 1. Oral contraceptive use in the United States, 1964-1977.
Source: Piotrow and Lee (1974,Table 4,A-8);Rinehart and Piotrow
(1979,Table 1).

relationship between oral contraceptive use and cancer is unclear. Cigarette smoking acts synergistically with oral contraceptives to produce an even higher risk of blood clots, particularly in women over 35. In short, most of the risks of oral contraceptives accrue to women over 35 who smoke (Kols et al. 1982,A-189).

Intrauterine Devices (IUDs)

Whereas the concept of uterine/cervical devices to prevent conception spans several thousand years, modern IUDs placed solely in the uterus are the descendants of devices developed during the early 1930s. By the end of that decade, IUDs had lost favor because of insertion difficulties and high infection rates, but they reemerged in the early 1960s with new designs made of flexible plastic that facilitated insertion (Huber et al. 1975;FDA 1977;FDA 1978b;Piotrow, Rinehart, and Schmidt 1979;Liskin 1982). The mode of contraceptive action is uncertain, with no single theory presently accepted, but is commonly ascribed to an inflammatory cell response (or infection) in the uterus, due to the presence of a foreign body (the IUD), which in turn prevents implantation of the fertilized ovum. IUDs apparently do not inhibit ovulation or fertilization, as ectopic pregnancies among IUD users show.

Except for those containing biologically active agents (such as copper or progesterone), IUDs were classified as inert medical devices, and, as such, they were largely unregulated until 1976. A shift in usage during the late 1970s reflects a trend toward copper devices and to a lesser extent, progestin-releasing IUDs (Liskin 1982). Accounting for 8.7 percent of all couples practicing contraception in 1975, and even more popular among those intending future pregnancies (Westoff and Jones 1977), IUDs are probably used by 2 to 3 million women in the United States today, with a total of about 9 million users between 1966 and 1974 (Huber et al. 1975,B-21;Mosher 1981;Liskin 1982).[1]

IUDs are a highly reliable method of contraception, with a failure rate of less than one to six pregnancies per 100 women-years of use. Major adverse effects include implantation in the uterus upon insertion, a higher incidence of pelvic inflammatory disease (PID), occasional heavy bleeding (which is unacceptable to many users and a cause of anemia in some), and a greater risk of miscarriage if pregnancy should occur and the device is allowed to remain in place.

Injectable Contraceptives

Longer-acting hormonal-based injectable contraceptives are an offspring of the 1950s contraceptive experiments that led to the daily-dosage Pill. A three-month, 150-mg injection of progestogen works in a manner similar to oral contraceptives, by inhibiting ovulation, impeding sperm movement, changing cervical mucus conditions, and impeding implantation of the fertilized ovum. The level of protection is at least comparable to the Pill (less than one pregnancy per 100 woman-years of use, and the injection eliminates failure to take daily medication as a cause for method failure.

Although the leading product, Depo-Provera® (Upjohn Company), is available in over 70 countries [with about 1 million users worldwide (Rinehart and Winter 1975) compared to 50 million for the Pill], it has not gained approval for contraceptive use in the U.S. because of concern over progestogen as a cause of breast cancer (in beagles) and possible cervical cancer. FDA has approved its current use as a palliative for advanced endometrial (uterine) cancer as well as its former use to prevent miscarriage (a use later found ineffectual).

Critics of FDA's decision to ban Depo-Provera after more than 10 years' investigation and every indication that approval was forthcoming cite the drug's good record and lack of observed side effects in its use in other countries. Proponents view the drug as especially attractive for women who, because of mental impairments, might neglect a daily pill and thus be candidates for unwanted pregnancies or sterilization. Thus FDA is seen as denying these high-risk users an opportunity for safe, effective, reversible contraception. The counterargument is that injectable contraceptives would place poor, minority, and institutionalized women at a disadvantage with unscrupulous physicians or overzealous family-planning programs. More than any other birth-control method except abortion, the use of injectable contraceptives has become a political issue (Maine 1978;Rinehart and Winter 1975). Internationally, the United States Agency for International Development is placed in the awkward position of subsidizing a method of contraception whose sale its own government has forbidden.

In addition to being nonreversible in the short run, injectables disrupt the menstrual cycle, particularly during the first months of use and may cause fetal abnormalities if mistakenly administered early in a pregnancy. Of greater concern are possible long-term effects, including cancer, and a possible period of infertility after treatment is discontinued. These concerns have prompted prohibition in some countries and restrictions in others. In Thailand, for example, a user must have at least two living children and consider her family complete (Rinehart and Winter 1975,K-12).

In 1973, even as FDA was considering a favorable ruling on Depo-Provera, it was proposing restrictions in use: the drug would be limited to women for whom other methods proved unacceptable; the manufacturer would have to maintain a patient registry; the patient would be required to provide her "consent" after reading an information brochure detailing known and suspected risks and side effects.[2]

Diethylstilbestrol (DES) as a Postcoital Contraceptive

Since its synthesis in 1938, the versatile estrogen Diethyl-stilbestrol (DES) has had many uses: treatment for threatened miscarriage (later found ineffective and a cause of transplacental cancer in female offspring); suppression of lactation; control of postmenopausal symptoms and cancer of the breast; treatment of cancer of the prostate; food additive and growth promoter (now prohibited) in poultry, sheep and cattle.[3] Accounts of DES's complex history abound (Noller and Fish 1974;Hartman and Segel 1979;Shapo 1979). Postcoital application, first suggested in 1938 (Parkes, Dodds, and Noble) did not undergo investigation until the second half of the 1960s, when Morris and van Wagenen (1966) found it effective in primates. The first significant study in humans reported no pregnancies among 1000 subjects at high risk (midcycle, ovulation determined likely to have occurred) following unprotected intercourse, a condition in which 20 to 40 pregnancies might have been expected (Kuchera 1971). A subsequent study of several thousand patients reported only three pregnancies. To be effective, DES requires prompt administration, within 72 hours of intercourse. The standard dosage is 25 mg every 12 hours for five days.

Information about DES as a morning-after pill circulated in medical journals and by word of mouth. Concerned over its apparently growing use (particularly in college medical facilities) without controlled studies of efficacy or safety, FDA invited manufacturers to submit modified drug applications. To date, none have accepted, and, consequently, postcoital use of DES is not advertised, although FDA has prepared and placed in the federal Code the patient insert warning that will be required in the event of approval.[4] Confirmation of Herbst's 1971 finding (Herbst, Ulfelder, and Poskanzer 1971) of rare vaginal cancer in young women whose mothers were administered the drug during pregnancy has placed DES use in an awkward position.[5] In an attempt to control its "misuse," FDA has prohibited the manufacture of the 25-mg tablet. Because use is not approved, and customary dosage thus does not appear on prescriptions, FDA and others have no satisfactory way of monitoring or even estimating use, effectiveness, or safety.

Locus of Responsibility
for Managing Contraceptive Technologies

There are four primary classes of managers for contraceptive technologies: consumer; medical provider; technology sponsor; regulatory agency. Extensive media coverage of the contraceptive debate during the 1960s and 1970s, a legacy of litigation over consequences of using the IUD, the Pill, and DES, and the efforts of consumer and feminist groups all contributed to a heightened public awareness about the benefits and risks of the new contraceptives. It is in the context of media exposure, the courts, and the consumer and feminist movements that responsibility for and limitations upon management at each level must be considered.

The "sexual revolution" of the 1960s was accompanied by an increasingly frank public discussion about contraception. Oral contraceptives, which overshadowed all other methods in extent of public media and medical journal coverage (Figure 2), illustrate the expanded interest in new contraceptive methods, first on social grounds and subsequently because of risk. Media coverage of the Pill rose and fell in response to each new risk assessment.

Early coverage of oral contraceptives by print media had three focal points: the Pill as a miracle drug that would revolutionize contraception; the issue of religious conscience; and the moral climate most likely to follow easy access to effective control over unwanted fertility. The Pill was assumed to be quite safe. As reports to the contrary began to surface during the second half of the 1960s, the Pill slowly lost its "diplomatic immunity" (Mintz 1968/69 and 1969). Because contraception was a popular news topic in newspapers and lay journals and by no means confined to the "women's pages" or "ladies' magazines," prospective users in the late 1960s were more or less continuously exposed to conflicting evaluations of

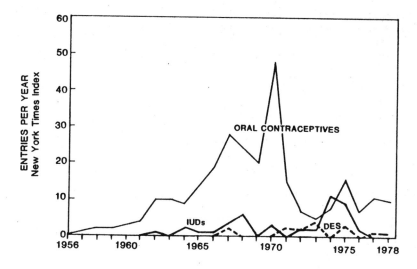

Figure 2. New York Times coverage of contraceptive technologies.
Source: New York Times Index (1956-1979).

risk, and repeated warning about the Pill's possible adverse effects. User uncertainty over the Pill's safety, changing demographics of the user population, and the growth of alternative contraceptive methods are all reflected in the leveling off and brief decline
in the use of the Pill during this period. Publicity over the Pill
peaked in 1970 with the televised and extensively reported hearings
before the Nelson Committee, a forum that was popularly credited at
the time with causing many women to discontinue oral contraceptive
use.[6] Because of this unprecedented media coverage of medical
matters, the time was ripe for FDA to get information about risk out
into the open, while presenting it in a manner designed to reassure
users about the Pill's overall safety record.

Oral contraceptive and IUD users (or their surviving spouses)
initiated a number of lawsuits against drug manufacturers for harm
inflicted as a result of contraceptive use. Not infrequently, these
suits charged the manufacturer with failing to provide warnings of
known adverse effects. The success of this litigation undoubtedly
sensitized manufacturers to the hazards of concealing information
about possible adverse effects of a drug's use, and, to a lesser
extent, alerted FDA and individual physicians who were occasionally
named as co-defendants in the lawsuits.

By the early 1970s, the consumer movement had come of age, as
witnessed by the emergence of consumer alliances and advocacy
groups, passage of the Truth-in-Lending Act (July 1969), establishment of the Consumer Product Safety Commission (1972), and other
consumer-oriented actions. Reinforcing this general climate of
increased consumer consciousness was the feminist position that for
women to be free, they must first regain control over their bodies.
This referred, of course, to reproduction and its control, a power
previously held predominantly by male physicians. Groups such as

the Boston Women's Health Book Collective in the highly successful Our Bodies, Ourselves,[7] a variety of feminist journals, and even more staid, traditional "women's" publications joined in advocating this position. It encouraged women to question the correctness and wisdom of deferring to their physicians' opinions on questions of reproduction and risk.

In this climate of public interest over contraceptive risks, what do the different classes of risk managers actually know? Consumers are best informed about important but nonmedical aspects of alternative contraceptive technologies, such as ease of use, aesthetic appeal, moral acceptability, and overall user satisfaction. These qualities are important because they affect method acceptability and therefore effectiveness of use. Consumers generally have not been well informed about medical aspects of drug use although some information about method reliability and adverse side effects has been available through the media, and later through patient information inserts, to prospective contraceptive users for about a decade. There is no evidence to indicate whether patients are directly benefiting from these efforts at consumer education.

Medical providers, in this case principally private-practice physicians and family-planning or hospital-clinic staffs, have generally better opportunities to be informed about health risks associated with alternative contraceptive methods once such information becomes part of the medical literature. They also manage the technology on the basis of personal clinical observations and patients' own reports, which reinforce particular ideas about efficacy and overall acceptability. Neither consumers nor medical providers are in a favorable position to identify systematically the hazards of technology, although they bear much of the responsibility for weighing its consequences.

As technology sponsors, pharmaceutical firms make the initial decisions about acceptable levels of safety and efficacy by screening products prior to seeking FDA approval. These firms are party to a conflict of interest: corporate goals of sales expansion and profit compete against goals of safety and efficacy for patients, for whom the regulatory agency has been the primary advocate. Past practices of some firms raise serious questions about their willingness to disclose unfavorable drug-use reports.

Primary regulatory responsibility for managing contraceptive technology rests with the FDA. In theory, FDA has the best access to information about risks and the greatest capability for evaluating such reports. In practice, FDA's access to data is constrained by budget considerations, an ineffective system for receiving reports of adverse drug reactions from physicians and pharmacists, and contrary motives of pharmaceutical firms. On the other hand, FDA has strong incentives to make "correct" judgments about risk, since failure to do so incurs considerable blame. FDA's functions are perceived (and to a considerable extent, defined) as "watchdog" actions, so institutional success is measured by its handling of risk and particularly by the avoidance of risk. Since in most instances the benefits of contraceptive technologies are only secondarily their life-saving capabilities,[8] FDA's ability and incentive to conduct benefit analysis is considerably diminished. How this impinges on FDA performance in managing the contraceptive hazard

requires some understanding of the history of FDA's regulatory
responsibilities for contraceptives.

Regulatory History of Contraceptives

When the first oral contraceptives were approved for marketing
in the United States, efficacy requirements were minimal and safety
standards considerably less than those acceptable today. The Pill
very quickly demonstrated extraordinary success in reducing incidence of unintended pregnancies, and FDA assumed a highly supportive, even defensive, posture with regard to safety questions. When
inquiries regarding the Pill's safety surfaced periodically, FDA's
Advisory Committee on Obstetrics and Gynecology would review the
available but limited data and conclude, without requiring additional investigations or data-collection efforts, that the drug was
safe. Only late in the 1960s did FDA qualify this position, when it
accepted the content and recommendations of the first British risk
assessments on oral contraceptive safety.[9] As a result, FDA began
to issue warnings to physicians about possible adverse effects of
blood clotting.

Throughout the first 10 years, regulatory actions were minimal,
comprising two quite different efforts. First, several "Dear Doctor" letters from FDA and manufacturers informed physicians of some
observed cases of thromboembolism and visual disorders among Pill
users, although the reports carried no implication of causality in
these early instances. Second, the discovery that an experimental
oral contraceptive formulation produced precancerous symptoms in
experimental animals prompted the institution in 1968 of long-range
cancer screening in animals. Clinical tests were also halted for
this product. During the 1970s, only two regulatory actions initiated by FDA stand out: the preparation of patient-warning labels
and their subsequent upgrading during the second half of that decade; and, based on British warnings first issued in 1969, a 1975
investigation of sequential oral contraceptives which resulted
in three U.S. manufacturers' "voluntarily" withdrawing their products in 1976.[10]

In contrast to oral contraceptives, which as prescription drugs
have always been vulnerable to FDA scrutiny prior to and following
marketing, as late as 1977 IUDs enjoyed considerable freedom from
regulation because of their classification as medical devices. Thus
FDA approval for safety and efficacy considerations was not required
prior to marketing, and FDA could initiate actions regarding IUDs
"only when there was either evidence of false or misleading claims
in the labeling or a health hazard associated with their use" (FDA
1978b,38). Individual designs but not the generic product were subject to postmarket regulation. FDA initiated only two actions
against specific products: a 1973 seizure of Anka Research Ltd.'s
Majzlin Spring IUD from its warehouse on grounds of probable harmful
complications (excessive bleeding, cramping, uterine perforations)
and an investigation into charges that A. H. Robins' Dalkon Shield
IUD was responsible for septic abortions resulting in uterine infec-
tions and some deaths (Levinson and Richardson 1976;Cates et al.
1977;Morrison 1975a,1975b,1975c). In the Dalkon Shield case, FDA
first suspended sales pending an investigation, then required redesign of the removal string believed responsible for introducing

bacteria into the uterus, and finally insisted upon patient registry as a condition for remarketing. Robins refused to accept the registry notion and never remarketed the device, but since the product was also never recalled, an unknown number of Dalkon Shields are in place today.

What is significant in both the Majzlin Spring and Dalkon Shield incidents is that FDA was precluded from taking action until it could demonstrate probable harm had already occurred. This impediment to insuring safety was quite directly responsible for the subsequent extension of FDA's premarketing authority to cover inert medical devices. FDA's only other significant regulation of IUDs was an extension (effective 7 November 1977) of the patient labeling requirement to include IUDs, although as in the case of the Pill, some questions have arisen about the effectiveness and timeliness of the process for informing patients about risks.

FDA has failed to secure regulatory control over DES in its use as a postcoital contraceptive precisely because no manufacturer openly promotes that use. This is a good example of FDA's inability to control the use of a drug that has been approved for a specific application. Although the use of DES in animals has been forbidden, it is approved for noncontraceptive uses in humans. Attempts to curb its use by prohibiting production of the 25-mg dosage have resulted, paradoxically, in greater difficulties in monitoring the extent of its use.

Perhaps because 20 years' experience with oral contraceptives provided some nasty surprises and left some issues (cancer, for example) still unresolved, FDA has exercised greater caution toward a longer lasting, less easily reversed, inexpensive injectable contraceptive. This is possible because the Pill and the IUD, although imperfect, constitute substitutes already approved for use, with known risk factors. FDA has elected to manage injectable contraceptives by banning them.

Issues in Hazard Management

FDA's struggles in managing newer contraceptive technologies reflect at least three generic issues of hazard management: How are hazards best identified? How should the adverse consequences of a technology's use be weighed against foregoing the technology's benefits? Whether, and if so, how should exposed populations be informed about risk? What is instructive from the contraceptive management experience is the existence of innovative solutions, either in place or under consideration, which hold promise for other technologies.

Identifying Hazards

A regulator's failure to anticipate risks and take corrective action all too often invokes editorial charges of negligence. That anticipation and recognition of hazards seem—at least after the fact—unforgivably delayed makes the accusations all the more compelling.

The past 20 years' experience with innovative contraceptive technologies has not found the FDA immune to these charges. The Pill, IUDs, and DES in its various uses all testify to an atmosphere of recriminations and presumptive negligence on the part of technology regulators. FDA's decision **not** to allow sale of injectable contraceptives has been widely criticized for neglecting needs of potential contraceptive users for whom existing methods are unsatisfactory.

Hazard identification, the first step in the hazard assessment process (see chapter 3), usually entails some combination of screening, monitoring, and diagnosis (chapter 11). The course of drug development and utilization offers a number of separate, distinct opportunities for hazard identification: screening through preliminary animal experiments and pre- and postmarket clinical studies; diagnosing through voluntary reporting of clinicians' observations and consequence-associated searches back to common drug causes; and monitoring through postmarket surveillance. To a degree, the management of contraceptive hazards has employed all of these identification techniques,[11] but premarket screening has been the preferred technique in the United States, whereas the United Kingdom has relied more on monitoring and diagnosis.[12] These differences in detection techniques are due in part to different philosophies of animal versus human testing, in part to the organizational abilities of a national health service. Whatever the cause, an international experiment of sorts has been conducted around a central issue of management: what is the best means of identifying hazards before consequences become widespread?

Current U.S. drug development practices (U.S. Congress 1980) call for **in vitro** and animal studies to determine a drug's toxicological and pharmacological properties prior to human testing. Provided the metabolism of experimental animals appropriately imitates that of humans, adverse effects are easily simulated and readily detected, and **in vitro** tests offer a valuable tool for prescreening drugs and avoiding unnecessary human exposure. Drugs may be administered to animals in high doses that can simulate long-term exposure, and autopsies allow for the study of organs not normally examined in humans. Extrapolation from animals to humans, however, has its own pitfalls, as seven recent National Academy of Sciences committee deliberations demonstrate: three Academy committees supported the use of extrapolations from animals tests; three did not; one was uncertain (National Research Council 1981,14).

Following animal studies comes human testing, first in normal volunteers to establish dosage tolerance, then in clinical studies of affected populations to determine efficacy and provide further safety evaluation. Such studies are limited—between 500 and 3000 patients typically participate in premarket studies—and for ethical reasons, usually exclude certain vulnerable populations—infants, children, pregnant women (U.S. Congress 1980,9 and 12). Such restrictions necessarily constrain opportunities for early detection of adverse effects, particularly when these effects seldom or never occur in nonuser populations, or have a delayed appearance. Searle's Enovid®-10 trials were virtually certain not to detect heart attacks as one consequence of using the drug. This effect is uncommon in young women, whether or not they use the Pill, and the trials employed only 132 subjects who used the drug for more than 12 consecutive months (Silverman and Lee 1974,63). Yet oral contraceptive users face two to 12 times higher risk of fatal heart attacks.[13] DES was in use several years as a postcoital contraceptive before its transgenerational consequences became apparent from an entirely different application. The link was established only upon the reproductive maturity of the women whose mothers had been administered DES to prevent miscarriage.

The U.S. model of drug management allocates resources primarily to these preclinical animal studies and premarket clinical studies in human populations. Burden of proof for safety and efficacy claims rests upon the manufacturer. FDA evaluates the results and may require further tests or halt experimentation altogether. After 10 years of extensive animal and clinical testing in the United States and abroad, the injectable Depo-Provera failed to secure expected FDA approval, on grounds of potential carcinogenic effects of progestogen-based drugs, and clinical trials in certain progestogen-only oral contraceptives were also halted. An experimental male contraceptive was abandoned by the manufacturer when clinical trials revealed adverse reactions among drinkers of alcoholic beverages.

In addition to requiring premarket studies, FDA may require further postmarket animal or clinical studies, as it did in 1967 for all oral contraceptives, following reports of precancerous tissues in animals administered the experimental MK-665 (New York Times 1967). Once FDA approves a drug for sale in the United States, however, physicians are free (subject to threat of malpractice suits or violation of the Controlled Substances Act) to prescribe a drug for any patient, in any dose, for any therapy. Manufacturers, however, are prohibited from advertising uses, doses, or populations other than those included in the drug application. Instead, this information circulates through medical journals and by word of mouth, as in the case of DES as a "morning-after pill," a use of the drug that has never been officially approved.

Support for cautious, premarket screening to detect potential drug hazards remains strong in the United States because of the 1961 experience with the tranquillizer Thalidomide. Side effects reported in experimental animals caused FDA to delay approval for Thalidomide, just as spontaneous reports from abroad began to focus attention on serious birth defects among infants whose mothers had taken the drug during pregnancy. FDA's averting of the Thalidomide tragedy through time-consuming analysis of animal data persists as a major argument supporting careful premarket scrutiny.[14]

Manufacturers' premarket studies on oral contraceptive safety apparently revealed no adverse health effects of Pill use. Instead, the United States was alerted to the Pill's possible risks by studies undertaken through the British Medical Research Council (Royal College of General Practitioners 1967;Inman and Vessey 1968;Vessey and Doll 1968) in response to voluntary reports in British medical journals, citing thromboembolism among Pill users in the United Kingdom. These British studies provided FDA with sufficient cause to initiate its limited warning actions in the late 1960s, well before comparable American studies of the Pill were under way. The British system of early warnings relies in large measure upon a tradition of published clinical surveillance reports of seemingly isolated or rare events, which in turn frequently undergo evaluation through more substantive, retrospective risk assessments. The British approach to hazard detection, relying as it does upon postmarket surveillance, voluntary reporting, and prompt risk assessment, permits more rapid, less costly drug approval mechanisms than does the more time-consuming, premarket screening approach employed in the United States. These benefits must be weighed against the costs, both to individuals and to society, of granting premature approval to drugs whose adverse consequences might have been detected and hence avoided, by means other than extensive human exposure, as in the Thalidomide case.

Detection of hazards associated with the IUD followed a pattern quite different from that of the Pill. IUD designs introduced during the 1930s had been abandoned during that same decade because of a high incidence of pelvic infection, a factor that caused physicians 25 years later to view the new flexible plastic designs with certain reservations. Because at this time FDA had little jurisdiction over (and consequently little interest in) premarket demonstrations of medical device performance, the Population Council established the Cooperative Statistical Program (CSP)[15] to referee the debate over IUD safety and efficacy through an analysis of data from various clinical studies. Researchers at CSP developed a standard method based on the idea of a life table, so that important indicators of IUD performance such as pregnancy, expulsion, and removal rates were comparable from one study to another. Over the period 1963 to 1969, CSP issued nine reports on IUD performance. From these emerged a picture of IUDs as generally effective contraceptives, comparing favorably with oral contraceptives (whose risks were just becoming apparent) but possessing certain unique problems only partly explained by factors such as design, experience of the inserter, and patient history. The reports provided a number of suggestions for improving IUD design and provided a standard of performance against which subsequent IUDs could be assessed.

A buildup of voluntary reports about septic abortions among Shield users triggered events leading to the demise of a specific product, A. H. Robins Company's Dalkon Shield. Discovery of problems with the Shield was facilitated by voluntary reports that physicians submitted to Robins and FDA, the requirement that manufacturers inform FDA of reported adverse effects, and a standard of performance for IUDs established through the CSP data base.

Sometimes, adverse effects are discovered only after an unusually high incidence of rare outcomes is traced back to a common drug source. Such was the case when, over a five-year period, a single Boston hospital admitted eight patients with rare adenocarcinoma never previously noted in young women. The common factor in the eight medical histories was maternal use of DES to prevent miscarriage. Years after such use had been abandoned as ineffective, the link provided the clue necessary for initiating retrospective studies among former DES users who produced daughters (Stolley 1974).

Which method of identifying drug hazard works best? Successful detection of consequences depends upon the frequency with which adverse effects become evident and how carefully adverse effects are monitored and reported. Animal experiments help screen against obvious flaws and provide an important safeguard against indiscriminate testing among clinical populations, provided consequences are not delayed (as in cancer) and animal metabolism mimics human in essential features. These conditions frequently can not be met. Because clinical studies are limited in size, duration, and types of users sampled, failure to detect adverse effects at this stage transfers the burden to postmarket studies, provided there is sufficient interest in the drug to warrant conducting these, or to astute clinical observers who may voluntarily report drugs they suspect of causing adverse effects. The United Kingdom benefits from a tradition of voluntary reporting by physicians; the United States lacks such a history. More systematic market surveillance systems, such as the Boston Collaborative Drug Surveillance Project or the postmarketing surveillance (PMS) program proposed by the Joint Commission on Prescription Drug Use,[16] require greater societal investment but offer another opportunity to detect adverse effects when user populations are dispersed among many medical practitioners and consequences are rare. Some drug effects remain undetected until cases with similar, rare outcomes are successfully linked to a common cause, but many side effects, particularly those which are infrequent or delayed, are unrecognized as drug-related, and will remain so.

Considering Consequences

Technology sponsors frequently lament that risk assessments that consider only the unwanted consequences of that technology's use fail to acknowledge that nonuse of a technology also carries with it undesired consequences. Thus, a risk assessment of nuclear power typically ignores the risks associated with employing alternative energy sources or limiting energy growth. In addition, discussions of risk often suffer from the flaw of noncomparability, as when the risk of dying from oral contraceptive use is compared with the risk of dying in an automobile accident.[17] Because driving a car is not a substitute for using the Pill, such comparisons are of little use except to create a sense for the magnitude of the risk factor. They are a way of placing the risk of Pill use in the context of life's more familiar risks.

Relative risk is one means of comparing the risks associated with genuine alternatives, such as the decision to use or not to use the Pill. Pill users experience 2 to 12 times as many heart attacks, 4 to 9.5 times the incidence of stroke, and 4 to 6 times greater risk of postsurgical thromboembolitic complications compared to similar nonuser populations (FDA 1978a).

Another way of expressing relative risk is through **excess mortality**, an absolute measure of deaths beyond what is normal for nonexposed populations. Estimated mortality from heart attacks is 5.7/100,000 in women ages 40-44 who are light smokers (less than 15 cigarettes per day) and do not use oral contraceptives. Pill users who are also light smokers should experience a comparable level of risk from heart attack due to causes unrelated to the Pill, so any risk beyond that is attributed to the Pill. Thus an estimated mortality of 28.6 deaths per 100,000 Pill users amounts to an **excess** mortality of 22.9/100,000 as a consequence of Pill use (FDA 1978a, 4225). Because excess mortality is an absolute value rather than a ratio, we can distinguish between low- and high-frequency events. Relative risk for users may be twice that of nonusers, but the absolute severity of the event—whether it involves 2 or 20 or 200 deaths per 100,000—may be of more consequence than relative risk.

Direct comparison of the hazards of alternative methods of contraception appears in Table 1. Overall mortality risk associated

TABLE 1

Annual number of birth-related, method-related, and total deaths associated with control of fertility per 100,000 nonsterile women, by regimen of control and age of woman.

REGIMEN OF CONTROL AND OUTCOME	AGE GROUP					
	15-19	20-24	25-29	30-34	35-39	40-44
No control						
Birth-related	5.6	6.1	7.4	13.9	20.8	22.6
Abortion only						
Method-related	1.2	1.6	1.8	1.7	1.9	1.2
Pill only/nonsmokers						
Birth-related	0.1	0.2	0.2	0.4	0.6	0.5
Method-related	1.2	1.2	1.2	1.8	3.9	6.6
Total deaths	1.3	1.4	1.4	2.2	4.5	7.1
Pill only/smokers						
Birth-related	0.1	0.2	0.2	0.4	0.6	0.5
Method-related	1.4	1.4	1.4	10.4	12.8	58.4
Total deaths	1.5	1.6	1.6	10.8	13.4	58.9
IUDs only						
Birth-related	0.1	0.2	0.2	0.4	0.6	0.5
Method-related	0.8	0.8	1.0	1.0	1.4	1.4
Total deaths	0.9	1.0	1.2	1.4	2.0	1.9
Traditional methods only						
Birth-related	1.1	1.6	2.0	3.6	5.0	4.2
Traditional methods, plus abortion						
Method-related	0.2	0.2	0.3	0.3	0.3	0.2

Source: C. Tietze (1977,75). Reprinted with permission from Family Planning Perspectives, Volume 9, Number 2, 1977.

with the effort to control fertility consists of two independent measures: method-related mortality and birth-related mortality. Method-related mortality reflects all deaths attributed to the method itself. For the Pill, this would include excess deaths due to stroke, heart attack, and the like; for the IUD, excess deaths associated with causes such as septic abortion. Method-related deaths are zero for those who practice no method of contraception or employ traditional methods such as the diaphragm. Birth-related deaths are those associated with pregnancies resulting from a method's failure to prevent pregnancy. Because the Pill and IUD have low failure rates, birth-related deaths are correspondingly low, whereas traditional methods, with their higher incidence of failure, carry a higher risk of birth-related death. Use of "no method" of contraception results in the greatest risk of birth-related death.

A comparison of overall risk of dying, according to method of contraception employed and age group (Figure 3) enables us to draw several important conclusions about contraceptive risk: (1) irrespective of age, annual mortality is consistently lowest (0.3/100,000 or less) among those who employ traditional methods with abortion as a backup; (2) for all other methods except **no method**, risk is low (2/100,000 or less) for women under 30; (3) after

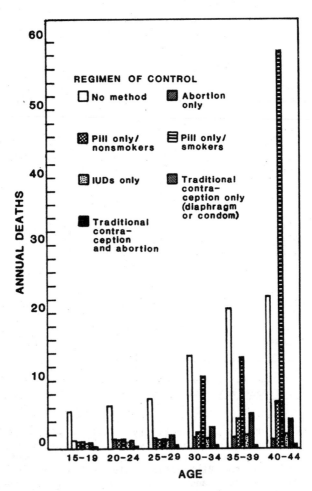

Figure 3. Estimated annual number of deaths associated with control of fertility and no control per 100,000 nonsterile women, by regimen of control and age of woman. <u>Source</u>: After Tietze (1977).

30, risk of dying from traditional methods, abortion, and the IUD (5/100,000 or less) still falls short of risks from no method (over 20/100,000 after age 34) and the Pill; (4) among Pill users, risk is particularly high among those who smoke (58.9/100,000, ages 40-44).

Illustrations such as Figure 3 routinely appear in patient package inserts and physician labeling and provide both prospective users and their physicians with a guide to the risks associated with different approaches to fertility control. Of course, the guide is only a general one, since mortality is estimated from sample populations, and small differences may be due to error. More important, although stratified by age groups, the sample neglects socioeconomic differences and other attributes that may be reliable predictors of mortality, if they result in differential access to medical care during the reproductive period.

A more significant flaw in the use of such mortality estimates results from the fact that contraception is not generally undertaken solely or even predominantly in order to preserve the user's life. Thus the relevant criterion, at least from the user's perspective, is not the risk of dying from fertility-related causes. Instead, it is the risk of the unintended pregnancy, or the rate of method failure (Table 2), which is the salient consideration. When safe means "safe from pregnancy" rather than "safe from dying," a rather different ranking of methods emerges. Oral contraceptives offer the greatest freedom from unintended pregnancy, followed closely by IUDs. Using no method of contraception confers significantly fewer benefits, because it involves a substantial risk of pregnancy, on the order of 60 to 80 pregnancies per 100 fertile women-years. In practice, Table 2 provides only rough guidelines regarding method failure. As the spread of values for some methods indicates, different studies have considerable variations in failure rates.

As this discussion suggests, the prospective contraceptive user must weigh two basically incommensurate measures: risk of dying against the benefit for which contraception is undertaken, namely, avoiding the unintended pregnancy. Comparing mortality consequences of contraceptive use against the likelihood that pregnancy can be

TABLE 2
Method failure: pregnancies per 100 women per year

METHOD	
Oral contraceptives	Less than 1
IUDs	Less than 1-6
Diaphragm with spermicidal products	2-20
Condom	3-36
Aerosol foams	2-29
Jellies and creams	4-36
Periodic abstinence (rhythm)	Less than 1-47
No contraception	60-80

Source: FDA (1978a,4231).

avoided meets only with frustration, however, because the risks and benefits being compared are so dissimilar. Death and pregnancy have no common denominator to which both can be reduced.[18]

Recognizing that dissimilar benefits and risks complicate the choice among contraceptive methods represents a step forward in answering the question: which method is safest? Because safety is necessarily ambiguously defined and represents a highly personal evaluation, choice of method is most properly made by the user herself in consultation with her physician and family. Easy access to such information enhances the prospective user's ability to weigh incommensurate benefits and risks. How contraceptive users come to be informed about risks (and benefits) is discussed in the section that follows.

Informing about Risks

A third area of innovation in the management of technological hazards is illustrated by the development of prescription warning labels and subsequent, more detailed patient package inserts (PPIs). PPIs were designed to provide directly to individual patients information supplemental to that provided by (and at the discretion of) the attending physician. PPIs originated in two different, potentially contradictory objectives: on the one hand, to encourage greater patient compliance with physicians' instructions for use; and on the other hand, to inform patients more fully about the risks as well as the benefits of a particular medical regimen (Institute of Medicine 1979). PPIs as devices for enhancing patient compliance were designed to reassure patients about the need for treatment, method of drug use, and possible "normal" side effects of a particular drug. Patient compliance objectives envisioned a more cooperative, less demanding patient. In contrast, informing patients about possible risks and comparing risks of alternative treatments were intended to encourage patients to assume a more active role in the management of their own health, by bearing more personal responsibility for choosing among alternative methods of treatment.

Historically, a paternalistic shouldering of risk decisions by regulators, technology sponsors, and professional risk managers has been the **modus operandi**, whereas the informed-use principle is of more recent origin. Contraceptive technologies are among the earliest cases where patient labelling has provided a direct means of informing potentially exposed populations about risks.

In the case of prescription drugs, the presence of third-party decisionmakers complicates the informing of users about risks. As professional risk managers, physicians have traditionally made drug risk/benefit comparisons on behalf of their patients. FDA has supported this right as consistent with good medical management practices. As lay people, patients could claim little competency to weigh medical benefits and risks, and any notion of their "right" to do so was tantamount to an abrogation of the physician's traditional responsibilities. Thus, as FDA required pharmaceutical companies to inform physicians (and subsequently, pharmacists) of adverse drug reactions, it allowed physicians full discretion in the matter of sharing this information with prospective users. Some did; most did not.

Further complicating the issue of a user's right to know about risks of medical treatment is the fact that, in sharp contrast to most patient populations, contraceptive users are usually young, healthy adults. The benefits they seek when using contraceptives are not entirely, and often not even substantially, medical in nature, except insofar as the avoidance of pregnancy is considered a medical issue. Users choose among several competing methods of contraception and are in a position (unlike the sick for whom few alternatives may exist) to exercise this choice in a "rational" fashion. Recognition of these unique circumstances encouraged users to pursue information about the nature of risks they faced and ultimately helped justify FDA's acceptance of the informed-use principle.

The thought of directly informing users of contraceptives about the hazards of such regimens did not come about easily or quickly. Not until 1969, in the wake of extensive, ongoing public discussion over oral contraceptive safety, did FDA specifically request physicians to warn their own patients about the risks of using the Pill. Perhaps this historic request represents a turning point, for within four months FDA drafted the first compulsory warning label aimed at any sizeable patient population. The first draft was made public during the final days of testimony before the widely publicized Senate Nelson Committee hearings on oral contraceptive safety.

The idea of bypassing physicians to warn users directly met stiff opposition from drug manufacturers, physicians, and the American Medical Association, which threatened legal action against the proposed package label. FDA prevailed in principle only: the final version of the warning bore little resemblance to the fairly informative original. Short and severely edited, the version finally approved read as follows:

> Do not take this drug without your doctor's continued supervision.
>
> Oral contraceptives are powerful and effective drugs which can cause side effects in some users and should not be used at all by some women. The most serious known side effect is abnormal blood clotting, which can be fatal.
>
> Safe use of this drug requires a careful discussion with your doctor. To assist him in providing you with the necessary information (pharmaceutical firm's name) has prepared a booklet written in a style understandable to you as the drug user. This provides comprehensive information on the effectiveness and known hazards of the drug including warnings, side effects, and who should not use it.
>
> Your doctor will give you this booklet and answer any questions you may have about the use of this drug. (New York Times 10 September 1970,25).

Physicians controlled distribution of the 800-word brochure written by the American Medical Association. Given the dilution of the 1970 warnings, the first patient labeling requirement for oral contraceptives was remarkable not for what it said—the warning was considerably less substantial than those available through contemporary news sources—but simply because it assigned to drug manufacturers a responsibility for informing users about product risks, and thereby diminished somewhat the physician's role as intermediary in monitoring risk information.

In 1978, patient package inserts for oral contraceptives underwent extensive revisions. These PPIs now detail known and suspected risks of the drug's use and compare the Pill's effectiveness with that of alternative contraceptive methods. Figure 4 shows a portion of a typical PPI, representative of those in use today. Specific requirements governing all PPIs are detailed in the **Federal Code.** PPIs closely resemble information sheets whose distribution to physicians and pharmacists is also required by law. Dissemination requirements have also been expanded to include a PPI with every prescription refill and PPIs have been extended to the IUD and many estrogen-based drugs, including DES.

Several conditions undoubtedly contributed to FDA's, physicians' and manufacturers' gradual, grudging acceptance of the consumer's right-to-know about medical risks. Extensive public media discussion of oral contraceptives, IUDs, and DES expanded the population knowledgeable about the existence of risks. Several lawsuits successfully established manufacturer negligence in failing to provide patient warnings about possible consequences of drug use. The contraceptive user's near unique status in the acceptable risk debate became more widely acknowledged. The consumer movement's general interest in product safety and consumer protection, and the feminist movement's more specific interest in asserting informed control over reproductive processes, helped generate an audience that was better informed, more articulate, and more demanding than in the past. In 1975, an alliance of consumer and feminist groups petitioned FDA to establish written warnings for certain drugs, including those (such as oral contraceptives) that posed dangers when taken by pregnant or nursing women.[19] Collectively, these conditions helped to mold a climate in which advocating the rights of vulnerable populations became an increasingly acceptable posture for regulatory agencies to assume. FDA has become increasingly supportive of patient warnings for prescription drugs, and in mid-1979 proposed regulations that would extend the use of PPIs to most prescription drugs in order "to promote the safe and effective use of prescription drugs by patients and **to ensure that patients have the opportunity to be informed of the benefits and risks involved in the use of prescription drug products"** (FDA 1979,40018).

The objective of informed use, which lies at the heart of the debate over PPIs, is not entirely fulfilled by current measures. On the positive side, users of oral contraceptives and IUDs are now guaranteed access to information about product risks and benefits and are in a better position to compare the consequences of alternative methods, however difficult these tradeoffs may be personally. But several important issues still remain regarding the efficacy of PPIs in creating an informed user population. First, do current PPIs provide the "right" kinds of information, under conditions that are appropriate for "informed judgments"? An informed decision of necessity includes the option of rejecting one method in favor of another. Has the choice already been made by the time the information becomes available? The law requires that PPIs be issued each time a Pill prescription is filled, but not before; PPIs for IUDs are provided in physicians' offices shortly before insertion is

DETAILED PATIENT LABELING

What You Should Know About Oral Contraceptives

Oral contraceptives ("the pill") are the most effective way (except for sterilization) to prevent pregnancy. They are also convenient and, for most women, free of serious or unpleasant side effects. Oral contraceptives must always be taken under the continuous supervision of a physician.

It is important that any woman who considers using an oral contraceptive understand the risks involved. Although the oral contraceptives have important advantages over other methods of contraception, they have certain risks that no other method has. Only you can decide whether the advantages are worth these risks. This leaflet will tell you about the most important risks. It will explain how you can help your doctor prescribe the pill as safely as possible by telling him about yourself and being alert for the earliest signs of trouble. And it will tell you how to use the pill properly, so that it will be as effective as possible. There is more detailed information available in the leaflet prepared for doctors. Your pharmacist can show you a copy; you may need your doctor's help in understanding parts of it.

Who Should Not Use Oral Contraceptives

A. If you have any of the following conditions you should not use the pill:

1. Clots in the legs or lungs. 2. Anging pectoris. 3. Known or suspected cancer of the breast or sex organs. 4. Unusual vaginal bleeding that has not yet been diagnosed. 5. Known or suspected pregnancy.

B. If you have had any of the following conditions you should not use the pill:

1. Heart attack or stroke. 2. Clots in the legs or lungs.

> C. Cigarette smoking increases the risk of serious adverse effects on the heart and blood vessels from oral contraceptive use. This risk increases with age and with heavy smoking (15 or more cigarettes per day) and is quite marked in women over 35 years of age. Women who use oral contraceptives should not smoke.

D. If you have scanty or irregular periods or are a young woman without a regular cycle, you should use another method of contraception because, if you use the pill, you may have difficulty becoming pregnant or may fail to have menstrual periods after discontinuing the pill.

Deciding To Use Oral Contraceptives

If you do not have any of the conditions listed above and are thinking about using oral contraceptives, to help you decide, you need information about the advantages and risks of oral contraceptives and of other contraceptive methods as well. This leaflet describes the advantages and risks of oral contraceptives. Except for sterilization, the IUD and abortion, which have their own exclusive risks, the only risks of other methods of contraception are those due to pregnancy should the method fail. Your doctor can answer questions you may have with respect to other methods of contraception. He can also answer any questions you may have after reading this leaflet on oral contraceptives

Figure 4. A portion of a typical patient package insert accompanying each prescription refill of oral contraceptives. The substance of the text is detailed in the federal Code and thus does not vary significantly from manufacturer to manufacturer.

scheduled. Second, are PPIs read and understood by the target populations?

Preliminary studies of PPI effectiveness suggest that patients fall into one of two classes: **information seekers** and **information avoiders** (Institute of Medicine 1979,26). Information seekers search out and read all available information; they also tend to be the more highly educated, affluent clients with better access to follow-up medical consultations. Information avoiders avoid exposure to potentially disturbing information or information that they cannot understand. Whereas users have a legitimate right to reduce their own anxieties by deferring to their medical advisors for judgments about acceptable risk, those with limited reading skills are at a distinct disadvantage with respect to informing themselves about risk. They defer by default, not choice.

Interpreting and Extending the Managerial Lessons Derived from Contraceptives

Remarkable changes in control over unwanted fertility have occurred in the United States during the past twenty years, thanks in part to the development of oral contraceptives, IUDs, and more acceptable sterilization and abortion procedures. Such technologies diffused rapidly and in the United States have resulted in a total exposure of perhaps 20 to 25 million women to oral contraceptives and roughly 9 million to IUDs. Despite the drawbacks associated with these and other methods of fertility control, the consequences for fertility reduction in the U.S. are indisputable.

Insights from the managerial experience with the newer contraceptive technologies have broad application to numerous nonmedical technologies. The contraceptive experience offers lessons for at least three issues in hazard management: hazard identification; consequence evaluation; and informing about risk.

Lesson 1: Increasing Opportunities for Identifying Hazards

The management histories of the Pill, the IUD, DES in its postcoital uses, and injectable contraceptives illustrate the futility of relying upon a single or even a few events in the drug development process in order to detect adverse effects of new technologies. Awareness of hazards often emerges slowly, particularly when technologies are viewed as benign and their causal mechanisms are not well understood. Extrapolation from experiments with laboratory animals often produces results of uncertain application to humans, as was the case for IUDs. Research on human subjects of necessity utilizes experimental populations that may be quite different from the ultimate, intended users, thus delaying the appearance of adverse effects until observed in vulnerable populations. Establishing of appropriate human dose levels proceeds slowly, and in the case of oral contraceptives, resulted in early users' being exposed to much higher dosages than ultimately proved necessary for effective therapy. Some drug reactions are unlikely to become apparent through animal or limited human testing, requiring instead broad exposure over large, diverse patient populations before consequences become apparent in a statistically significant fashion. For some drugs, such as DES, consequences are delayed, displaced to others than the original user, or discovered by chance. Individual stages of drug testing contain built-in liabilities, but each stage added to the evaluation process increases the likelihood that adverse effects will be detected in a timely manner.

Because utilizing multiple opportunities to detect hazards adds to the direct costs of producing a product, producers resist expanding opportunities for hazard detection. In the United States, arguments against formally extending the requirement for drug hazard research into the postmarket phase must be weighed against the system's current failure to internalize the hidden costs of drug failures. The costs of death or pregnancy as a result of the method's failure are borne almost exclusively by the victims or their families, unless through litigation they seek compensation from the drug manufacturer.

The first lesson from the contraceptive management experience affirms the need for a more systematic approach to postmarket surveillance of drug effects in the United States, a proposal that has already been made to FDA. In addition, it has direct applications to technologies quite unrelated to medicine. Blind spots exist in our perception of new technologies as potentially harmful. Hazard detection occurs slowly and erratically and, unlike prescription drugs, many technologies are not characterized by any multiple-phase search for adverse effects. Thus the technology is rapidly diffused and consequences become widespread before the causal links are drawn and action can be taken. Technologies as diverse as the automobile and DDT show the effects of delayed hazard recognition. The management of prescription drugs in the United States offers a model, although an imperfect one, for a more systematic approach to hazard detection, and, as a result, an opportunity to allocate more equitably the costs of hazard detection while achieving a more timely recognition of risks. The contraceptive experience also shows why we ought not to be repeatedly surprised when bright new technologies, which seem to offer such great promise, subsequently become tarnished by links to unanticipated side effects. What should amaze us is that such occurrences continue to take us by surprise.

Lesson 2: Choices Involving Incommensurate Risks and Benefits

Weighing the risk of dying from the Pill's side effects against that of dying in an automobile accident serves a useful but limited purpose: it places the Pill in the context of life's more familiar risks. Such a comparison, however, falls short because it provides a prospective user with no information about the safety of the Pill as a method of birth control, the reason, after all, for which the user risks exposure. The IUD, diaphragm, or even the "no-contraceptive-method" alternatives are among the appropriate controls against which the Pill's risks should be compared. The medical community has taken this message to heart: compare the medical regimen with its alternatives, including that of foregoing treatment entirely. Discussions of contraceptive risk in the media, professional journals, and PPIs suggest that this lesson has been quite broadly applied. It is a lesson that consumers, regulators, and technology sponsors would do well to apply to a wider range of technologies, so that the risks of nuclear power plants are appropriately compared with genuine alternatives, rather than the familiar but not equivalent automobile or the eating of charcoal-broiled steaks (Wilson 1979,45).

The question, "which method of contraception is safest?" must inevitably be followed by the query, "safest from what?" Contraception is employed principally to prevent pregnancy, yet pregnancy differs fundamentally from death, the traditional common denominator to which risks are reduced. For most women in the United States, the risk of dying from birth-related or method-related causes is a low-probability event that seldom reflects the reason for practicing contraception: for its benefit, the avoidance of pregnancy. The literature on contraceptives recognizes that the choice of method involves two fundamentally different but measurable qualities of risk, and hence contraceptive methods are routinely evaluated for efficacy (benefits) as well as their life-threatening (risk) properties.

The acceptance of incommensurate benefits and risks is transferable to medicine and beyond: in medicine where the illusive "quality of life" issue is raised with increasing frequency in evaluating life-saving techniques; beyond medicine in children's viewing of television (chapter 18) where benefits such as entertainment value and child care must contend with the quite different risks of behavioral effects. Life's choices often require that we weigh incommensurate risks and benefits. The desire to reduce all to some common unit of measurement tempts us to ignore those effects that most defy the reduction process.

Lesson 3: Informing about risk

The contraceptive management experience in informing about risk has proved successful by one criterion; pleased with the results and supportive of the principle, FDA has proposed extending PPIs to most prescription drugs. The first 10 years of experience with PPIs have provided some interesting findings. First, consumers are generally receptive to the notion of being directly informed (FDA 1980,60754, 60756, 60759-60). Second, groups originally opposed—manufacturers, physicians, and pharmacists—constitute the reservoir of continued opposition, although a softening of this line is apparent. Third, initial fears of the effects PPIs have proved unfounded (FDA 1980,60754, 60764). PPIs have generated no obvious increase in symptoms through suggestion; no widespread, abrupt discontinuation of contraception out of fear; no marked increase in patient demands on physicians' time; no exorbitant increase in the cost of prescription drugs. These observations suggest that the principle of informing about risk could be extended to other hazardous technologies in consumer markets and in the workplace without meeting significant consumer or worker opposition. In fact, both consumer and worker groups have already initiated efforts in these areas just as feminist and consumer groups disseminated information about the Pill. If the PPI experience proves typical, fears about the consequences of informing these populations will probably prove largely unfounded and may well result in more risk-conscious, careful workers and consumers.

What is unclear from the PPI experience for prescription contraceptives is the extent to which this approach has actually produced a better-informed user population. Nor is it clear that the

present distribution methods provide information in the most timely
manner possible. The issue of legal liability arising from PPI
warnings has not been tested. FDA believes that informing about
risk produces a better-informed, hence less adversely affected user,
resulting in fewer occasions when product liability lawsuits are
appropriate. FDA prefers to view PPIs as divorced from the question
of informed consent. "Patient labeling [is not] intended to serve
as a vehicle for obtaining informed consent of patients for the use
of a drug product. Patient labeling will be required solely because
of its positive effects" (FDA 1979,40023). Whether intentionally
informing consumers, workers, or other exposed populations about
risks will shift the liability burden is hard to predict, but surely
the issue will not be laid to rest this easily.

Because contraceptive users may choose from a range of methods,
consistent with their own levels of pregnancy aversion, medical circumstances, and personal preferences, they are quite different from
many medical patients, for whom such a range of regimens is seldom
available. They are distinctly different from workers in hazardous
occupations, such as coal mining, lead manufacturing, or pesticide
applications, who face severe economic consequences of acting upon
risk information and are much more constrained in their opportunities to alter exposure levels.

The significance of the third lesson from contraceptive management is that adverse information is preferable to uncertainty or
ignorance, and fears about the adverse effects of increased access
to risk information are likely to be unwarranted. Equity considerations plus a current social climate that encourages access to information favor a broadening of the principle behind PPIs. Today's
workers and many consumers are in a position similar to that of contraceptive users 15 years ago: through the print media and television they are exposed to a considerable volume of information about
hazards, but such information is scattered, of varying reliability,
often sensationally presented, and not usually directed toward those
who have the greatest need to know. Under these circumstances, the
contraceptive management experience with informing about risk provides some encouragement about the prospects for expanding, in a
nonthreatening manner, the scope, quality, and availability of
information about hazards to populations at risk.

NOTES

1. Unlike consumable drugs or devices whose monthly or quarterly
    sales indicate fluctuations in user acceptance, IUD use is more
    difficult to ascertain because a device may remain in place for
    several years or more. Neither FDA nor other data sources are
    able to estimate extent of use with the level of accuracy pre-
    sumed correct for oral contraceptives or other prescription
    drugs.

2. The proposed patient leaflet for injectable contraceptives, as
    described in the provisions of the Code of Federal Regulations
    (21 CFR 510:301a), contains the following statement: "(Trade

name of drug) should be used only if: other methods of preventing pregnancy have failed; or you are not able to use other
methods of preventing pregnancy such as the "pill," an IUD, or
a diaphragm **and you accept** the small possibility of permanent infertility (inability to have children) **and you understand and accept the risks and drawbacks described below
. . . .**" The patient, having familiarized herself with the
insert or having had it explained to her, was then to give her
"informed consent."

3. DES is among the hazardous substances monitored by the Interagency Regulatory Liaison Group (IRLG), which comprises the
Consumer Product Safety Commission (CPSC), the Environmental
Protection Agency (EPA), the Food and Drug Administration (FDA)
and the Occupational Safety and Health Administration (OSHA).

4. 21 CFR 310:501. Reasons for changes appear in the Federal
Register 40 no. 25 (5 February 1975):5351-5355.

5. The concern over DES as postcoital contraceptive is not that
its users will develop cancer. Rather, if an unsuspected
pregnancy has already occurred and is not terminated, then a
female embryo over six weeks is at risk from the DES therapy,
which will be ineffective in its contraceptive application.
See Rinehart (1976,146).

6. The Subcommittee on Monopoly, Senate Select Committee on Small
Businesses, chaired by Senator Gaylord Nelson, conducted an
inquiry from 14 January to 4 March 1970, into "the question
whether users of birth control pills are being adequately informed concerning the Pill's known health hazards" (Statement
from the Committee's press release, as quoted in Djerassi
1979,92). The popular joke at the time suggested that extensive publicity about the Pill's adverse effects might result in
widespread, abrupt discontinuation of pill use and a crop of
unplanned babies, named "Nelson" by their mothers in "honor" of
the Senator.

7. See 1971 and subsequent editions. The chapters entitled "Birth
Control" and "Women and Health Care" are particularly critical
of the traditional treatment of women patients at the hands of
drug companies and physicians.

8. At least this is so for most women in industrialized societies,
where maternal mortality rates are low. Where maternal mortality is high, consequences of unintended pregnancies are
proportionally more severe, as measured by the criteria of
mortality.

9. Consider the following brief chronology:

In August 1963, the Wright Committee (FDA Advisory Committee on Obstetrics and Gynecology) found **"no significant increase** in thromboembolitic disease" (Silverman and Lee
1974,99).

In 1965, FDA aide Sadusk stressed that a "**cause-and-effect relationship**" between oral contraceptives and clotting **had not been established** (New York Times, 18 November 1965,1), and several days later a quickly formed Advisory Committee on Ob/Gyn reported "**no evidence**" of eye disorders due to clotting associated with Pill use (New York Times, 25 November 1965,60).

In 1966, the FDA Advisory Committee on Ob/Gyn found **no evidence** that oral contraceptives were **unsafe**, and recommended removing all time limits on their use, while urging further study of Pill safety (New York Times, 15 August 1966,1).

In 1968 the Ob/Gyn Advisory Committee reported a "**small element of risk**" associated with Pill use (New York Times, 22 January, 1968,1), and in November of that year, the committee found the pill/cervical cancer linkage "inconclusive" (New York Times, 3 November 1968,118) but supported the World Health Organization committee's recommendation that users undergo regular medical examinations.

In 1969, FDA's Second Report on Oral Contraceptives, citing British risk assessments, found a **definite cause and effect relationship existing** between pill use and deaths due to blood clotting disorders. It labelled the Pill "**safe**," however (New York Times, 5 September 1969,1).

This brief chronology suggests how the assessment of oral contraceptive safety evolved over a decade of warnings and subsequent risk assessments.

10. Late in 1975, FDA threatened to remove the drugs from the market if manufacturers could not demonstrate the existence of a population for whom they were the preferred medical treatment. By that time, sequentials held a relatively small portion of the entire oral contraceptive market, down to 5.4 percent of sales volume from a high of 22 percent in 1967. This fact, and the desire to avoid unfavorable publicity for their other, non-sequential contraceptives, prompted the three U.S. manufacturers to withdraw their products from distribution early in 1976, although several months of already distributed stock was allowed to remain on pharmacy shelves. Ironically, the reason that the manufacturers gave for their "voluntary" action was new evidence suggesting a sequentials/cancer link (New York Times, 23 December 1975,57; 24 December 1975,14; 29 December 1975,1; 26 February 1976,23; and FDA Consumer, April 1976,25).

11. For a review of premarket and postmarket sources of information about drug efficacy and safety specifically applied to oral contraceptives, see Ortiz (1978).

12. As a result, there are considerable differences in the average length of time required for each government to complete the approval process for new drugs. A GAO study initiated in response to the "drug lag issue" found that FDA took an average of twenty-three months to grant approval for new drugs, whereas the United Kingdom averaged five months (General Accounting Office 1980). For an examination of the postmarket phase of

drug analysis in the U.S., Great Britain and eight other countries, see Wardell (1978).

13. For these and other comparative risk estimates, see FDA (1978a).

14. Paradoxically, Searle Company blamed initial concern over its Enovid product on the interest in drug risks stimulated by media coverage of FDA's success in protecting the U.S. from the drug Thalidomide, whose harmful consequences had not been detected in early trials abroad. (New York Times, 4 August 1962,20).

15. CSP resulted from the First International Conference on IUDs, held in New York City in 1962 under the auspices of the Population Council. The first of a series of statistical evaluations conducted by CSP was presented at the Second International Conference in 1964 (Tietze and Lewit 1965).

16. The Boston Collaborative Drug Surveillance Project employs a computer to record reactions reported among hospitalized patients in several countries. A more comprehensive, government sponsored PMS for the U.S. has been suggested by the Joint Commission on Prescription Drug Use, in its final report to the Senate Committee on Labor and Human Resources (U.S. Congress 1980). Intended to complement rather than replace current premarket screening mechanisms, the proposed PMS would give highest priority to new chemical entity drugs, commonly used drugs, and those with delayed effects.

17. Consider estimated mortality from the Pill and the automobile, per 100,000 women (nonsterile in the case of Pill estimates):

ages	the Pill	the automobile
15-19	1.4	23.6
40-44	24.8	12.4

It is apparent that young women are at greater risk from the automobile, whereas older women face greater risk from Pill use. But what is the meaning of such conclusions, when the choice is **not** that of the Pill **or** the automobile? Data source: Tietze, Bongaarts, and Schearer (1976,9, Table 3).

18. At a societal level, one could presumably estimate the "costs" associated with death and pregnancy outcomes, but this exercise is of doubtful value to an individual user concerned with personal, not societal, consequences of her actions.

19. On 31 March 1975, the Center for Law and Social Policy filed the petition, on behalf of itself and other groups such as Consumers Union, Consumer Action for Improved Food and Drugs, the National Organization for Women, the Women's Equity Action League, and the Women's Legal Defense Fund [Federal Register 44 no. 131 (6 July 1979):40018].

REFERENCES

Boston Women's Health Book Collective. 1971. Our bodies, ourselves: A book by and for women. New York: Simon and Schuster.

Burke, Laurie B., Dianne L. Crosby, and Chang S. Lao. 1977. Drug use analysis of oral contraceptives (mimeo). Rockville, Maryland: Drug Use Analysis Branch, Food and Drug Administration (December).

Cates, William, Jr., David R. Grimes, Howard W. Ory, and Carl W. Tyler, Jr. 1977. Publicity and public health: The elimination of IUD-related abortion deaths. Family Planning Perspectives 9 no. 3 (June):138-140.

Davis, Kenneth S. 1978. Story of the Pill. American Heritage 29 (August/September):80-90.

Djerassi, Carl. 1979. The politics of contraception. New York: W.W. Norton.

FDA (Food and Drug Administration). 1977. Patient labeling requirements for IUDs. Federal Register 42 (10 May):23777.

FDA (Food and Drug Administration). 1978a. Oral contraceptive drug products: Physician and patient labeling. Federal Register 43 no. 21 (31 January):4223-4234.

FDA (Food and Drug Administration). 1978b. Medical Device and Drug Advisory Committees on Obstetrics and Gynecology. Second report on intrauterine contraceptive devices. Rockville, Maryland: FDA.

FDA (Food and Drug Administration). 1979. Prescription drug products: Patient labeling requirements. Federal Register 44 (6 July):40016-40041.

FDA (Food and Drug Administration). 1980. Prescription drug products: Patient package insert requirements. Federal Register 45 no. 179 (12 September):60754-60784.

General Accounting Office. 1980. FDA drug approval: A lengthy process that delays the availability of important new drugs. HRD 80-64. Washington: GAO.

Hartman, Sarah, and Vikki A. Segel. 1979. Diethylstilbestrol (DES) drug controversy. Issue Brief IB75023 (updated 19 December). Washington: Congressional Research Service, Library of Congress.

Herbst, A.L., H. Ulfelder, and D.C. Poskanzer. 1971. Adenocarcinoma of the vagina: Association of maternal stilbestrol therapy with tumor appearance in young women. New England Journal of Medicine 284 (22 April):878-881.

Huber, S.C., P.T. Piotrow, B. Orlans, and G. Kommer. 1975. IUDs reassessed: A decade of experience. Population Reports Series B, no. 2 (January). Washington: Population Information Program, George Washington University.

Inman, W.H.W., and M.P. Vessey. 1968. Investigation of deaths from pulmonary, coronary and cerebral thrombosis and embolism in women of child-bearing age. British Medical Journal 2 (27 April):193-199.

Institute of Medicine. 1979. Evaluating patient package inserts. Washington: National Academy of Sciences.

Kols, Adrienne, Ward Rinehart, Phyllis T. Piotrow, Louise Doucette, and Wayne F. Quillin. 1982. Oral contraceptives in the 1980s.

Population Reports Series A, no. 6 (May/June). Baltimore: Population Information Program. The Johns Hopkins University.

Kuchera, L.K. 1971. Postcoital contraception with Diethylstilbestrol. JAMA 218:526-563.

Levinson, Carl J., and David C. Richardson. 1976. The Dalkon Shield story. Advances in Planned Parenthood 11:53-63.

Liskin, Laurie. 1982. IUDs: An appropriate contraceptive for many women. Population Reports Series B, no. 4 (July). Baltimore: Population Information Program, The Johns Hopkins University.

Maine, Deborah. 1978. Depo: Debate continues. Family Planning Perspectives 10 no. 6 (November/December):342-345.

Mintz, Morton. 1968/1969. The Pill: Press and public at the experts' mercy. Columbia Journalism Review 7 no. 4 (Winter): 4-10.

Mintz, Morton. 1969. The Pill: Press and public at the experts mercy (second part). Columbia Journalism Review 8 (Spring): 28-35.

Morris, J.M., and G. van Wagenen. 1966. Compounds interfering with ovum implantation and development, III: The role of estrogens. American Journal of Obstetrics and Gynecology 96:801-813.

Morrison, Margaret. 1975a. Contraception with IUD's. FDA Consumer 9 no. 5 (June):15-20.

Morrison, Margaret. 1975b. The Dalkon Shield. FDA Consumer 9 no. 1 (February):21.

Morrison, Margaret. 1975c. Dalkon IUD won't be returned to market. FDA Consumer 9 no. 7 (September):25.

Mosher, W.D. 1981. Contraceptive utilization, United States, 1976. Vital and Health Statistics, Series 23. Data from the National Survey of Family Growth, no. 7 (March). Hyattsville, Maryland: National Center for Health Statistics.

National Research Council. 1981. Governing Board Committee on the Assessment of Risk. The handling of risk assessments in NRC reports. Washington: [National Academy of Sciences], March.

New York Times. 1967. Birth pill tests ordered by FDA. 13 August, p. 64.

Noller, Kenneth L., and Charles R. Fish. 1974. Diethylstilbestrol usage: Its interesting past, important present, and questionable future. Medical Clinics of North America 58 (July): 793-810.

Ortiz, Edwin M. 1978. The role of the regulatory agency in drug development. In Risks, benefits and controversies in fertility control (Workshop, Arlington Virginia, 1977), ed. John J. Sciarra, Gerald Zatuchni, and J. Joseph Spidel, 12-16. New York: Harper and Row.

Parkes, A.S., E.C. Dodds, and R.I. Noble. 1938. Interruption of early pregnancy by means of orally active oestrogens. British Medical Journal 2:557-559.

Piotrow, Phyllis T., and Calvin M. Lee. 1974. Oral contraceptives: 50 million users. Population Report Series A, no. 1 (April). Washington: Dept. of Medical and Public Affairs, George Washington University.

Piotrow, Phyllis T., Ward Rinehart, and J.C. Schmidt. 1979. IUDs: Update on safety, effectiveness, and research. Population Reports Series B, no. 3 (May). Baltimore: Population Information Program, The Johns Hopkins University.

Planned Parenthood Federation of America. 1977. Within our reach:
Annual report. New York: The Federation.
Rinehart, Ward. 1976. Postcoital contraception: An appraisal.
Population Reports Series J, no. 9 (January). Washington:
Dept. of Medical and Public Affairs, George Washington University.
Rinehart, Ward, and Phyllis T. Piotrow. 1979. Oral contraceptives:
Update on usage, safety, and side effects. Population Reports
Series A, no. 5 (January). Baltimore: Population Information
Program, The John Hopkins University.
Rinehart, Ward, and Jane Winter. 1975. Injectable progestogens:
Officials debate but the use increases. Population Reports
Series K, no. 1 (March). Baltimore: Population Information
Program, The Johns Hopkins University.
Royal College of General Practitioners. 1967. Oral contraception
and thrombo-embolic disease. Journal of the Royal College of
General Practitoners 13:267-279.
Shapo, M.S. 1979. A nation of guinea pigs, 163-190. New York:
Collier Macmillan.
Silverman, Milton, and Philip Lee. 1974. Pills, profits and poli-
tics. Berkeley and Los Angeles: University of California
Press.
Stolley, Paul D. 1974. Assuring the safety and efficacy of thera-
pies. International Journal of Health Services 4 no. 1:131-
145.
Tietze, Christopher. 1977. New estimates of mortality associated
with fertility control. Family Planning Perspectives 9 no. 2
(March/April):74-76.
Tietze, Christopher, John Bongaarts, and Bruce Schearer. 1976.
Mortality associated with the control of fertility. Family
Planning Perspectives 8 no. 1 (January/February):6-14.
Tietze, Christopher, and Sarah Lewit. 1965. Intra-uterine contra-
ception: Effectiveness and acceptability. In Proceedings of
the Second International Conference on Intra-Uterine Contra-
ception (2-4 October 1964), 98-110. International Congress
Series, no. 86. Amsterdam: Excerpta Medica Foundation.
U.S. Congress. 1980. Senate Committee on Labor and Human Re-
sources. Subcommittee on Health and Scientific Research. The
final report of the Joint Commission on Prescription Drug Use,
Inc. (April). Washington: Government Printing Office.
Vessey, M.P., and Richard Doll. 1968. Investigation of relation
between use of oral contraceptives and thromboembolic disease.
British Medical Journal 2 (27 April):199-205.
Wardell, William M. 1978. Controlling the use of therapeutic
drugs: An international comparison. Washington: American
Enterprise Institute for Public Policy Research.
Westoff, Charles F., and Elise F. Jones. 1977. Contraception and
sterilization in the United States, 1965-1975. Family Planning
and Perspectives 9 no. 4 (July/August):154, Table 1.

18
Television:
A Social Hazard

Julie Graham and Roger E. Kasperson

The possibility that television poses significant risks to society has fanned scholarly and popular debate since the advent of the technology over 30 years ago. Critics of television have alleged a wide variety of ill effects, ranging from increased heart disease among children to the erosion of the family and traditional values. A flood of research on these questions has followed: in the past 25 years some 3000 books, articles, reports, and documents have examined the effects of television on human behavior (NIMH 1982,1:87); and major risk assessments have addressed televised violence, advertising to children, and social stereotyping.

Despite this effort, the risks of television remain elusive and shrouded in dispute. As of the end of 1984, no comprehensive risk assessment has been undertaken for the technology as a whole, and no generally accepted quantitative risk assessment links types of exposure to specific effects. Nevertheless, the scientific and regulatory community abounds with appeals exhorting researchers to new and larger efforts, and popular concern over television appears to be growing.

Meanwhile, the regulatory agencies charged with the responsibility for managing television have an unblemished record of impotence. Constrained by the First Amendment and a lack of congressional support, the agencies have been unwilling, perhaps unable, to challenge the networks and the television industry. When the new chairman of the Federal Trade Commission (FTC) in 1978 decided to propose regulations for televised advertising to children, he quickly found himself in a bitter fight for his job. Television today stands almost unique among modern high technologies, its risks unassessed and uncontrolled.

Why has the system for managing the technology so failed the public interest? The analysis to follow examines the social risks posed by television, the adequacy of assessments to date, and the effectiveness of control efforts. Attention is also devoted to changes on the horizon: the likely impacts of new technologies, pending reforms, and the deregulation actively in progress. These issues are viewed not only through the lens of risk management but that of political economy as well. But the necessary components of sound risk assessment—information on dose, dose/response relations, the various adverse consequences which may occur, control opportunities—receive their due. Finally, a number of new initiatives for improved risk management are proposed.

428

Growth of the Technology

Although initially developed in the 1920s and 1930s, television is basically a postwar technology. The Federal Communications Commission approved commercial television broadcasting in 1941. After the war, which interrupted the development of the technology, television quickly reached maximum market penetration. Whereas only 9 percent of American households owned television sets in 1950, 30 years later that figure has risen to 98 percent (Figure 1). The 1979 Washington Post poll of 1693 persons found that only 1 percent had no working television sets in their homes, over 50 percent had at least two working television sets, and only 2 percent said they never watched television at all (Sussman 1979).

Time allocated to television viewing has grown steadily since the introduction of the technology, though the rate of growth has slowed since the mid-1960s. The average daily household use of television in 1979-1980 was six hours and 35 minutes (Figure 2). Since 1970, the average individual in a television household has spent approximately three hours viewing per day (Roper 1979), more time than is spent on anything but work or sleep.[1] Longitudinal time budget studies indicate that television consumes daily more than 90 minutes of time that was formerly devoted to other activities, such as sleep, visiting, and the use of other mass media (Robinson 1972). By 1978, Americans allocated one-half of all their free time to the mass media, and television received more of this time than all other mass media combined.

Disaggregation of the gross figures on television ownership and viewing reveals several interesting trends. According to the 1977 Nielsen report (Nielsen 1977), television viewing varied by sex and age but not significantly by income level. Women watch television more than men; men and women over 65 watch more than any other age group; and children 2-5 years old watch more than older children or

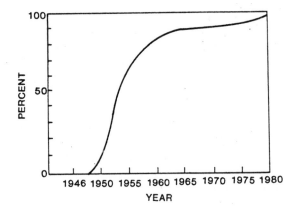

Figure 1. Percent of U.S. homes with one or more television sets, 1946-1980. Source: Adapted from Liebert et al. (1973) and Nielsen (1981).

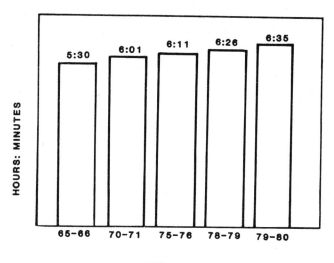

Figure 2. Average hours of household TV usage per day. <u>Source:</u>
Nielsen (1981,6). Reproduced by permission.

teenagers. Other studies (Greenberg and Dominick 1969;Bogart 1972)
have found that viewing is higher among minorities and varies inversely with socioeconomic status and education. The trend over
time, however, has exhibited convergence of viewing times for these
subpopulations toward the higher end of the scale. Though there may
be a saturation point in quantity of viewing time, the increase in
multiset households and the related increase in the number of hours
that individuals spend watching television suggest a qualitative
change in the viewing experience.

This remarkable growth of the technology is due in no small
part to its varied roles, as companion, consoler, eye on the universe, diverter, educator, entertainer, babysitter, filler of un-
structured time, and addicter. Prominent among these roles may well
be television's importance as "electronic fireplace," the modern
equivalent to the campfire (Sneed 1979). Indeed, the particular
information transmitted by television is often incidental, perhaps
even obstructive, to the relaxation it affords from the hectic pace
of everyday life. For many individuals, television is a pleasant
source of background noise, and, for the lonely, it offers electronic companionship and serves as a primary source of recreation (Davis
and Kubey 1982).

Whatever its role, television is indisputably central to American society. Whereas television in 1960 was seen as the most enter-
taining of the mass media (including newspapers, radio, and magazines), by 1974 the Roper poll revealed that the public also accorded television news two and one-half times the credibility accorded
to newspapers (Roper 1979). A 1979 survey of public attitudes
toward some 50 types of leisure activities found that watching
television ranked at the very top (Brown 1980).

Although the general public attitude toward television has been favorable over time (Comstock 1978), nagging concerns persist about the role of television in American life and about its impacts on society. Television in the United States is largely an entertainment rather than an educational medium. Viewers voice concern about the medium's unfulfilled educational potential; perhaps they even feel a little guilty that a technology with such compromised virtues commands so much of their time (Opinion Research Corporation 1978). Over one quarter of the respondents to a recent Washington Post poll indicated frequent disappointment with the shows they watch (Sussman 1979). There is evidence of substantial resentment over television commercials (Roper 1979). Finally, there is the understandable fear that a relatively new technology with such wide acceptance poses unknown but potentially serious risks to this society and its members, especially children.

The Social Risks of Television

The risks of television are associated with the release of information rather than the release of materials (as with toxic materials,) or energy (as with radiation or collisions), both of which receive substantial attention in chapters 2-4. Such information risks reflect par excellence a newly emerging class of societal hazards, one which includes such technologies as electronic surveillance, computer records, CB radios, photocopying, and electronic funds transfers.

The problems in assessing television risks resemble those involved with toxic substances such as PCBs (chapter 15) and mercury (chapter 9): one usually encounters a substantial time lag between exposure and the onset of effects; it is difficult to isolate the "toxic" information received from television from the "doses" received from other sources (in this instance, other socialization agents); current theory is inadequate to establish the precise mechanisms by which exposure converts into a particular consequence; and there is no general agreement about the best indicators for dose or for the various effects. Since the poorly understood processes of human development mediate between exposure and consequences and since control groups untouched by television are virtually unavailable, it is not surprising that the risks of television have generally resisted attempts to define their nature or magnitude. Most of the effort expended on the study of television risks has centered upon specific "messages" or program content, rather than on the medium itself (Figure 3). In particular, televised aggressiveness and violence have been subjected to lengthy research; to a lesser extent, advertising to children and sexual, racial, and occupational stereotyping have also received attention. For such risks, major state-of-the-art assessments have been completed. The following brief exploration of the risks of the medium and of violence and advertising to children affords a glimpse into television risk assessment and the difficulties it presents.

The Hazards of the Medium

Most research on television risks has been devoted to exploring the relationship between specific types of programming content and

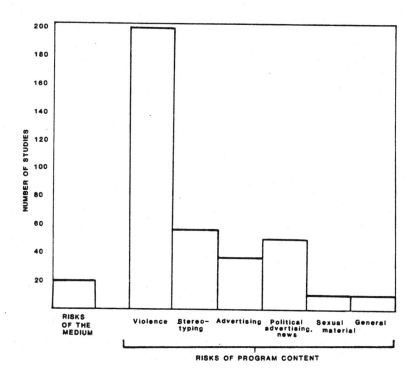

Figure 3. Distribution of risk assessment effort, as indicated by a 1977 ERIC search.

various undesirable social consequences. The hazards of exposure to the medium itself--that is, the risks of watching television regardless of content--have rarely been the subject of intensive investigation. And yet the average American experiences more than three hours per day of exposure to the television medium; and the average high school graduate will have spent approximately 22,000 hours in front of the set, as opposed to 11,000 hours in school (Adler et al. 1980). Such widespread and intensive exposure carries obvious potential for a serious hazard. Yet a major study by the National Institute of Mental Health (NIMH 1982,1:8) turned up no experimental study relating television viewing to IQ.

Critics of television have attempted to attach a wide range of negative consequences to exposure to the television medium. Television has been accused of contributing to the disintegration of family life and the breakdown of community; it has been blamed for increasing the passivity of viewers, for undermining their creativity, and for impeding the development of children. Some critics claim that television creates a psychological dependency, or addiction.

Such claims are not buttressed by any substantial body of empirical evidence, in part because of insufficient concern on the part of the public and policy makers. Research funds have flowed in the direction of risks with more visible social consequences or more clearcut policy implications. In addition, the alleged negative consequences are often hard to define. Researchers have not established a consensus about what constitutes passivity or creativity, and there is no generally accepted understanding of the processes of human development. Many of the questions and criticisms directed at the medium are concerned with holistic phenomena, such as breakdown of community. Such issues thwart attempts at risk estimation; they remain unresolved, as nagging concerns.

Research on television and passivity provides an interesting example of the primitive state of hazard assessment for the television medium. Some investigators have observed a high level of alpha activity in the brain waves of television viewers. The alpha state is associated with relaxed and unfocused behavior and with limited mental activity (Krugman 1971). But brain-wave studies that show minimal brain activity do not necessarily imply that viewers are more passive overall than are nonviewers who may simply seek another source of relaxation. Other studies question the contribution of television to passivity. Platt (1975) emphasizes the degree to which television provides knowledge directly relevant to value choices and policy decision, whereas Lyle and Hoffman (1972) show that viewing by young people is highly active and discontinuous, not passive.

Televised Violence

Past research has dealt at considerable length with children's heavy viewing of televised violence and subsequent aggressiveness and antisocial behavior. Indeed, the research in this area clearly dominates all risk-related research on television. Much of the research has occurred in laboratory settings and has generally indicated an increase in short-term aggressive behavior among young viewers. The extent to which such results are generalizable to real-life settings is questionable; they tend to be suggestive rather than confirmatory, reflecting short-term responses to the viewing of violence under artificial conditions.

The Surgeon General's assessment, which included some 25 individual research projects, concluded that there was evidence of a causal relationship between exposure to television violence and aggressiveness. It also concluded, however, that the impact was likely to be greatest on children who were already predisposed to behave more aggressively. Other overall assessments have led their authors to similar judgments. A "headcount" of some 67 studies linking exposure to televised violence to aggression, divided between those reporting negative or inconclusive results and those with positive results, found that the majority fall into the latter group (Andison 1971). The 1978 state-of-the-art assessment by Comstock and associates concluded that:

- studies have consistently found a relationship between viewing of violence and various measures of aggression;
- the way in which television violence is portrayed (e.g., as punished or as rewarded) can inhibit or enlarge the probability of subsequent aggression by the viewer;

- those most influenced by televised violence are boys, younger children, and more aggressive children;
- heavy exposure to television violence conceivably may desensitize children to the negative consequences of real-life violence (Comstock et al. 1978).

Over the years, well-publicized incidents of violence have allegedly been imitated from television or films:

- In 1956, a Chicago boy jumped to his death trying to imitate Superman
- A girl in England died from injuries resulting from a fall from a three-story building while she was playing Mary Poppins
- In 1978, a court case failed to resolve whether a bizarre sexual assault on a nine-year-old girl was imitated from a 2.5-minute scene broadcast some four days before the assault
- Two nights after the airing of Fuzz on television, a youth gang re-enacted the dousing with gasoline and burning to death of a young woman
- A 1979 movie called The Warriors, which graphically portrayed the violent subway journey of a gang in New York City, was accompanied by such widespread outbreaks of violence that Paramount suspended advertising and offered to pay for extra security at the 670 movie theatres showing the film

The risk of imitative violence is relatively amenable to assessment because of the short time between exposure and the onset of consequence. It is, in this sense, an **acute** hazard. Nevertheless, research on television has generally neglected the study of incidences of violence and has concentrated instead on laboratory simulations or studies of long-term socialization processes, both of which present serious methodological difficulties. A careful "epidemiological" study that might relate television viewing to the social incidence of violence remains to be undertaken.

Studies that attempt to measure long-term socialization effects are more difficult to control but produce findings that bear more directly on actual behavior. The most ambitious study of television's long-term effects on violent behavior is William Belson's (1978) six-year investigation of 1,565 teenage boys in England. The boys, aged 12 to 17, were paid to record how many specific episodes from 68 television series they had watched between 1959 and 1971. A 50-person panel drawn from ex-members of the BBC viewing panel rated each of the programs for **amounts** and **kinds** of violence. The boys were split into two groups (highly exposed and less exposed) and evaluated according to the acts of violence they had committed during the six months prior to the interview. Then came a painstaking effort to sift out television's impact from 22 other stable correlates of violent behavior and from possible inverse relationships (e.g., predisposition to violence causes more viewing of violence). A number of important findings emerge from the study, including some that discriminate among types of dose:

- Portrayal of violence that occurred in the context of close personal relationships and presented very realistic violence with which the boys could identify was particularly linked to subsequent violent acts.
- Forms of program violence that did not appear to promote subsequent serious violence are comedies featuring slapstick violence, violent cartoons, sports other than boxing or wrestling, and science fiction shows.
- Whereas heavy exposure to televised violence appeared to relate to subsequent violent behavior, it did not apparently change boys' conscious attitudes to the acceptability of violence or create a more callous mindset for acts of violence.

Interestingly, the study also provides some evidence of a recurring phenomenon in hazard management, namely, the presence of a threshold limit value (TLV) for dose; the level of serious violent behavior apparently increases only after an accumulation of a substantial amount of viewing of televised violence.

Another ambitious effort is the statistical treatment by Susan Hearold (1979), of some 230 experiments and surveys on the effects of televised violence. This assessment analyzed the degree as well as the direction of reported relationships and covered more than 100,000 individuals included in the studies. The results largely confirm the findings of the Surgeon General's report, which the NIMH (1982) has recently validated. The viewing of violence was positively associated with antisocial behavior and negatively associated with prosocial behavior.

Not all studies point to a positive correlation between television viewing and aggressive and violent behavior. One large-scale, technically sophisticated study (Milavsky et al. 1982) found no such evidence of a causal link to implicate television in the development of aggressive and violent behavior patterns in children and adolescents. This research included the compilation of data on several hundred elementary-school boys and girls and teen-age high-school boys. Levels of aggression--measured at multiple points in time-- were related to television program content, assessed independently for level and type of violence.

In all these studies, the magnitude of the dose is a critical issue. Unfortunately there are no generally accepted indicators of "violence" (is a pie in the face or a verbal insult a violent act?), and dose trends are quite sensitive to the indicator chosen. The most widely used (though not accepted by the networks) indicator-- that of George Gerbner and associates--centers on physical violence or the "credible threat" thereof. With this rather broad definition, Gerbner's data indicate that some 85 percent of all television shows and some 65 percent of all characters were associated with violence in 1978 (Gerbner 1978). Despite a general belief that violent programming has declined, time trends using this measure indicate little change in the ten years since the Surgeon General's report (NIMH 1982,1:37), except perhaps a reduction in the killings portrayed (Figure 4).

We conclude that there is enough accumulated evidence to suggest that televised violence contributes (probably in conjunction with other sources) to increased aggressiveness, although the exact

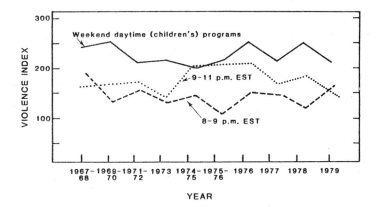

Figure 4. Violence in children's and prime-time programming, 1967–1979. Source: Gerbner et al. (1980,13). Reproduced by permission.

nature of dose/response relationships and necessary conditions are not now known. In this respect, we concur with the NIMH finding that "television violence is as strongly correlated with aggressive behavior as any other behavioral variable that has been measured. The research question has moved from asking whether or not there is an effect to seeking explanations for the effect" (NIMH 1982,1:6). Except for imitative (i.e., acute) effects, further research is unlikely to clarify in more than a marginal way the magnitude and causality of the risk. But, as the saccharin case suggests, even "weak" toxicity can constitute a significant hazard if the dose is sufficiently widespread. The management problem appears to lie primarily in determining risk tolerability rather than in further risk estimation—is there a sufficiently compelling reason on the benefit side to continue a dose of potentially toxic information at current levels of magnitude? Since programming violence is largely a by-product of network efforts to capture a larger share of the market, it appears quite similar to other industrial "externalities," such as chemical wastes and air pollution.

Televised Advertising

As with other U.S. mass media, television exists not only to inform and entertain the public but to sell products. In theory, the technology should be able to accommodate these various functions. In a capitalist society, advertising is a way of life, and the American public has adjusted readily to its ubiquitous nature. In practice, however, no other communications technology has joined advertising with the pervasiveness and intrusiveness of television.

Public disenchantment with television advertising is increasingly apparent. A 1979 Washington Post poll found an 18-year trend of growing public resentment of television advertising (Sussman 1979). Asked whether they agreed or disagreed with the statement "Commercials are ordinarily in poor taste and very annoying," 54 percent of those interviewed agreed as compared with 40 percent in 1960 and 43 percent in 1970. The sources of public disenchantment lie not only in the fragmentation of programs and the intrusiveness of the sales pitch, but in unease over possible social hazards. Unlike televised violence, advertising has only recently come under close scrutiny in the publicly available literature. That litera- ture has noted a wide variety of possible impacts, including distortions in consumer choice (such as in dietary habits and nutri- tional criteria), accidents attributable to advertising unsafe products, increased child-parent conflict stemming from denial of children's wants, and long-term societal value change in the direc- tion of greater materialism and consumption. The studies to date, however, have generally succeeded in relating only intermediate variables (attention, cognition, affect), and not behavior itself, to exposure to television advertising.

The sketchy evidence that does exist suggests that most adults are capable of evaluating advertisements and selecting among them. Not so with young children, however, for research indicates clearly that they do not comprehend the purposes of television commercials, are unable to distinguish between progamming and advertising, and tend to believe and trust the commercial message indiscriminately. A state-of-the-art risk assessment (Adler et al. 1978), supported by the National Science Foundation (NSF) included among its major findings that

- television advertising does influence children. Chil- dren attend to and learn from commercials; advertising is at least moderately successful in creating positive attitudes toward and desire for products advertised;
- the strongest determinant of children's perception of television advertising is the child's age. Children under eight years of age have substantial difficulty in comprehending the difference in purpose between commer- cials and programs;
- heavy television viewing seems neither to retard nor accelerate children's understanding of commercials, though it does seem to produce more favorable attitudes toward advertising and advertised products; and
- relatively little mediation of children's exposure and reactions to commercials occurs in most households.

It is for these reasons that most industrialized nations have his- torically forbidden television advertising directed specifically at preschool children.

Given the susceptibility of young children to adverse conse- quences, the nature and extent of the dose assumes considerable importance. The NSF-sponsored assessment estimated that children on the average are exposed to some 20,000 commercial messages each year, or slightly more than three hours of television advertising each week (Adler et al. 1978). The sheer volume of televised adver- tising viewed by children acquires greater significance when the content of the messages is considered. The NSF-sponsored assessment (Adler et al. 1978) found that of the total advertising aimed at children 25 percent is for **candy/sweets** (e.g., cakes, cookies, fruit drinks), 25 percent is for **cereal** (mostly presweetened), and another

10 percent for eating places (principally fast-food restaurants). The remaining major category of products is toys, for which advertising is heavily concentrated just before Christmas. In its analysis of some 7,515 food commercials aimed at children during weekend daytime hours in 1975, the staff of the U.S. Federal Trade Commission (FTC) estimated that only 4.3 percent were not for sugared products (FTC 1978a). Basic foods (e.g., fruit, vegetables, fish, and meat), by contrast, are rarely advertised nationally, whether to children or to general audiences.

Quantitative assessments of the health risks involved are unavailable. The FTC's rule-making procedure in the late 1970s (FTC 1978b) focussed on the contribution of television advertising to the increased consumption of sugar (estimated in 1977 as 128.1 pounds per person) and its role in the rising incidence of tooth decay, obesity, coronary heart disease, and diabetes. Although these proceedings spawned no specific statement on the risks of televised advertising, they left little doubt that television advertising was a significant contributing agent to nutritional problems—and this at a time when the Director of the National Cancer Institute had issued nutritional guidelines for improved national health.

Managing the Risks

To understand the history and potential of television hazard management, it is necessary to examine the political economy of television, including the structure and power of the television broadcasting industry, and the regulatory context, including the statutes and institutions that bear on the technology. Within these general contexts, an assessment of the managerial performance to date may be made.

The Political Economy of Television

The television broadcasting industry is oligopolistic in structure, characterized by concentration of revenues and centralization of control in three national networks—CBS, NBC, and ABC. These networks produce most programming and provide local affiliated stations with programming that carries advertising. The industry also involves smaller networks or group owners, such as Westinghouse (Group W) and Metromedia. Of the more than 700 commercial stations now operating, 85 percent, including almost all VHF stations in major markets, are affiliated with networks.

The production companies that create television's dramatic content are ostensibly independent of the broadcast industry. Yet one of three commerical networks usually provides the funds for program creation and acts as the original distributor. Thus the creators of dramatic fare must satisfy the production companies, which in turn must satisfy the networks, before the program reaches an audience. The networks then transmit the programs, replete with commercial advertisements, to network-owned or affiliated local stations (Cantor 1982,351).

In 1952, the U.S. Federal Communications Commission (FCC) established the present VHF-UHF television system, which allocates VHF channels in a manner that limits actual and potential competition. Many large-market areas were allowed only three commercial

VHF stations, resulting in the fact that only 30 percent of all television homes can receive four or more commercial VHF stations. The FCC does not acknowledge that it deliberately created monopoly power in television broadcasting, but it has, over the years, either failed to act upon or has rejected outright all proposals aimed at increased competition (Crandall 1978). The rationale for this protection has been that "discretionary" revenues from entertainment have been needed to subsidize allegedly unprofitable programs deemed to be in the public interest.

In fact, the television industry is highly profitable. Even during the 1974-1975 recession, television broadcasting outperformed the rest of the economy, and pre-tax profits currently stand at record levels. The distribution of revenues reflects the concentration of the industry--the three networks, the five stations each is allowed to own, and the network affiliates account for 91 percent of the annual profits of the entire television industry (FCC 1979).

Centralization of control over programming in the industry is largely related to the costs of producing the types of programs currently broadcast: one half hour of prime-time programming costs approximately $150,000 to produce, but occasionally reaches as high as the $660,000 per episode required for **Battlestar Galactica.** Local stations cannot offer advertisers an audience large enough to make such expenditures worthwhile. This means that community-based attempts to control television hazards will tend to be ineffective. Though a television station may seem to be part of the local business community and therefore susceptible to local pressure, it has little actual control over programming, and seldom exercises whatever control it has. The FCC study found, for example, that affiliates gave clearance to 95 percent of network prime-time shows and 87 percent of network offerings during off-peak hours (FCC 1979).

In this highly concentrated industry, the driving force in choice of programming is not a conception of the public interest (such as meeting diverse needs) but the competitive struggle among networks to maximize profits over the short term. The keys to success are performance in the ratings sweeps held three times a year when advertising rates are set. Programs that fail to attract 30 percent of the audience are dropped, regardless of merit. The result is a competitive struggle in which networks program in lockstep, aiming for habitual, heavy rather than selective viewers, cloning shows from successful predecessors and lifting out successful characters from existing shows to seed new ones. Experimentation, alternative programming, and public-service-oriented programming are avoided on the grounds that the audience cannot be guaranteed.

The economic power of the television industry is manifested not only in its balance sheets but also in its relation to Congress and potential hazard managers. Television broadcasting exerts an extremely powerful influence in Congress, one of its prime hazard managers. Most legislators take seriously the power of local broadcasters in their home states to promote or damage their chances of reelection through selective reporting of their stands on particular issues. A large majority regularly use free air time that broadcasters provide under the guise of "public affairs" programming. Thus, the industry not only resembles other industries in maintaining a powerful lobby in Washington, but television also has at its disposal the means of punishing or rewarding individual legislators for their expressions of concern or for their active intervention.

The FCC suffers from the "revolving door" problem that characterizes many regulator-regulatee relations. Most FCC commissioners upon retirement become affiliated with the industry, either as employees or as communications lawyers practicing before the Commission. Between 1945 and 1970, 33 commissioners retired from the FCC; 21 of these found industry-related employment and most of the other 12 retired from professional life. From 1970 to 1977 nine commissioners left the FCC, and five of them became affiliated with the communications industry (Cole and Oettinger 1978). Though the prospect of future employment would not necessarily render the commissioners sympathetic to the industry, it is clearly not in their interest to confront the industry in an aggressive or offensive manner.

A Self-Regulatory System

The First Amendment dominates the regulatory framework of television hazard management. This amendment, which guarantees individual freedom of speech, guarantees within certain limitations the same right to broadcasters and other media. Thus, the television hazards of most concern to the public—particular types of program content—enjoy explicit protection from intervention. An exception to this rule is the content of advertising.

Other constraints on hazard management emanate from the statutory mandates of agencies, such as the FCC and the FTC, which have responsibility for television. Not only is the FCC constrained from intervening in programming content, but its jurisdiction is limited to stations. The Commission was created primarily to allocate spectrum space among competing uses. Since the networks themselves are not broadcasters operating over the airwaves, they elude the FCC's direct regulatory grasp.

The FCC's principal regulatory activity is licensing of stations. All stations must apply for license renewal every three years, undergoing an essentially pro-forma process characterized by self-monitoring and extensive paperwork. Predictably, the procedure seldom yields license denials. Only the statutory requirement that stations must operate in the public interest allows the FCC's regulatory apparatus to approach programming content, and at best the requirement means that broadcasters must provide each week a certain amount of public-affairs and community-oriented programming.

In the case of the FCC, the regulatory structure incorporates a curious contradiction. Television regulation rests on a principle of decentralization, focussing on stations and their local markets, whereas the industry and the production of programming are highly centralized. Because the regulatory framework tends to channel effort in the direction of local stations, citizens concerned about television hazards have concentrated many dollars and hours of effort on "petitions to deny" station licenses. The individual station, however, is only nominally the locus of control over programming; such local attacks at best constitute weak signals to the center of control that certain outposts have exceeded appropriate bounds.

In the case of the FTC, a similar problem has scattered and vitiated regulatory energy. Though the FTC has jurisdiction over television advertising, it did not engage in industry-wide rulemaking until the 1970s because it was unclear under the Federal Trade Commission Act whether the Commission had rulemaking authority. As a consequence, the FTC pursued a case-by-case approach or issued industry-wide guidelines that did not have the force of law. Considering the volume and diversity of television advertising, such an approach is a highly inefficient and patently indirect route to the elimination of alleged hazards. In 1975, however, Congress enacted the Magnuson-Moss FTC Improvements Act to guarantee the Commission broader regulatory powers. Those powers were tested in the FTC investigation and the proposed rulemaking on advertising to children, and, one must conclude, found wanting.

Given these various constraints on public regulators and the political clout of the industry, it is not surprising that self-regulation has emerged as the characteristic means of hazard control. In other words, the television industry itself has the primary responsibility for hazard management. The self-regulation system is embodied in the "broadcast standards" departments of the network production units, which review and edit all network programming according to guidelines based on criteria of public acceptability, and in the National Association of Broadcasters (NAB) Code Authority. Approximately two-thirds of all television stations are members of the NAB and they subscribe voluntarily to the NAB Code, which provides guidelines on both programming content and advertising. Failure to adhere to the guidelines can result in expulsion from the organization but carries no direct economic sanctions.

Public regulators, confronted by ill-defined risks and frustrated by legal and political obstacles, have tended to deal case-by-case with risks, rather than adopting a comprehensive approach to the array of risks that television presents. The selection of what risk to work on has been guided more by a hope of assembling sufficient political clout to have some impact than by a need to reduce those risks that pose the greatest dangers. These characteristics of the self-regulatory systems are quite apparent in efforts to cope with televised violence and advertising to children.

Managing Television Violence

From the early days of the medium, managerial attention has focussed on televised violence. Congress, spurred by widespread concern over the social problems of violence and crime, has conducted at least eight major investigations of televised violence. Congressional hearings have consistently sought more information about the nature of the hazards and produced repeated admonitions to the industry to regulate itself.

Attempts at improved understanding of the risks have emerged from these investigations. In 1963, the Joint Committee for Research on Television and Children, composed of representatives from The Department of Health, Education, and Welfare (HEW) and the three major networks, was established at the behest of the Senate Judiciary Committee. As late as 1969, however, no reports or recommendations were forthcoming. In 1968, in the wake of the Kennedy and King assassinations, President Johnson appointed the National

Commission on the Causes and Prevention of Violence and that Commission in turn set up a Task Force on Media. After reviewing all existing research on the effects of violence on behavior, the Task Force concluded that violence on television indeed encouraged violence in society. In 1969, Senator Pastore, Chairman of the Senate Commerce Committee's Subcommittee on Communications, called for a major program of research on the subject of television and violent behavior, resulting in the ambitious $1.8-million Surgeon General's assessment <u>Television and Social Behavior</u>, completed in 1971 (Comstock and Rubinstein 1972).

The 1972 Senate Subcommittee on Communications hearings on the report's findings eventually led to the recommendation, a familiar one, that the industry should regulate itself in this area (U.S. Congress 1972). The subcommittee also recommended that an outside research organization monitor television content to produce a violence index that would keep Congress and consumers informed about the success of industry self-regulation. During the hearings, FCC Chairman Dean Burch agreed that the FCC should take some action, the nature of which was only vaguely specified but which included establishing a children's unit.

Over the next two years, despite much conferring and correspondence among Pastore, HEW, and the FCC, little happened. The "violence index" was not developed, although in 1974 the National Institute of Mental Health (NIMH) promised to produce such an index within a year or two. The FCC established a children's television task force but delayed producing a policy statement on televised violence. The industry, for its part, gave signs that it would address the problem: industry-sponsored research increased, the networks promised to reduce violent programming and established children's programming advisory boards, and the 1973-1974 television season spawned new family-oriented programs that achieved high ratings. These signs of change proved misleading, however, for televised violence quickly climbed to pre-1974 heights and remained there for the rest of the decade.

Sporadic congressional attention continued during the 1970s. In 1974, the House and Senate Appropriations Committees, concerned over the FCC failure to act, directed the FCC to report on actions it had taken or planned to take by 31 December 1974. This ultimately led to the institution of the family hour in 1975. According to this provision of the NAB code, the first hour of prime time, as well as the access hour before it (i.e., between 7:00 and 9:00 p.m., Eastern Standard Time), should contain programming appropriate for general family viewing. The provision was struck down in 1976 by the Federal District Court in Los Angeles because the restrictions were partially imposed through threats by the government. Judge Ferguson in his decision, however, noted the provision was largely a "public relations gimmick," since it covered only a fraction of children's viewing. Congress again held hearings in 1976 and in a subsequent report <u>Violence on Television</u> (U.S. Congress 1977) recommended rather weakly that

- research on amounts and effects of television violence should be deposited with the Library of Congress;
- NSF and NIMH should continue to support pertinent research;

- parents should be encouraged to participate positively in their children's viewing and to write to local broadcasters about offensive programs;
- the FCC should consider rulemaking as to whether television sets should carry locks and whether an unobtrusive program guide (such as a white dot on the screen) should be transmitted by broadcasters of adult material; and
- the FCC should complete rulemaking regarding increased opportunity for local network affiliates to prescreen network programming.

Confronted by continuing inaction by the managers of television hazards, various public and professional groups have begun to assume stronger stands. In 1974 the National Parent Teachers Association initiated a monitoring program at 25 television stations throughout the nation in an effort to force stations to reduce programming with excessively violent or sexual content. Later it turned its attention to publicizing the advertisers who buy time on such programs and to teaching parents how to challenge the licenses of network-owned TV stations. In 1976 the American Medical Association (AMA) adopted a resolution stating that televised violence is a risk factor threatening the health and welfare of young Americans (Washington Post 1976), and AMA President Palmer in 1977 called on advertisers to reduce their sponsorship of violent programming content (Bogart 1980). A major public-interest group, the National Citizens' Committee for Broadcasting, adopted the strategy of linking advertisers to particularly violent programming, and several major advertisers (as well as the J. Walter Thompson Advertising Agency) voluntarily refused to sponsor programs that contained excessive levels of violence. Despite these efforts, televised violence has not subsided, and it is uncertain whether the 1980s will see increases or declines.

Managing Advertising to Children

Television advertising to children differs from televised violence in several significant ways: as an issue, it is relatively recent in origin; it developed as a part of the management agenda through the persistent leadership and prodding of public-interest groups rather than through legislative and executive directive; and it is more susceptible to regulatory action, since advertising content is not protected by the First Amendment. Two very active public-interest groups, Action for Children's Television (ACT) and the Council on Children, Media, and Merchandising, generated activity in this area in the early 1970s. Primarily as a result of their endeavors, the FTC became involved in a rulemaking proceeding to consider (1) banning all television advertising directed at audiences comprised of a significant proportion of children who are under eight, (2) banning televised advertising--directed to or seen by audiences composed of a significant proportion of older children--of sugared products, and (3) requiring nutritional and/or health disclosures funded by advertisers (FTC 1978a).

Public-interest groups involved in this proceeding expressed concern over using children as tools to promote adult purchasing and over selling products to children who have not reached the age at which they understand the function of advertising. Though these groups had presented their objections to the FCC, the FTC, the NAB, and advertisers, a serious response was not forthcoming until 1978 when the FTC staff issued its report and recommendations.

As with televised violence, management of advertising risks has occurred through self-regulation, despite the existence of rulemaking authority vested in the FTC and FCC. The NAB's Television Code Review Board responded to pressures from FCC and public-interest groups by issuing in 1974 voluntary guidelines that reduced from 16 to 9.5 minutes per hour the limit on commercials during weekend children's programming and prohibited "host selling" by a figure who was an object of children's respect. The FCC buttressed NAB self-regulatory attempts with its 1974 <u>Report and Policy Statement on Children's Television</u> (FTC 1974), a belated response to an ACT petition of 1971. This statement applied the NAB time standards to all stations and stipulated a clear separation between program content and advertising on children's programs. In 1976, the National Advertising Division of the Council of Better Business Bureaus issued comparable guidelines for advertisers.

ACT had originally petitioned the FCC to issue a rule with the force of law prohibiting advertising to children, but the FCC instead merely brought forth a policy statement that attempted to formalize and generalize self-regulation. The results have not been encouraging. Code subscribers and other stations violate with impunity the NAB time standards for commercials (Barcus 1978), and the restrictions on advertising apply in any case to no more than 20 percent of children's viewing time. It is likely that children during the second half of the 1970s were exposed to as much or more commercial materials than they were prior to the compliance date of 1 January 1976. It is not surprising, then, that ACT and the Center for Science in the Public Interest (CSPI) petitioned the FCC in 1977 to initiate rulemaking to extend the types of programming covered by the guidelines, reduce commercial minutes on weekday programs, and phase out commercialization on all children's programming.

Until recently, public-interest groups dealing with the FTC have achieved only qualified successes. In 1971 and 1972, ACT filed petitions with the Commission to make Trade Regulation Rules prohibiting the selling of toys and edibles on children's commercial television programs. By 1975, when ACT had received no response from the FTC on either of the petitions, they sued the Commission for inaction on the edibles petition, and the Commission denied the petition without a public hearing. ACT has also pursued the course of filing against individual companies specific complaints for hazardous advertising, such as vitamin commercials directed at children. By February, 1976, ACT had submitted three petitions and at least seven specific complaints to which the FTC had not responded in any way, except to issue a negative ruling when threatened with a suit. With this record, the 1978 rulemaking on advertising to children constituted a clear reversal of direction. Chairman Michael Pertschuk, appointed by President Carter in 1977, adopted the issue to inaugurate his new leadership of the Commission, responding positively to the ACT and CSPI petitions.

The history of that rulemaking venture indicates the industry's impermeability to regulatory attempts to control television hazards.

The stakes are high. Television advertising to children brings broadcasters annual revenues of some $600 million, with the breakfast cereal industry's contributing $172.5 million to that figure (McDowell 1978,D1). Broadcasters argue that networks would experience difficulty recouping losses if advertising to children were eliminated entirely, a conclusion confirmed by a 1972 FCC study (Pearce 1972); such a regulatory structure, it is claimed, would have the incidental effect of eliminating programming produced specifically for children. A later report by the FCC argued, however, that a reduction of advertising on children's programming from 16 to six minutes per hour would not materially alter network finances (Pearce 1974).

Despite this apparent flexibility in network budgets, the industry mounted a powerful campaign to oppose the proposed rule. The sugar and food industries joined with the broadcast industry in an intense lobbying effort with the members of the congressional oversight committees. Three national advertising trade associations (the Association of National Advertisers, the American Association of Advertising Agencies, and the American Advertising Federation), the Kellogg Company, and the Toy Manufacturers of America (a $3.5-billion industry with some 250 members) asked FTC Chairman Pertschuk to disqualify himself from participating in the inquiry on the grounds of fairness and, when he refused to do so, brought the case to a federal court.

This campaign brought quick results. In November 1978, U.S. District Judge Gerhard Gesell ordered Chairman Pertschuk to disqualify himself on the grounds of bias. Since another commissioner had previously disqualified himself (because of a prior association with one of the original petitions) and yet another had retired in March 1979, the decision left the FTC with only two members available to participate in the inquiry.

Meanwhile in Congress, a House subcommittee recommended that Congress prohibit the FTC from spending any money on the television advertising inquiry, and the Senate also expressed concern over the various FTC proposed bans. In September of 1979, the House State, Justice, Commerce and Judiciary Subcommittee approved a proposal prohibiting spending on a series of FTC planned and current actions, including the television advertising procedure. The Senate delivered the **coup de grace** two months later when the Senate Commerce Committee voted overwhelmingly to prohibit the Commission's acting against "unfair" advertising, an action that effectively killed the FTC investigation and was widely interpreted as the worst congressional defeat for the consumer movement since the 1977 failure to approve the proposed Consumer Protection Agency (Kramer 1979). In April 1981 the FTC staff recommended that the Commission drop its investigation, holding that "...there do not appear to be, at the present time, workable solutions which the Commission can implement" (Sulzberger 1981).

Continuing its well-honed tactic of meeting public or regulatory pressure with token self-regulatory responses, the industry announced several new measures. Effective in January 1980, ABC unilaterally instituted a two-step, 20-percent cutback in the amount of advertising aimed at children during Saturday- and Sunday-morning programming. The resulting free time has been used for special messages on nutrition, health, public service announcements, or general program information. ABC also proposed a standardization in the use of **separators** (usually lasting between three and 12 seconds) between children's advertising and programming. These modifications leave intact the basic structure that creates risks, however, and will have little impact upon the dose to young children. As with televised violence, efforts to control risks have failed, and there is every reason to expect the current hazard to children to continue unabated or even to increase.

Our analysis of television hazard management provides little ground for optimism that the risks will soon be controlled. The extant system of self-regulation has functioned primarily to resist regulatory intervention and to ensure continued profits rather than to reduce or mitigate hazards. This failure in hazard control lies partly in the poor understanding of the risks, partly in the various constraints on regulation, partly in the political power of the constellation of economic interests in the broadcast industry and those it serves, and partly in the ambivalence of the public and the lack of a powerful consumer lobby. Given this reality and the deregulatory efforts of the Reagan administration, future efforts to engage television hazards may well have to utilize the market rather than the Federal Register.

The Coming Revolution

There is widespread consensus that television has encountered "the kind of storm that swept Dorothy off to Oz" (Smith 1979) a technological revolution that will outmode the present experience of programming risks as well as existing institutional structures. The technological developments now under way are quite remarkable both in pace of change and potential impact:

- **Communications satellites:** Since 1972, the FCC has had an **open skies policy** that has permitted the operation of domestic communications satellites. By 1983, 15 North American satellites were beaming television programs 24 hours per day to cable systems distributed throughout the United States, competing directly with offerings from the three national networks. In 1979, Comsat launched a $600-million program to provide direct satellite-to-home pay programming--a transmission system that would offer to consumers three new original programming networks and inexpensive receiving dishes--and promises to outmode existing television program delivery (Brown 1981).
- **Growth of cable television:** In 1970, 2,490 cable television systems served some 4.5 million American homes, and the growth prognosis was not bright. But this was before the linking of cable to satellites, a development that made pay-TV movies and entertainment programs available nationally. Cable television mushroomed, reaching some 2.9 million households by 1983, and new subscribers were joining at the rate of 300,000 per month (Marcus 1983). Approximately 34 percent of U.S. residents received cable in 1983, with half the homes having access to cable subscribing to it, and overall

industry revenues reached $436 billion by the end of 1982 (Kahn 1983). Offering a choice upwards of 100 channels, satellite-linked cable television can offer narrowcasting--highly diverse programs to specialized interests, a capability of great interest both to the public and to the advertising industry but one whose promise is yet to be realized (Marcus 1983).

- Space antennae and super stations: When communications satellites were first orbited, earth stations--the dish-shaped space antennae that receive satellite signals--were huge (350 tons and 100 feet in diameter) and cost millions of dollars. In 1980 these antennae cost only from $25,000 to $37,000, measured a scant 15 feet in diameter, and over 2,000 existed in the U.S. Sophisticated mobile earth stations, which can both broadcast and be rotated to tune into different satellites, are now appearing. In 1979 RCA startled the television industry by offering to build and maintain at its own expense earth stations at all 725 U.S. commercial television stations (Brown 1979). By 1982, the National Cable Television Association could identify 51 national satellite-distributed cable networks, including three all-news channels, two all-sports channels, four religious networks, three movie channels, a health network, a Spanish-language network, and numerous others (Kahn 1983,49).

- Home video: It is perhaps video technology, however, which promises the greatest change in televisiion. An estimated 4.6 million video cassette recorders were already in use in the U.S. by the end of 1982, and the cost per unit has dropped from about $1,000 to $400 in 1984. The video recorder provides owners with options, thus freeing the viewer from network schedules and irritating interruptions.

- Videodisks: Equally significant are the emerging video-disk systems, which play back prerecorded video records. Videodisks enable viewers to select from a limitless universe of options, ranging from first-run movies to computer games and tennis instruction (New York Times 1980). In 1981, RCA began marketing a $500 system with disks priced at $15-$20, projecting that videodisks would be in 30-50 percent of all U.S. households and become a $7-billion industry in annual sales by 1990 (Pollack 1981). For the projected 1985 market of $4.2 billion, a competitive struggle is currently under way among RCA, Magnavox, and Mitsubishi of Japan.

What does all this mean for the average American? One suggestion is contained in the interactive cable television system known as Qube. In Qube, cable television equips each home television set with a return line that connects to a computer. Using a lap console, viewers can send back a variety of digital responses to the central computer, thereby making "talking back" (such as for public opinion polls or shopping) possible. In Columbus, Ohio, some 38,000 households have access to ten channels of commercial and public television programming, ten special pay-per-view channels with movies not yet shown on commercial television, self-help courses (How to Prepare for College Entrance Examinations), Ohio State University football games, and ten "community" channels that utilize the interactive capability of the system. Experiences with Qube suggest possible use of interactive television:

- A talk by Ralph Nader on one of these channels produced a computer printout of names and addresses of some 700 volunteers which could be handed to Nader as he left the studio.
- Rock musician Todd Rundgren had the audience evaluate his concert performance and used the responses to re-structure his format.
- The planning commission of a Columbus suburb sponsored a 2.5-hour "hearing" on a draft renewal plan, and the computer was programmed to "narrowcast" to the subscribers who lived within the suburb. Although previous public meetings had failed to attract more than 125 residents, 2,000 participated in the televised hearing, registering their preferences on their home consoles.
- American Express has experimented with a shopping-at-home program using the company's mail-order catalog of products. Planned is the integration of credit cards into the shopping system.
- A fire-alarm option for subscribers combines sensors installed in the home with a print-out of information on the location of the house, flammables inside it, and the nearest fire hydrant. An alarm in the house, meanwhile, rouses sleepers if fire occurs.

Implications for Hazard Management

It is not only the technology sponsors who espouse the "revolu-tion," but the regulators, Congress, and many of the critics as well. Frustrated by two decades of unsuccessful jousting with the industry, and viewing the current antiregulatory mood in Washington, it is understandable that the managers of television hazards see technological innovation and associated changes in the market for video products as their best bet. The best encouragement of such changes, it is argued, is to reduce rather than increase regula-tions, in short to deregulate the industry.

In 1979, the FCC issued the results of its **Network Inquiry**, a detailed five-volume analysis of network performance in 1977. It concluded that "We can expect major changes in broadcaster behavior only if broadcasters' incentives change due to an alteration of the overall structure of the industry" (FCC 1979,1:47). In 1979, the House Subcommittee on Communications also issued its long-awaited plan to deregulate telecommunications. The bill died quickly, however, the victim of resistance to change in the status quo by the broadcast industry, citizen groups who feared the abolition of the public trustee function, and indifference in the Senate. It appears that institutional reform will continue to lag behind technological change and industrial developments.

Yet, although a number of benefits of technological innovations and increased product diversity are apparent, the impact upon social risks is much less clear. At what price will these increased benefits be obtained? There has been no overall assessment of this question although the scope of impacts could well dwarf existing television hazards. It is likely, in fact, that a complex pattern of change will occur, reducing some risks, enlarging others, and creating entirely new, unanticipated risks. Consider, for example, that

- A videodisk system will increase parental control over children's viewing while potentially eliminating the barrage of television advertising directed at children.
- Prerecorded videodisks or pay-TV programming could lead to a remarkable growth of violence and pornography, increasing the societal dose of violence and exploitive sex. A 1982 study by the National Coalition on Television Violence, for example, found pay-television programming three times more violent than comparable network programs and ten times more violent than fare offered by the Public Broadcasting System (Prial 1983).
- Ghetto television may well emerge in the 1980s, as television becomes a two-tiered system of quality, based on ability to pay, in which the privileged will have greater diversity and the poor greater homogeneity and lower-quality programming.
- The growth of interactive television, linking television and computers, could lead to new abuses and erosion of privacy. Interactive television introduces capability to integrate computer records on individuals and to monitor behavior and preferences, threatening new infringements of civil liberties and more manipulative social control of individuals.
- The ability to conduct more activities, ranging from shopping to entertainment, in the home could further isolate the individual from society.
- As the risks of specific programming content (violence, sex) are either reduced or made more voluntary, the risks of the medium may well expand. Enlarged benefits inevitably will mean increased exposure to the medium—more time spent before the television set, with a reduction in the role of other media and uses of leisure.

Conclusions

Our analysis leads us to a number of conclusions regarding the assessment and management of television risks. In our view, both processes have been relatively unsuccessful. Despite the potential magnitude of television risks, no comprehensive risk assessment of the technology has been conducted to date. Instead, the societal effort to understand the risks has been seriously imbalanced, preoccupied with those risks (such as violence) provoking media attention and public concern, whereas other potentially more serious risks (impact on learning, passivity) have remained relatively unassessed and neglected. Furthermore, the history of television

risk management to date has been a record of failure—one of Johnson's "tales of woe" (chapter 13). A number of factors have con-
spired to thwart risk-management efforts: the poorly understood
nature of the risks; the constitutional and regulatory constraints
that limit the intervention of public managers; the political power
of a constellation of economic interests composed of the broadcasting industry and the consumers of its services; the lack of a well-
developed consumer movement; and public willingness to tolerate the
situation.

Despite the discouraging record, it is still possible to make a
strong case for the need for increased control over television
risks. Although the risks are poorly understood and managers are
probably condemned to operate in an environment of scientific uncertainty for the indefinite future, there are clear grounds for managerial action:

- Studies of violence and advertising to children have
  identified effects which are certainly cause for concern, though causality cannot be established and the
  risks cannot be stated in precise quantified terms.
- Even if the toxicity of televised information is weak,
  exposure is nearly universal and the average dose is
  very substantial. This means that the sum total of
  social harm could be very great, though the harm to the
  average individual may be small (or at least difficult
  to establish).
- Children are a particularly vulnerable group, for whom
  the risks are largely involuntary and for whom, in other
  risk contexts, society is particularly protective.
- Finally, the risks—whatever their precise nature and
  magnitude—are not counterbalanced by significant social
  benefits. Television is potentially a technology that
  could serve society in a number of beneficial ways; at
  the moment, however, its use is governed by a private
  industry that acknowledges no obligation to serve the
  larger social good. A significant reduction of the
  known risks appears entirely achievable with little or
  no reduction in social benefits.

Recent history suggests that risk-management initiatives of the
types already attempted are unlikely to be successful. Television
violence continues to thrive under the umbrella of the First Amendment, and Congress has made it clear, even before the Reagan administration, that it was unwilling to attack advertising to children.
Other social risks of television are only beginning to be assessed
or are still being neglected. We see no set of changes that would
make the existing self-regulatory system more responsive to the
public interest, nor is Congress, in its current mood, likely to arm
the regulators. Recognizing that television will probably, as
suggested by a 1983 FCC policy weakening its commitment to children's educational programming (Smith 1983), continue to be conceived as primarily an entertainment rather than an educational
medium, that increasing reliance upon market forces will intensify
the profit factor, and that the American public will be left to cope

with the fallout of risk, we suggest three specific steps and one major study that need to be undertaken:

1. <u>Internalizing Social Costs</u>: In the course of increasing its profits, the current industry knowingly exposes the American public to harmful information. A decade of public pressure has failed to reduce this trend. As with other polluters, the television industry should be made to bear its true costs. Qualitative indicators and monitoring of violence, sexual exploitation, stereotyping, and advertisements to children should be formulated and broadcasters taxed according to the level of dose. The resulting funds should be used to support public broadcasting, particularly high-quality children's programming free of advertising, to compete with the current network fare.

2. <u>Protecting a Vulnerable Group</u>: Unlike the general experience of television risks, children are particularly sensitive to toxic information and require special protection. Television is also for the young child, more than the adult, an educational medium. All children's programming should be taken out of the hands of commercial television and made the responsibility of public television.

3. <u>Arming the Public</u>: Since the locus of television hazard management will continue to be in the home, steps should be taken to enhance the power of the consumer to make risk judgments and to make the experience of risk more voluntary. Two areas of improvement are needed: better information as to what exposure is likely to be experienced, and enhanced means to use this information (as through locking devices) to control the use of the television in the home.

In regard to improved information, the development of the qualitative indicators suggested above would permit the characterization of forthcoming programs in television guides, much as newspapers increasingly do for movies and as some pay-television companies do for subscribers. In addition, a symbol could be displayed in the corner of the screen to indicate, for example, when a program with excessive violence was playing. Robert Choate has suggested several means to help parents and children evaluate advertisements: placing a special border around the picture when commercials are playing; revealing product costs when the price is over $2.00; indicating potentially hazardous products through the use of graphic symbols; and using graphic symbols to indicate the nutritional value of advertised food (<u>New York Times</u> 1979).

In addition, parents need increased control over children's television viewing, much of which is distributed over programs not aimed at children. All new television sets should be equipped with plug-locks that permit the locking of the television set. As cable and pay television expands, it should include the use of selective locking devices, as currently exist for Qube and some pay television, that regulate which channels can be received.

4. <u>Risks of New Technologies</u>: The lack of a technology assessment of the new television technologies is an outstanding societal need in 1980. A task for the Office of Technology Assessment (OTA), such an assessment should be thorough and comprehensive.

These steps will not, of course, eliminate the social risks that television presents. A large-scale reduction of risks awaits a new conception of the medium in American society.

NOTES

1. There are, of course, problems associated with the various assessments of viewing time. The Nielsen reports are considered the most reliable estimates of viewing because they are based on the largest sample. But the measurement device is a mechanical audimeter which simply registers when the set is turned on and not whether anyone is watching it. Viewing diaries and survey instruments that attempt to record individual viewing time may also produce unreliable results. On the average, people tend to record four hours for every three spent in actual viewing. Some evidence is available concerning the actual viewing behavior of children. See Anderson et al. (1979).

REFERENCES

Adler, Richard, et al. 1980. The effects of television advertising on children. Lexington, Mass.: Lexington Books.

Anderson, Daniel R., et al. 1979. Watching children watch television. In Attention and cognitive development, ed. Gordon A. Hale and Michael Lewis. New York: Plenum Press.

Andison, F. Scott. 1977. TV violence and viewer aggression: A cumulation of study results. Public Opinion Quarterly 41:314-331.

Barcus, F. Earle. 1978. Commercial children's television on weekends and weekday afternoons: A content analysis of children's programming and advertising broadcast in October 1977. Newton, Mass.: Action for Children's Television.

Belson, William. 1978. Televised violence and the adolescent boy. London: Saxon House.

Bogart, Leo. 1972. Negro and white media exposure: New evidence. Journalism Quarterly 49:15-21.

Bogart, Leo. 1980. After the Surgeon General's report: Another look backward. In Television and social behavior: Beyond violence and children, ed. S.B. Whitney and R.P. Abeles, 103-133. Hillsdale, N.J.: Erlbaum Associates.

Brown, Les. 1979. Satellite offer may aid tv stations' programming. New York Times, 20 March, C11.

Brown, Les. 1980. Viewers' dissatisfaction with tv programs found increasing. New York Times, 3 January, C18.

Brown, Merrill. 1981. The brave new world of television. Washington Post, 11 January, A1 and A18.

Cantor, Muriel G. 1982. The organization and production of prime time television. In NIMH (1982) 2:349-362.

Cole, B., and M. Oettinger. 1978. The reluctant regulators. Reading, Mass.: Addison-Wesley.

Comstock, George A., et al. 1978. Television and human behavior. New York: Columbia University Press.

Comstock, George A., and Eli A. Rubinstein, eds. 1972. Television and social behavior: A technical report to the Surgeon General's Scientific and Advisory Committee on Television and Social Behavior. 5 vols. Rockville, Maryland: Dept. of Health, Education, and Welfare.

Crandall, Robert W. 1978. Regulation of television broadcasting: How costly is the public interest? Regulation 2 (January/ February):31-39.

Davis, Richard H., and Robert W. Kubey. 1982. Growing old on television and with television. In NIMH (1982) 2:201-208.

FCC (Federal Communications Commission). 1974. Report and policy statement on children's television. Washington: Federal Communications Commission.

FCC (Federal Communications Commission). 1979. Television programming for children: A report of the Children's Television Task Force. 5 vols. Washington: Federal Communications Commission.

FTC (Federal Trade Commission). 1978a. FTC staff report on television advertising to children. Washington: Federal Trade Commission.

FTC (Federal Trade Commission). 1978b. Proposed trade regulation rule on food advertising, 16 CFR Part 437, Phase 1: Staff report and recommendations. Washington: Federal Trade Commission.

Gerbner, George. 1972. Violence in television drama: Trends and symbolic functions. In Television and social behavior, ed. George A. Comstock and Eli A. Rubinstein. Vol. 1, Media content and control. Washington: Government Printing Office.

Gerbner, George, et al. 1980. The "mainstreaming" of America: Violence profile no. 11. Journal of Communication 30 no. 3:6.

Greenberg, B.S., and J.R. Dominick. 1969. Racial and social class differences in teen-agers' use of television. Journal of Broadcasting 13:331-344.

Hearold, Susan L. 1979. Meta-analysis of the effects of television on social behavior. Ph.D. dissertation. University of Colorado.

Kahn, Robert D. 1983. More messages from the medium. Technology Review 86 no. 1 (January):49-51.

Kramer, Larry. 1979. Senate panel votes stiff curbs on FTC powers. Washington Post 21 November, A1.

Krugman, H.E. 1971. Brain wave measures of media involvement. Journal of Advertising Research 11:3-9.

Liebert, Robert M., et al. 1973. The early window: Effects of television on children and youth. Elmsford, New York: Pergamon Press.

Lyle, J., and H.R. Hoffman. 1972. Children's use of television and other media. In Television and social behavior. Vol. 4, Television in day-to-day life, ed. E.H. Rubinstein, G.A. Comstock, and J.P. Murray, 129-256. Rockville, Maryland: Dept. of Health, Education, and Welfare.

McDowell, Edwin. 1978. Storm ahead on TV ads for children. New York Times, 8 August, pp. D1, D5.

Marcus, Steven J. 1983. Cable TV: Competing in the wrong place. Technology Review 86 no. 3 (April):72.

Milavsky, J. Ronald, Ronald C. Kessler, Horst H. Stipp, and William S. Rubens. 1982. Television and aggression: A panel study. New York: Academic Press.

NIMH (National Institute of Mental Health). 1982. Television and behavior: Ten years of scientific progress and implications for the eighties. 2 vols. DHHS Publication ADM 82-1195. Rockville, Maryland: U.S. Dept. of Health and Human Services.

New York Times. 1979. Label on TV children-ads urged. 13 March, A16.

New York Times. 1980. The TV set: Growth of tapes and disks. 17 February, p. 1.

Nielsen, A.C. 1977. The television audience: 1977. Chicago: A. C. Nielsen Company.

Nielsen, A.C. 1981. Nielsen report on television 1981. North brook, Illinois: A.C. Nielsen Company.

Opinion Research Corporation. 1978. Public attitudes toward advertising. ORC Public Opinion Index 36 (Mid-October):1-12.

Pearce, Alan. 1972. The economics of network children's television programming: Staff report submitted to the Federal Communications Commission, July.

Pearce, Alan. 1974. The economics of children's television: An assessment of the impact of a reduction in the amount of advertising. Staff report for the Federal Communications Commission, June.

Platt, J. 1975. Information networks for human transformation. In Information for action, ed. M. Kochen. New York: Academic Press.

Pollack, Andrew. 1981. RCA's big videodisk campaign. New York Times, 26 February, A16.

Prial, Frank J. 1983. Cable TV is said to top networks in movie violence. New York Times 22 January, 46.

Robinson, John W. 1972. Television's impact on everyday life: Some cross-national comparisons. In Television and social behavior: A technical report to the Surgeon General's Scientific and Advisory Committee on Television and Social Behavior, ed. George A. Comstock and Eli A. Rubinstein. Vol. 4, Television in day-to-day-life, 410-431. Rockville, Maryland: Dept. of Health, Education, and Welfare.

Roper, Burns W. 1979. Trends in attitudes toward television and other media: A twenty year review. In Public perceptions of television and other mass media: A twenty year review, 1959-1978, ed. The Roper Organization, Inc. New York: Roper Organization.

Smith, Desmond. 1979. Television enters the 80's. New York Times Magazine, 19 August, 16-21, 64-67.

Smith, Sally Bedell. 1983. F.C.C. alters policy on TV for children. New York Times, 23 December, C25.

Sneed, Laurel Crone. 1979. Peter Crown: Tending tv's fireplace. American Libraries 10 (September):492-495.

Sulzberger, A.O., Jr. 1981. FTC staff urges end to child-tv ad study. New York Times, 3 April, D6.

Sussman, Barry. 1979. Less tv for more? A majority say they watch less than before. Washington Post, 28 February, B1 and B11.

U.S. Congress. 1972. Senate Committee on Commerce. Subcommittee on Communications. Surgeon General's report by the Scientific Advisory Committee on Television and Social Behavior: Hearings . . . March 21-24, 1972. Washington: Government Printing Office.

U.S. Congress. 1977. House Committee on Interstate and Foreign Commerce. Subcommittee on Communications. Violence on

454

television, 95th Cong. 1st sess. Washington: Government Printing Office.

Waters, Harry F., et al. 1981. Cable tv: Coming of age. _Newsweek_, 24 August, 44-49.

19
Congress as Hazard Manager

Branden B. Johnson

Since 1838, when President Van Buren signed into law provisions for the prevention of explosions on steamboats (Burke 1966), the United States Congress has been a major policy maker in the area of technological hazards. The hazard management activities of Congress call forth a somewhat awesome interplay of its legislative, budgetary, and oversight functions. Insofar as analysis of all three congressional functions would require a prohibitively massive undertaking, however, the present author has elected to focus on a single portion of that activity--legislative policy-making on hazard management. Analysis of the statutes alone will not itself explain the overall role of Congress as hazard manager, but the study of legislative activity is crucial to an understanding of congressional attention to technological hazards. Thus this chapter offers a brief analysis of that attention, as reflected in the congressional hazard agenda and in the hazard controls mandated by Congress. Out of the analysis comes a set of findings that carry implications for federal hazard management and suggest avenues for further research.

The study confines itself to federal nonappropriations legislation enacted during the 85th through the 95th Congresses, 1957-1978. That period witnessed the enactment of 7909 laws, from which were culled 179 laws addressing some 36 distinct hazards. The sample includes only laws that provide for substantive and direct federal control of nationally important technological hazards. Table 1 lists the 36 hazards and the number of laws relating to each. A detailed account of the selection and hazard identification processes comprises the appendix at the end of this chapter.

The Congressional Agenda, 1957-1978

What is the place of technological hazards on legislative agenda? This section of the chapter explores three aspects of that question as revealed by the statutory output of the 85th-95th Congresses (1957-1978):

1. What proportion of the agenda comprised hazard laws?
2. Which hazards commanded, or failed to command, legislative attention, and why?
3. Has hazards agenda-setting in Congress dealt with the items on society's hazard agenda?

TABLE 1
Hazard listing and number of sample laws per hazard

automobile collisions	19
nuclear fuel cycle	16
operation of nuclear plants	
waste disposal/storage	
nuclear incident liability	
ship accidents	15
railroad accidents	15
food additives	14
occupational safety and health risks	13
oil discharges to water	13
water pollution	12
airplane crashes	11
consumer products	9
transfer of nuclear materials	9
motor vehicle air pollution	7
noise pollution	7
explosives	7
pesticides	6
stationary source air pollution	5
air piracy	4
firearms	4
highway environmental effects	4
dam failure	3
small boat accidents	3
drugs	2
lead-base paint	2
pipeline breaks	2
hazardous materials and solid waste	2
miscellaneous radiation	2
surface effects of mining	2
poisons	2
treated seeds	1
switchblade knives	1
medical devices	1
cargo transport containers	1
Antarctic environmental effects	1
Corps of Engineers projects	1
airport environmental effects	1
chemicals	1

Overall Hazard Statute Output

The overall output of hazard laws has remained remarkably stable (Table 2), belying the impression that the number of such statutes enacted per Congress has been snowballing (Sittig 1979). During most of the study period the number of hazard laws enacted per Congress varied between a low of 11 or 12 and a high of 19, 20, or 21. Exceptions occurred in the early 1960s, when this number fell to eight for two Congresses and in 1977-1978 when there was a dramatic increase to 35 hazard laws. The passage during the 96th Congress of about 21 laws meeting the study criteria (Congressional Research Service 1957-1980) may presage a rise in hazard laws per Congress (the impact of the Reagan administration's initiatives to reduce government regulation on statutory output in this field is not yet clear).

To be sure, the number of hazards dealt with (i.e., a statute passed) per Congress has shown a fairly constant rise. Yet the number of "new" hazards (those not previously the subject of federal legislation) does not exceed four per Congress. Only a third of the 17 hazards legislated upon in the 94th Congress were not among the 15 acted upon in the 85th Congress; even in the extremely active 95th Congress, only half of the hazards were not part of the 85th Congress's legislation. During this period the sample hazard agenda has thus been fairly stable and focused on hazards already federally controlled.

Such stability is unusual in the context of two opposite trends: the long-term decline in overall federal enactments over this period and the increase (often 100 percent) in several measures of hazard risk, public concern, or hazard management activity.[1] If technological hazards were agenda items in the same way as other issues, one would expect a similar decline in hazard statute output;

TABLE 2
Frequency of selected federal technological hazard legislation, by congress 1957-1978

CONGRESS	TOTAL LAWS*	HAZARD LAWS	HAZARD/ TOTAL (%)
85 (1957-1958)	936	20	2.1
86 (1959-1960)	800	11	1.4
87 (1961-1962)	885	8	0.9
88 (1963-1964)	666	8	1.2
89 (1965-1966)	810	21	2.6
90 (1967-1968)	640	12	1.9
91 (1969-1970)	695	19	2.7
92 (1971-1972)	607	12	2.0
93 (1973-1974)	649	21	3.2
94 (1975-1976)	588	12	2.0
95 (1977-1978)	633	35	5.5
TOTAL	7909	179	2.3

*U.S. Bureau of the Census 1975,1081;U.S. Bureau of the Census 1978, 514.

if they were completely insulated from whatever factors have resulted in the overall reduction in federal laws passed, it is reasonable to expect a rise in the hazard law output equivalent to other measures of technological hazardousness. The absolute stability of statutory output may indicate that these trends are not only opposite but countervailing: if it were not for strong pressures upon Congress to act on this particular set of issues, the proportion of overall legislation devoted to hazards would have remained constant rather than increasing; but the overall agenda-setting process has acted as a brake on the rate of such an increase.

If strong pressure is being put upon Congress to act against technological hazards, what sorts of pressure and how does Congress respond? The legislature may well be responding in varying degrees to public opinion—as the theory of speculative augmentation in policy-making suggests (Jones 1975)—to the apparently growing number of potential hazards (rational choice theory), to improved monitoring and regulatory capability, and to the battles between bureaucracies to maintain and extend their authority and power (Downs 1967,216). Congress can deal only to a limited extent with the details of hazard management. Under these various pressures to act on technological hazards, it has kept its output (and presumably, therefore, its workload) from increasing exponentially by passing to the executive branch responsibility for dealing with individual hazards. Both the overall number of regulations (Heclo 1978) and the percentages of the federal budget devoted to technological hazards (see chapter 7) have increased. One means Congress has used to transfer responsibility is by enacting broader laws. A single statute may cover thousands of actual or potential hazards, as do (for example) the Consumer Product Safety Act, the Occupational Safety and Health Act, and the Toxic Substances Control Act. Increases in the comprehensiveness of hazards legislation during the study period were found for 10 of the 27 hazards subject to more than one law; there was no evident change for the others.

Hazards on Congressional Agenda

Which hazards prompt federal legislation, and why? To answer this question, the study first compared the bills introduced and laws enacted in the 95th Congress (a not necessarily typical example, of course).[2] Of the 48 separate hazards that were the subjects of bills, 19 eventually became the subjects of statutes enacted by the 95th Congress. The mean of 24.7 bills submitted per legislated hazard was about three times that per nonlegislated hazard, and most of the laws enacted by this Congress were distilled from ten or more bills apiece (food additives, for example, were the subject of 67 bills that produced one law). Congress appears to legislate on those hazards for which there is already considerable interest, reflected in bill introductions ($r^2 = 0.6$, between numbers of bills introduced and laws enacted). Yet some laws were based on relatively few bills. For example, two laws—to prevent Antarctic pollution, and to insure the safety of cargo transport containers, respectively—were the results of one bill each; two laws on dam failures came from only seven bills. The 95th Congress demonstrates that exceptions are likely when hazards are perceived to be uncontroversial or minor (Antarctic pollution and cargo containers), have recently contributed to a disaster (dam collapses and airplane crashes), or have been the subject of considerable congressional effort to obtain legislation (surface effects of mining).

These results leave open the question of why only 19 of the 35 sample hazards subject to 95th Congress bills prompted statutory action: legislators were capable of introducing dozens of bills on the others as well. It turns out that Congress is more likely to enact legislation for a particular hazard if legislation already exists, particularly if the existing legislation is of recent vintage. In any given year, Congress deals largely with old hazards and adds a few new ones. Of the 36 sample hazards, 15 had been the subject of legislation prior to 1957, the starting date for the present analysis. These 15 "old" hazards took up 72 percent of all hazard laws before 1966 and 60 percent thereafter (through 1978). Thus the addition of some new issues had only moderate effects on the level of attention to older issues. As for the recency of such enactments, 59 percent of the hazard control laws passed by that Congress concerned hazards already legislated upon in the 94th Congress (excluding hazards acted upon only in the 95th Congress). For the 1957-1978 period under study, the proportion per Congress of laws that pertained to hazards already legislated upon in the immediately preceding Congress ranged from 38 percent to 100 percent, with seven of the 10 relevant Congresses exhibiting above 50 percent of their enactments in this condition.

The development of Congress's formal hazards agenda thus appears to be a two-step process. Technological hazards are more often the subject of bills when they have previously been covered by federal statutes. Once they have gotten on the agenda, hazards bills are more likely to be enacted into law when there are large numbers of bills introduced, providing the momentum (by expressing, or merely suggesting, a widespread interest in legislative action) to get candidate hazards "through the gate." This result coincides with the observation that it is very difficult to get new issues on the agenda, since legislators' time is limited and their agenda is overloaded; they "presume that older problems warrant more attention because of their longevity and the greater familiarity officials have with them" (Cobb and Elder 1972,89). Elsewhere in this volume (chapter 6) Harriss et al. note the progress in controlling well-known, well-understood acute hazards. Given the political need for legislators to demonstrate successful action on their part, it is scarcely surprising that Congress should direct more attention to such older hazards than to less tractable hazards with chronic and delayed consequences.

Congress and the Societal Hazards Agenda

How does the congressional hazards agenda relate to that of society? The societal hazards agenda, which has been termed **systemic** (Cobb and Elder 1972,86-87), includes those hazards commonly perceived as meriting public attention and as coming under government jurisdiction. The hazards on Congress's formal agenda (i.e., bills and laws) were at one time, and may still be, on the systemic agenda. If we presume that Congress has acted upon a given hazard because that hazard has appeared on the systemic agenda--and that, as a representative institution, it ought to do so—then it is appropriate to ask how well it has done in acting on the overall systemic agenda.

A surrogate for the systemic agenda was constructed by combining the hazards in this legislative sample with those from two other sources (a research project compilation and a news column from an environmentalist journal).[3] Of the 54 distinct hazards on the resulting list, the legislative sample includes 36 (two-thirds). Three of the remainder were subjects of 95th Congress bills that did not get enacted. And if the complete congressional agenda (bills plus laws) were known, all or almost all might have been included (with the exception of hazards traditionally managed by local governments or in the private sector). Congress thus appears to be covering the systemic agenda on technological hazards, as defined by this synthetic list, fairly well.[4]

Summary of Congressional Agenda Findings

In terms of both the number of statutes and the hazards concerned, the stability of the congressional hazards agenda from 1957 to 1978 has been little short of remarkable. The number of hazard laws per Congress has not declined as has the number of total enactments; neither, due to the increasing comprehensiveness of hazard laws, has this number shown the dramatic increase of other measures of hazard concern and action. Those hazards with which Congress has had most experience are the most likely to elicit legislation in a given Congress, particularly if they were the subjects of recent laws. Despite the difficulty of getting new issues on the legislative agenda, however, Congress appears to be acting upon most of the hazards that appear on society's agenda.

Legislated Hazard Controls

After its authorization for federal agencies to take action on technological hazards, perhaps the most important component of congressional legislation on such hazards is the mandate of specific controls to be implemented by those agencies. The significance of these controls lies in the differential impact they can have on risks, the economy, and the political system. This section of the paper examines three aspects of control legislation: (1) the nature and comprehensiveness of the hazard controls mandated by Congress; (2) the variations in application of enforcement techniques and in civil and criminal penalties for violations of these hazard controls; and (3) moves by Congress to restrict regulators' discretion in implementing hazard legislation.

Hazard Controls

The 12 categories of controls (Table 3) used in this research derive empirically from the sample legislation (see Johnson 1980, 99-123 for discussion of their derivation and comparison with the results of using the hazard evolution model, described in chapter 2, to classify statutory hazard controls). The most common control (in terms of the number of laws using it) was restriction on the design or manufacture of a technology (47 percent of the laws), with

TABLE 3
Hazard control modes

CONTROL MODE	EXAMPLE
1. Mandatory Ban	Prohibit addition of carcinogenic substance to food
2. Discretionary Ban	Food or fuel additive may be prohibited if air pollutant emission problems arise
3. Restriction of Uses or Users	License or permit required to buy, sell, transport, manufacture, etc. firearms
4. Restriction of Distribution or Handling	Regulations for transportation of dangerous explosives
5. Restriction of Design or Manufacture	Emergency locator beacons required on aircraft
6. Restriction of Releases or Emissions	Standards for emissions of particulates into the air
7. Restriction of Exposure	Require removal of oil after spill
8. Change in User Behavior	Driver performance improvement program
9. Informing Those at Risk	Inform purchaser or renter of federally supported dwelling of lead-based paint risk
10. Indemnifying Those Harmed	Mandatory repair, replacement, or refund of hazardous consumer products
11. Mitigation of Harm	Require airport firefighting and rescue safety equipment, and minimum standards for them
12. Support for an Alternative Technology	Provide instructions on integrated pest management to those who request it

restrictions on releases (34 percent) and uses or users (24 percent) also common. By contrast, support for an alternative technology—e.g., subsidizing mass transit as an alternative to automobile use, thus reducing the risk of collisions—occurs in only 2 percent of the laws (all but one, on pesticides, dealing with automobile alternatives). Given the economic commitment to existing technologies, it is scarcely surprising that the latter approach has been rare. Outright bans of specific technologies are similarly rare (mandatory bans occur in 1 percent of the laws, discretionary bans occur in 4 percent). Even that most famous of all mandatory bans—the Delaney Amendment, which prohibits use of carcinogens as food additives—has served "as the explicit basis for decisions to ban fully registered products on only two occasions, both involving compounds used in food packaging materials" (Lowrance 1976,82), as well as the ban on cyclamates in 1970. Such bans may bans may be less significant in hazard control than their proponents and opponents appear to believe (National Academy of Sciences 1974,18-20 and 25-26;Moore 1978, 1157).

Congress typically mandated a limited number of controls for a given hazard. Only five (food additives, pesticides, surface effects of mining, occupational safety and health risks, oil discharges to water) of the sample hazards were subject to more than six of the 12 sample controls (and these are not necessarily all the possible controls that could have been applied by Congress). By contrast, a quarter of the hazards were subject to **domination** by a specific control (i.e., one control was used more than twice as often as the next most frequent control for that hazard). Restrictions on design and manufacture of the hazardous technology dominated legislated controls on railroad accidents, airplane crashes, and automobile collisions (see chapter 14 for further discussion of controls on the latter hazard). Restrictions on emissions dominated controls on stationary-source air pollution and water pollution. Specific controls dominated certain other hazards: indemnification dominated the nuclear fuel cycle; restrictions on handling and distribution dominated explosives; restrictions on uses and/or users dominated firearms; and requirements for changes in user behavior dominated small-boat accidents. (It should be emphasized that this narrowness of mandated controls did not preclude major policy changes in the studied legislation, though most laws made only incremental changes.[5])

This lack of comprehensiveness[6] in the mandate of hazard controls by Congress defies ready explanation. Perhaps Congress's "narrow" approach actually complements control actions taken by other parts of the hazard management system, both outside the federal government and in legislation excluded from the study sample (e.g., non-hazard-specific risk mitigation programs such as Medicaid). The dominance of a hazard's statutory management by one control may simply reflect legislators' perception that that control was the "optimal" one (in terms of cost-effectiveness, ease of application, its acceptance by the regulated technology sponsor, etc.) for the hazard. Not all controls are reasonable for all hazards--banning of a technology, for an extreme example. On the other hand, such dominance might well signal nothing more than the ease of following precedents; once the initial hazard control is in place, an incrementalist approach to hazards policy-making would tend to foster a recurring mandate for that particular control.

It has been argued that a sign of "mature" hazard management is the use of a wide range of controls (Kasperson and Morrison 1980, 42). The assumption underlying this argument is that subject to constraints of ignorance, scarce resources, and limited political will, society will use every means at its disposal to reduce risks. If maturity is defined chronologically, one would expect that more controls would be legislated for hazards that had been the subject of many laws and/or had had the first national statute concerning them enacted long ago. A testing of this hypothesis for six "old" (subject to federal legislation before the 1957 beginning of the study period) and six "new" hazards unearthed no correlation (r^2 =

0.1) between years on the agenda, or number of laws enacted, and the total number of controls mandated.

Enforcement Options and Penalties. These do not reduce hazard risk directly but are intended by legislators to act as incentives for organizations and individuals to do what is required by law. If one assumes that such incentives are always useful, if not necessary, it is surprising that Congress has legislated their use for only 22 hazards and has applied any single incentive to fewer than half of the 36 sample hazards. Injunctions—judicial orders that prohibit any further action by the alleged violator until compliance can be achieved or it is determined that the law was not violated—were allowed for 14 hazards. Other mandated enforcement options included citizen-initiated lawsuits (eight hazards); seizure of the hazardous product (four); suspension or revocation of an offending individual's or facility's license to operate (two); and prohibition of federal contract awards to, or procurement of goods or services from, a violator of the law. Monetary civil penalties (for assessment of damages or liability for harm incurred) have been set for seven hazards, and criminal penalties—including prison terms and/or fines—for between three and 10 hazards for each of the following categories: giving false information to government officials; failure to correct a cited violation; tampering with a monitoring or control device; using a deadly weapon to resist enforcement of the law; causing injury; causing death. The hazards to which the largest number of such incentives were applied were oil discharges (seven of the 12 listed above), water pollution, and hazardous materials (five each). Although this skewed distribution is attributable in part to the inappropriateness of some incentives for some hazards—e.g., firearms and explosives rarely have monitoring or control devices attached so that fining those who tamper with such nonexistent devices will serve no useful purpose—one may still ask why they were not applied more broadly.

In addition to a noncomprehensive application of incentives for compliance with hazard controls, Congress has legislated civil and criminal penalties with a fine disregard for consistency in the level of penalties. A case in point is the set of laws pertaining to liability for removing oil from the environment after an illegal discharge. Maximum levels for such liability increased over the years since the incentive's first statutory appearance in 1970 with the level's remaining several times higher for discharges from vessels than for discharges from onshore or offshore (e.g., oil drilling) facilities. The maximum fine (set in 1972) for discharging oil and other substances that could **not** be removed from the environment was a third of that for removable discharges.[7] Criminal penalties provide further examples. If an injury was caused by—or a deadly weapon was used to resist enforcement of the law regarding—shipping, the penalty is up to $100,000 and/or up 10 years imprisonment, whereas the same penalty for oil discharges (and passed in the same year, 1978) is only $10,000/10 years. In 1977, failure to correct a cited violation became punishable by a civil fine of $1,000 per day for occupational hazards, but only $750 for surface mining. In 1970 the prospect of false statements to government officials brought simultaneously enacted fines and prison terms of $5,000/5 years (explosives), $10,000/6 months (occupational safety and stationary source air pollution), and $10,000/5 years (airplane crashes). Even when the penalized action was a constant—causing death—the prison term for such an offense was up to 10 years if explosives (1958, 1960) or transfer of nuclear materials (1960) were involved but no more than six months for occupational hazards (1970).

Similar inconsistencies showed up in comparing penalties for a single hazard over time. For example, the penalty for illegal discharge of oil was the same in 1966 as it was in 1961, rose 400 percent in 1970, stayed at that level for the next three laws (through 1975), and increased 500 percent in the 1978 statute. The fine for violating a legislative provision on railroad safety increased in 1970 to 10 times its 1957 level. Yet some fines did not change at all: $500 was the penalty for breaking a ship safety rule in 1977, as it was in 1958. In either case, inflation can erode the incentive power of such fines.

Congress has variously levied the same or different penalties for different hazards at the same time, for different hazards at different times, and for the same hazard at different times. Such behavior may be indicative of one hazard's being considered more important than others to control. But such priority-setting does little to explain differential penalties for one consequence—death—that spans a number of technological hazards. Erratic patterns suggest rather that legislators act according to a "principle of conservation of effort"—analogous to Lindblom's theory of disjointed incrementalism on policy-making (Lindblom 1959,1979). Such a principle implies that (1) when legislators again turn their attention to a hazard, they do not amend an entire law (even on an incremental basis) though previously enacted fines may thereby lose their power to inflation; (2) legislators may copy penalties from laws on other hazards, without considering whether the penalty is commensurate with the hazard (in fact, Congress may have spent no time as a deliberative body thinking about what penalties should be assessed); and (3) legislators do not appear to be interested in interhazard comparisons. The last implication, though as yet untested, is particularly interesting in light of Representative Ritter's bills in the 96th and 97th Congresses to promote use of comparative risk data in hazard management (Congressional Research Service 1979-82; cf. more general proposals in the same direction: Cohen and Lee 1979;Wilson 1979); it may augur that such proposals, if implemented, would exert little or no influence on congressional policy-making.

Statutory Constraints on Regulatory Discretion. When Congress passes laws to control technological hazards, further management is usually left to the regulatory and executive agencies. Although these laws do not grant unlimited discretion—they may limit control alternatives, factors that can be considered in setting standards, and the time period during which controls must be implemented, among other items (National Research Council 1977,3)—it has been generally assumed that Congress has neither the time nor the expertise to handle all the necessary details. This is particularly true when the scientific basis for regulation is uncertain or flexibility is necessary (National Research Council 1975,19), which is often the case for technological hazards.

Recently, considerable questioning has asked whether this delegation of authority (see Mitnick 1980,327-337) has gone too far, and some have wondered whether Congress's resultant tightening of statutory constraints on bureaucratic discretion might lower agency effectiveness. An example of the latter concern is the great controversy over the legislative veto [e.g., Sulzberger 1980; Roberts 1982; see Norton (1976) for an inventory of such veto enactments since 1932]. Certainly the number of statutory constraints on federal hazard management has increased drastically in recent years. Over the course of the 11 Congresses under examination, the number of laws that require the submission of proposed regulatory actions to Congress before their promulgation has increased--all but two of the 14 laws of this type were enacted in the last five Congresses. The same five Congresses (91st-95th) passed 18 of the 23 laws containing very specific instructions (e.g., maximum allowable level in micrograms of particulate air emissions). The number of laws containing instructions on how to set a standard or determine a level of acceptable risk (e.g., specifying factors that should enter into the decision or requiring risk/benefit comparisons) has also increased. Such "contextual risk evaluation criteria" (Johnson, 1980,161-164;Kasperson and Kasperson 1983,143-148) were applied to 17 hazards in some 15 percent of the sample laws. Congressional involvement in the details of hazard control has included 5 cases of orders to regulators not to require certain controls, primarily for hazards stemming from use of the automobile.

The net effect of such constraints on executive agency discretion is difficult to determine. The Environmental Protection Agency has avoided (at least during the early 1970s) asking that unworkable provisions of the hazards laws under its jurisdiction be changed, for fear that sound ones might also be altered by Congress (National Research Council 1977,2). A General Accounting Office study (1982) of the impact of congressional rule review requirements in the Federal Trade Commission Improvements Act of 1980 found little effect on the agency's decision-making processes up to that time, though Congress's ban on one proposed rule might change this. The very limited and inconclusive evidence from the sample hazard laws suggests that such constraints may not have altered federal hazard management activity significantly.

Summary of Hazard Control Findings

The type and number of controls per law and hazard have shown little change over time. Fines have not increased steadily to keep pace with inflation, nor have legislators maintained consistency either in assigning penalties for the same offense across hazards or in applying incentives for compliance. Legislated controls have strengthened Congress's hold on the regulatory agencies, but what little evidence exists does not suggest that such interventions have altered federal risk management significantly.

Research and Policy Implications

This study has portrayed congressional activity on technological hazards from 1957 to 1978 as dominated in a sense by precedent: hazards acted upon in the past maintain their prominence on the legislature's agenda, past hazard control mandates re-occur with little change, and penalties for violations of these laws change relatively infrequently. The evidence does not support wholesale labelling of Congress's policy-making on hazards as incrementalist in nature (Lindblom 1959,1979)—though the data on monetary penalties are persuasive (if not greatly significant for policy) in that regard—for two reasons: (1) only statutes were analyzed in this study—direct study of legislative behavior, or of budgetary and oversight activity, might provide different results; and (2) Congress has acted, in at least some hazard cases, in a nonincremental manner (for example, see note 5 and Jones 1975). Despite these caveats, however, the general stability of technological hazard management by Congress over the 22 years of the study period is quite striking, and it raises a number of questions that can be discussed only briefly here.

The relative stability of the hazards agenda may pose a number of problems for both federal bureaucrats and legislators. It has been suggested that Congress has kept its output (in number of laws enacted), and therefore perhaps its workload, from increasing drastically by passing responsibility for dealing with individual hazards to the executive branch. It might be asked whether the agencies can cope with an ever-increasing workload; some commentators (e.g., Wilson 1980,392-393) have suggested that the magnitude of agencies' hazard management responsibilities has put severe strains on their ability to be effective, beyond those imposed by the complexity of these problems.

Yet Congress has not escaped difficulties by awarding others primary decision-making authority, as indicated by the rising proportion of its statutory output devoted to technological hazards. These are issues for which information is at least as tentative, relevant values at least as difficult to identify, and concerned groups at least as much in conflict as for others, perhaps more so; it could be plausibly argued that these problems would make legislative oversight of federal hazard control actions extremely difficult. Since Congress has more than doubled the total number of hazards for which it has legislated controls between 1957 and 1978—though keeping the number per Congress constant—the difficulty of such activities as oversight becomes even greater. The legislature may become less and less able to cope with the burden of hazard management—e.g., the number of committee meetings and hearings more than doubled between 1957 and 1978 (Bibby et al. 1980,86-87).

A further issue concerns those hazards that get on that agenda: "old" hazards continue to dominate the legislative agenda even after they have come to total less than half of the total number of hazards acted upon by Congress. This proposition implies some inability of the legislature to respond quickly to the appearance of new hazards. A reduction in the difficulty of getting hazards on the formal agenda would mean that even less time (perhaps less than minimally necessary) would be available for legislators to weigh problems and make appropriate responses. In short, has Congress coped with some hazards at the cost of ignoring hazards of equal or greater importance?

The history of hazard management exacerbates this difficulty. Whereas the overall burden of risk has appeared to decline over the past decades, progress in hazard control has concentrated on acute hazards for which scientific knowledge is extensive and thus control is relatively easy (chapter 6). For controlling chronic hazards, the task ahead is of unknown magnitude but could require expenditure of a growing proportion of society's scarce resources. Competing demands for those resources make an all-out commitment to hazard management unlikely. A threshold value of knowledge, resources, and value consensus may exist below which Congress's current legislative processes cannot be effective; if it does exist, these "new" hazards may well fall below it. In which case Congress would seem to have two major options: (1) maintaining the present level of effort on the current hazard agenda and thus ignoring many chronic hazards (unless those are covered in existing comprehensive or generic hazard legislation),[8] or (2) setting priorities so that effort can be systematically focussed on the most important hazards (however "important" is defined). Neither option seems likely to be acted upon in the near future; thus future congressional success in managing technological hazards would be uncertain.

Further research is necessary to determine why Congress has mandated so few controls per hazard: because legislation and regulation complement hazard controls undertaken by other parts of society, because the favored controls are optimal ones for reducing risk to "acceptable" levels, because they are optimal from the viewpoint of the regulated technology sponsors, or because Congress is prone to maintain precedents? Each of these possible findings would have important normative implications for hazard management policy. The question of whether statutory specificity should be increased (e.g., National Research Council 1977,17;Fischhoff et al. 1980,274-275) or decreased (e.g., Marcus 1980), as well as the degree to which actual agency practice has been affected by statutory constraints, needs clarification as well. Finally comes the finding that Congress does not appear to amend laws consistently or to make interhazard comparisons while doing so. The main evidence cited in this study concerned penalties for violations of the hazard laws, though evidence also occurs in contextual risk evaluation criteria and other statutory language (Johnson 1980,152-164). If true on a wider scale, this conclusion could pose serious limits for any attempt to "rationalize" hazard management, such as reducing risks for all hazards to the same level or equalizing the marginal costs of hazard controls. Since rationalist decision-making techniques (most prominently, risk/benefit or cost/benefit analysis) are the most technically coherent, and among the most popular of all current legislative proposals for improving societal risk management, this finding could be particularly significant.

The ultimate question is whether all this activity by Congress has been effective in preventing, reducing, or mitigating technological hazard consequences. Since this study has not been concerned with measuring the efficacy of congressional policy, the most that can be said here is that the mix of policy approaches used by Congress has been at least partially successful (witness the stabilization or reduction of air and water pollution in many areas over the past decade—Council on Environmental Quality 1980,80,100,146).

The search for an answer to this question will be one of the more important future research tasks in the field of managing technological hazards.

APPENDIX

Methodological Issues of Law Selection and Hazard Classification

Population Definition

The population of legislation subject to study comprised public laws enacted by the 85th through 95th Congress, 1957-1978. The time period, which spans an interval long enough to reveal changes in patterns of legislative hazard management, includes the controversial Delaney Amendment (1958) regarding potential carcinogenic substances in food (included due to the vociferous debate over its powerful but rare hazard control technique of banning a hazardous technology).

The study population includes only laws of national scope that mandate various hazard controls or that alter the definition of a hazard subject to those controls. Although other classes of legislation may influence the implementation of hazard controls or may be more important than these "controls" laws for the management of a particular hazard, they require either a series of difficult assumptions about their effect on hazard risks or a review of regulatory activity that is outside the scope of this study.

The selected legislation was hazard- and technology-specific, directly related to federal hazard control, and relatively easy to analyze within limits of the study. The first criterion excluded such laws as the National Environmental Policy Act (NEPA), which is concerned with the environmental effects of undefined technologies in major federal projects, and statutes involving hazards not exclusively or primarily associated with the use of a technology (e.g., mortality and morbidity attributable to fire, cigarettes, and liquor).

The study excluded several categories on the grounds that they did not involve **control** of technological hazards by **federal** agencies. Laws that deal only with administrative or similar matters (e.g., the organization and staffing of a regulatory agency), or only with research on a hazard, do influence the efficacy of hazard controls through the marshalling of knowledge and other resources, but they do not themselves reduce, redistribute, or otherwise control risks. The same is true of laws on taxes and trade, and of such a "non-law law" (Nader 1965,342) as the one proclaiming **Safe Boating Week.** Statutes concerned with the local effects of hazards (e.g., protection against oil spills in Puget Sound) or with specific units of the technology (e.g., exemption of fishing vessels in Alaska, Washington, and Oregon from certain shipping regulations) were excluded because they appeared to reveal more about the power of congressmen to pass "special interest" (pork barrel) legislation than about national policy on the management of technological hazards. Laws that provided only technical or funding assistance to the hazard control activities of interstate, state, or local agencies (e.g., legislation that gave the "consent of Congress" to the signing of interstate highway safety compacts), **without** providing also for federal guidance of those activities, were excluded due to the research into state and local policy-making their inclusion would have made necessary.

Finally, the study excludes certain legislation that posed insurmountable analytical or identification problems. Although military use of military technologies is definitely hazardous (intentionally so), it was difficult to determine what controls, if any, were included in the pertinent legislation; in fact, such controls are more likely to be found in treaties. Ignorance about how one identifies the "toxic" consequences of using photocopying machines or electronic eavesdropping devices pointed up the lack of a consistent rationale for choosing laws on such "information" hazards and prompted their elimination from the study.

Laws enacted only to alter deadlines for regulatory actions (e.g., registration of pesticides) were also excluded (National Research Council 1977,13 and 68), though they are obviously important to the ultimate efficacy of federal controls. To avoid a mere accounting of these deadlines (often more than one in a single law, and sometimes contradictory) by asking which of them were significant, how they have influenced agency actions, and what proportion of promptness or delay in such actions was due to them, would go beyond the study focus on legislation.

The last exclusion involved appropriations laws, which are enacted at least once each Congress for each agency. Whereas the study focus on policy would exclude these laws as a class, budgetary provisions do mandate occasionally a substantive hazard control or an explicit deletion of funds. Moreover, the disappearance of a former line item may constitute, in effect, the ban of a hazardous technology. But the formidable task of reviewing hundreds of appropriations on the chance of detecting such provisions (not otherwise identified) loomed as too time-consuming for the results produced.

The categories of exclusions outlined above contain legislation requisite to a complete picture of congressional management of technological hazards. Nevertheless, the legislative population defined by this study—statutes of national effect and bearing direct and deliberate hazard controls to be implemented by federal agencies—is itself significant for hazard management and worthy of separate study.

Sample Identification

The basic tool used for identifying this population of technological hazard laws was the **Digest of Public General Bills and Resolutions** (Congressional Research Service 1957-1980), which provides a summary of each law enacted. Several other "comprehensive" lists of legislation (Bureau of National Affairs 1979-1980; Congressional Quarterly Service 1945-1976; Rodgers 1977) supplemented the **Digest.** **The United States Code, Annotated (USCA)** facilitated the follow-up of cross-references in the laws themselves and the identification of subsequent amendments.

The sample of hazard legislation was intended to be synonymous with the population. The final sample of 179 laws, enacted between 1957 and 1978, comprises 2.3 percent of the total number of laws (7909) enacted during that period. Comparison of the sample with lists of natural resources, energy policy, and environmental management legislation enacted in the 94th and 95th Congresses (Hughes, Caudill, and Yost 1977;Hughes, Bishop, et al. 1979) suggests that the sampling procedure identified at least 80-90 percent of the study population (those lists themselves did not contain several sample laws). The sample included 14 appropriations laws containing substantive hazards policy provisions, identified through the USCA.

Hazard Classification

The final methodological task entailed determining which technological hazard a given law intended to address and grouping together laws handling the same or similar hazards. This task was complicated, because statutes rarely specify a single technological hazard—for example, deaths from automobile collisions.

Of the 36 hazards identified in the 179 sample laws (see Table 1), 15 were easy to identify; they included such hazards as air piracy, airplane crashes, dam failure, explosives, ingestion of lead-based paint, and ship safety. A second group of hazards came from laws concerned with similar but usually not well-defined problems; occupational safety and health laws, for example, rarely pinpointed the specific hazard (falling machinery or inhalation of toxic fumes), only that its location was the workplace (the wharves, artistic productions, mines). Other such hazards included consumer products, noise pollution, the nuclear fuel cycle, and the transfer of nuclear materials from one location to another.

For other hazards, defined less clearly by the legislation itself, distinguishing one from another required a healthy dose of common sense and an eye for consistency. Laws on air pollution were divided according to the technology implicated (i.e., motor vehicles and stationary sources), even though certain controls (ambient air pollutant standards) could not be strictly allocated on this basis and were finally assigned to the stationary source hazard. By contrast, statutes rarely specified the technologies involved in pollution of water bodies. The final distinction between hazards of oil discharge to water and general water pollution came about because both hazards rarely appeared in the same law (10 on oil discharge, 8 on general pollution, only 3 laws on both hazards), and the controls mandated were largely distinct. The identification of three other hazards—miscellaneous radiation, surface effects of mining, and chemicals—was resolved on similarly pragmatic grounds.

NOTES

1. Between 1957 and 1978, Congress enacted 7909 public laws. The number of enactments per Congress (a high of 936 in the 85th, a low of 588 in the 94th) has declined over this period, though

the rate of decline appears to be stabilizing (Table 2); this is the longest, though not steepest, such decline in the 20th century. Against a background of declining enactments, the accompanying rise in the proportion of hazard laws has failed to keep pace with other rough measures of hazard risk, concern, or management. It is noteworthy, for example, that the number of research reports on "manmade environmental hazards" published in Science and Nature rose from 2 percent of the total in 1964 to 4-7 percent in 1971 (Halverson and Pijawka 1974,11). The estimated percentage of the federal budget devoted to hazard management doubled between 1964 and 1974 (see chapter 7).

2. Of the 18,045 bills introduced during the 95th Congress, 1039 (5.8 percent) met the study criteria (described in the Appendix). Only 680 of these were actually distinct bills; the rest were duplicates for introduction in other committees or the other chamber of Congress. Bills on technological hazards enjoyed the same chance of passage as those dealing with other issues: 5.3 percent of the Senate bills and 5.9 percent of the House bills concerned hazards, whereas 5.5 percent of all 95th Congress enactments concerned hazards.

3. A review of two lists of hazards provided some sense of the societal agenda on technological hazards. One list of 82 hazards was compiled by Clark University's Hazard Assessment Group from newspaper articles, scientific research papers, and government reports. The second list of 32 hazards was drawn from the "Spectrum" columns of Environment, a major environmentalist journal, from the issues of June 1976 to May 1980; this column discusses (among other items) the latest revelations of possible hazards and government action or inaction to control other hazards. The two lists together provide a reasonably comprehensive—though not exhaustive—survey of hazards of concern to a large proportion of the public. Seven of the hazards in the legislative sample do not appear on either list.

4. There is a caveat to this conclusion based on the classification of hazards on the two lists. The societal list is more specific than the congressional list—specific consumer products are found on the societal list, whereas generic hazards (e.g., occupational safety and health) are more common on the congressional list. For comparative purposes, specific hazards were combined into generic classes wherever possible. In practice, an agency created by Congress to regulate the generic group may not act upon all the specific hazards contained in the societal list.

5. Congress did effect massive policy augmentation at least once for about a third of the hazards. The indicator of augmentation was either a law that markedly altered previous policy, or the first law on a hazard if it entailed a major new departure in hazard management. This first group included the hazards of airplane crashes, consumer products, motor vehicle and stationary source air pollution, occupational safety and health, pesticides, and water pollution; the second category included

air piracy chemicals, hazardous materials and solid waste, medical devices, and the surface effects of mining.

6. Not only are controls applied narrowly to single hazards, but they are applied differently to domestic and imported technologies on the one hand and exported ones on the other. All imported hazardous technologies are subject to the same controls as domestic ones, with the exception (concerned with food additives and enacted in 1967) of less than 50 pounds of meat imported for one's own use. By contrast, the only absolute restriction on hazardous exports prohibited export of banned animal drugs or animal feed containing such drugs (1968). Some hazards have been controlled explicitly only for domestic uses (motor vehicle air pollution, 1965; automobile collisions, 1966; boat safety, 1971) or for use in American "installations" overseas (flammable fabrics, 1967, consumer products, 1972). Otherwise, hazard control has been limited to notifying foreign governments of the U.S. status of the hazard's management (pesticides, 1972, 1978; consumer products, 1978) and assuring that the proposed export meets the purchaser's specifications and complies with the laws of the importing country (household substances, 1960; pesticides, drugs, and food additives, 1976). The implicit justification for such noncomprehensive management of **all** U.S.-originating hazards is that this country has no right to impose on others its views regarding which risks are "acceptable" and which are not. If we accept this argument, rather than view the import-export difference as an instance of "double standards," we may still ask why requirements that information on hazardous technologies be provided to their recipients should not be extended to all hazards. (See Johnson 1980,137-138 for further discussion.)

7. Review of the legislative histories of these provisions shows that the difference in penalties for vessels vs. nonvessel polluters was justified as covering the worst-case disaster without imposing an unreasonable insurance burden, and the distinction between removable and irremovable substances may have been due to the observation that, by definition, fines levied on the latter type of pollution would function as penalties rather than as funding for cleanup activities. Neither explanation is completely satisfactory; see Johnson (1982) for further discussion.

8. This blanket approach constitutes "successful" hazard management only if the executive agencies have been able to cope with particular hazards on the systemic agenda as a result of their new responsibilities; otherwise, Congress has succeeded merely in "passing the buck."

REFERENCES

Bibby, John F., Thomas E. Mann, and Norman J. Ornstein. 1980. Vital Statistics on Congress, 1980. Washington: American Enterprise Institute for Policy Research.

Bureau of National Affairs. 1979-1980. Environment Reporter: Federal Laws Index 71:0001-2ff.

Burke, John G. 1966. Bursting boilers and the federal power. Technology and Culture 7 (Winter):1-23.

Cobb, Roger W., and Charles D. Elder. 1972. Participation in American politics: The dynamics of agenda-building. Baltimore: John Hopkins University Press.

Cohen, Bernard L., and I-Sing Lee. 1979. A catalog of risks. Health Physics 36 no. 6:707-722.

Congressional Quarterly Service. 1945-1976. Congress and the nation. Vols. 1-4. Washington: Congressional Quarterly Service.

Congressional Research Service. 1957-1982. Digest of public general bills and resolutions (85th-96th Congresses). Washington: Government Printing Office.

Council on Environmental Quality. 1980. Environmental Quality (11th annual report). Washington: Government Printing Office.

Downs, Anthony. 1967. Inside bureaucracy. Boston: Little, Brown.

Fischhoff, Baruch, Sarah Lichtenstein, Paul Slovic, Ralph Keeney, and Stephen Derby. 1980. Approaches to acceptable risk: A critical guide. NUREG/CR-1614, ORNL/Sub-7656/1. Oak Ridge, Tennessee: Oak Ridge National Laboratory.

General Accounting Office. 1982. Impact of congressional review on Federal Trade Commission decisionmaking and rulemaking processes. HRD 82-89. Washington: General Accounting office, 17 August.

Halverson, Bret, and David Pijawka. 1971. Scientific information about manmade environmental hazards. Monadnock (Bulletin of the Clark University Geographical Society) 48 (June).

Heclo, Hugh. 1978. Issue networks and the executive establishment. In The new American political system, ed. Anthony King, 90, Figure 3.1. Washington: American Enterprise Institute for Policy Research.

Hughes, Steve, Chris Caudill, and Spencer Yost. 1977. Appendix A: List of enactments of the 94th Congress pertinent to natural resources, energy policy, and environmental management. In Environmental protection affairs of the Ninety-Fourth Congress. Senate Report 95-3. Washington: Congressional Research Service.

Hughes, Steve, Susan Bishop, et al. 1979. Appendix A: List of enactments of the 95th Congress pertinent to natural resources, energy policy, and environmental management. In Environmental protection affairs of the Ninety-Fifth Congress. Senate Report 96-5. Washington: Congressional Research Service.

Johnson, Branden B. 1980. Congress as technological hazard manager: Analysis of legislation on technological hazards, 1957-1978. Ph.D. diss., Clark University, Worcester, Mass.

Johnson, Branden B. 1982. Statutory strategies for federal control of technological hazards (unpublished manuscript).

Jones, Charles O. 1975. Clean air: The policies and politics of pollution control. Pittsburgh: University of Pittsburgh Press.

Kasperson, Roger E., and Jeanne X. Kasperson. 1983. Determining the acceptability of risk: Ethical and policy issues. In Risk: A symposium on the assessment and perception of risk to human health in Canada, October 18 and 19, 1982, Proceedings, ed. J. T. Rogers and D. V. Bates, 135–155. Ottawa: Royal Society of Canada.

Kasperson, Roger E., and Murdo Morrison. 1979. A proposed program of energy risk research. Unpublished report for a Beijer Institute submission to the Swedish Energy Research and Development Commission. Worcester, Mass.: Center for Technology, Environment, and Development (CENTED), Clark University.

Lindblom, Charles. 1959. The science of "muddling through." Public Administration Review 19 (Spring):79–88.

Lindblom, Charles. 1979. Still muddling, not yet through. Public Administration Review 39 no. 6 (November/December):517–526.

Lowrance, William W. 1976. Of acceptable risk: Science and the determination of safety. Los Altos, Calif.: William Kaufmann.

Marcus, Alfred. 1980. Environmental Protection Agency. In The politics of regulation, ed. James Q. Wilson, 267–303. New York: Basic Books.

Mitnick, Barry M. 1980. The political economy of regulation: Creating, designing, and removing regulatory forms. New York: Columbia University Press.

Moore, George E. 1978. Carcinogens by fiat. Letter to Science 199 (March 17):1157.

Nader, Ralph. 1965. Unsafe at any speed: The designed-in dangers of the American automobile. New York: Grossman.

National Academy of Sciences. 1974. How safe is safe? The design of policy on drugs and food additives. Academy Forum, 15 May 1973. Washington: The Academy.

National Research Council. 1975. Committee on Principles of Decision Making for Regulating Chemicals in the Environment. Decision making for regulating chemicals in the environment. Washington: National Academy of Sciences.

National Research Council. 1977. Committee on Environmental Decision Making. Decision making in the Environmental Protection Agency. Analytical studies for the U.S. Environmental Protection Agency, vol. 2. Washington: National Academy of Sciences.

Norton, Clark F. 1976. Congressional review, deferral and disapproval of executive actions: A summary and an inventory of statutory authority. Washington: Congressional Research Service, 30 April.

Roberts, Steven V. 1982. Congressmen seeking turf the executive calls its own. New York Times, 28 March, E-5.

Rodgers, William H., Jr. 1977. Handbook on environmental law. St. Paul, Minnesota: West.

Sittig, Marshall. 1979. Legislation bearing on toxic hazards of industrial chemicals. Occupational Health and Safety 48 (October):64–65.

Sulzberger, A.O., Jr. 1980. Legal or not, Congress likes to have the last word. New York Times, 4 May 1980, F4.

U.S. Bureau of the Census. 1975. Historical statistics of the United States: Colonial times to 1970, part 2. Washington: Government Printing Office.

U.S. Bureau of the Census. 1978. Statistical abstract of the United States, 1976. Washington: Government Printing Office.

Wilson, James Q. 1980. The politics of regulation. In The politics of regulation, ed. James Q. Wilson, 357–394. New York: Basic Books.

Wilson, Richard. 1979. Analyzing the risks of daily life. Technology Review 81 no. 4:40–46.

Index